Biodiversity and Conservation along an East African Railway

Biodiversity and Conservation along an East African Railway

A Survey of the Dar es Salaam–Makutupora Standard-gauge Railway, Tanzania

Edited by

Catherine Aloyce Masao

Institute of Resource Assessment, University of Dar es Salaam, Tanzania

Joel Nobert

Institute of Resource Assessment, University of Dar es Salaam, Tanzania

Flora John Magige

College of Natural and Applied Sciences, University of Dar es Salaam, Tanzania

CABI

CABI is a trading name of CAB International

CABI
Nosworthy Way
Wallingford
Oxfordshire OX10 8DE
UK

Tel: +44 (0)1491 832111
E-mail: info@cabi.org
Website: www.cabi.org

CABI
200 Portland Street
Boston
MA 02114
USA

T: +1 (617)682-9015
E-mail: cabi-nao@cabi.org

A catalogue record for this book is available from the British Library, London, UK.

ISBN-13: 9781800626928 (hardback)
9781800626935 (ePDF)
9781800626942 (ePub)

DOI: 10.1079/9781800626942.0000

Commissioning Editor: Ward Cooper
Editorial Assistant: Helen Elliott
Production Editor: James Bishop

Typeset by Exeter Premedia Services Pvt Ltd, Chennai, India
Printed and bound in the UK by CPI Group (UK) Ltd, Croydon, CR0 4YY

Contents

Contributors

Edmond Alavaisha
Institute of Resource Assessment, University of Dar es Salaam

Zainabu S. Bungwa
Tanzania Forest Services Agency, Ministry of Natural Resources and Tourism

Deusdedith F. Bwenge
Tanzania Wildlife Research Institute, Arusha, Tanzania

Elikana John
Tanzania Forest Agency, Ministry of Natural Resources and Tourism

Jasson John
Department of Zoology and Wildlife Conservation, College of Natural and Applied Sciences, University of Dar es Salaam

Mohamed Julius Kibaja
College of Natural and Applied Sciences, University of Dar es Salaam

John Valentine Lyakurwa
College of Natural Resources and Applied Sciences, University of Dar es Salaam, Tanzania

Flora John Magige
Department of Zoology and Wildlife Conservation, College of Natural and Applied Sciences, University of Dar es Salaam

Catherine Aloyce Masao
Institute of Resource Assessment, University of Dar es Salaam

Baraka N. Mbwambo
Tanzania Wildlife Research Institute, Arusha, Tanzania

Heriel Moshi
Department of Botany, College of Natural and Applied Sciences, University of Dar es Salaam

Fatina Athumani Mturi
College of Natural and Applied Sciences, University of Dar es Salaam

Juma R. Mwangi
Tanzania Forest Services Agency, Ministry of Natural Resources and Tourism

Henry Ndangalasi
Department of Botany, College of Natural and Applied Sciences, University of Dar es Salaam

Wilirk Ngalason
College of Natural and Applied Sciences, University of Dar es Salaam

Joel Nobert
Institute of Resource Assessment, University of Dar es Salaam

Bruno Alberto Nyundo
Department of Zoology and Wildlife Conservation, University of Dar es Salaam

Anitha Byabato Philbert
Department of Zoology and Wildlife Conservation, University of Dar es Salaam

Felix A. Shayo
Tanzania Wildlife Research Institute, Njiro, Arusha, Tanzania

Dos Santos Silayo
Tanzania Forest Services Agency, Ministry of Natural Resources and Tourism

Chacha Werema
Department of Zoology and Wildlife Conservation, College of Natural and Applied Sciences, University of Dar es Salaam

Introduction

The Government of the United Republic of Tanzania, through the Tanzania Railways Corporation (TRC), embarked on a major standard-gauge railway (SGR) construction that runs from Dar es Salaam to Mwanza via Isaka. Such developmental structures are associated with modifying and changing the environment hence affecting the biodiversity and ecosystem functioning. This book intends to provide baseline information that will serve for monitoring and future reference. As part of its obligation to restore biodiversity within the SGR footprint areas, the TRC is implementing a Biodiversity Action Plan (BAP), within which one of the important activities is to undertake baseline studies on various areas around critical habitats along the SGR alignment. The idea of writing this book about the biodiversity status along the SGR that runs from Dar es Salaam to Mwanza is in compliance with the Environmental and Social Management Plan (ESMP), consolidating the material that was intended to be published.

The reasons for writing a comprehensive book were: (i) dissemination of scientific information about the biodiversity within the areas, especially those from Pugu and Ruvu forests that serve as critical habitats within the areas; (ii) putting together the literature concerning the biodiversity of the areas that were dispersed; and (iii) consolidation of information of species of conservation concern.

The book is needed and serves as a benchmark for potential book series on railway infrastructure in East Africa and beyond. It underscores the crucial role of infrastructure services – such as power, transportation, telecommunications, water and sanitation – in driving economic development and fostering urban and rural growth. However, the links between infrastructure and inclusive sustainable growth, including communities' livelihoods within the infrastructure corridors, remain a challenge to many developing countries and in most cases are developed with no, or minimal, biodiversity considerations.

The book is restricted to wetlands, terrestrial plants, some invertebrates, amphibians, reptiles, birds and mammals, since inclusion of the whole plant and animal kingdom would be very costly and superficial in terms of content.

Several researchers have conducted research on the biodiversity of Pugu and Ruvu protected areas; however, there have been no efforts to publish a biodiversity book along a railway alignment that mostly passes through areas that harbour sensitive wildlife species.

The book focuses on the status of wetlands, conservation status of various taxa, and the influence of developmental infrastructures on the biodiversity in Pugu Nature Forest Reserve and Ruvu South, all the way to Makutupora. It is chronologically structured along a theoretical sequence, following an evolutionary hypothesis; it begins with plants and moves to invertebrates, with emphasis on arthropods; it then proceeds to amphibians, reptiles, birds and mammals, with special emphasis on bats and galago. In addition, there are some species of both plants and animals that are of conservation concern; hence, their information needs to be assessed and compiled in a book for easy accessibility.

Finally, we offer concluding remarks and provide recommendations that will pave the way towards sustainable conservation of the biodiversity.

Benchmarking the biodiversity status prior to extensive construction activities prepares the ground for other detailed studies within or beyond these areas to include other relevant ecological and social issues alongside the biodiversity trends following construction of the projects.

We target national and international scientists, decision- and policy-makers, as well as ecologists and conservation managers, as our key audience.

Acknowledgements

This book is the result of consultancy and research activities undertaken in implementation of the construction of the standard-gauge railway (SGR) line from Dar es Salaam to Dodoma, funded by the Tanzania Railways Corporation (TRC) through the Government of Tanzania.

Special thanks are extended to the following colleagues from TRC: Director General, Masanja K. Kadogosa, for the full support to the team that executed the studies as part of compliance to IFC Performance Standards; Director of Civil Engineering Infrastructure, Machibya Masanja; Project Manager, Lot 1, Thadei Komu; Project Manager, Lot 2, Tito Mateshi; Environmental and Social Co-ordinator, Veronica Mirambo; and Environmental Safeguard Manager, Magdalena Kitila. All studies compiled in this book were guided and supervised by Dr. Catherine A. Masao who is the Biodiversity Advisor for TRC's SGR Environmental, Social and Health (ES&H) Team and a team leader in preparation of this book. Dr. Catherine Also played a major role in scrutinizing the various reports before they could be submitted to lenders for compliancy checks.

The various reports that were submitted to TRC further received professional feedback from the Lenders' Independent Environmental and Social Consultant (IESC-ARUP). Some IESC members that had direct encounter with the biodiversity surveys undertaken between Dar es Salaam and Makutupora include: Tom Coyne (Biodiversity Specialist), Colette Curran (Director) and Pete Gabriel (Environmental and Social Team Lead for Lots 1 & 2).

We also acknowledge the researchers from the University of Dar es Salaam, the Tanzania Forest Service Agency, and TRC for their tireless contributions and unwavering determination that made this publication possible.

Illustrations

All figures in the book belong to the authors unless otherwise stated.

1 Book Overview

Catherine A. Masao[1]*, Joel Nobert[1] and Flora John Magige[2]
[1]Institute of Resource Assessment, University of Dar es Salaam, Tanzania; [2]College of Natural and Applied Sciences, University of Dar es Salaam, Tanzania

Abstract

Infrastructure services are fundamental to economic growth. Railways are an important part of world transportation infrastructure and have several advantages. Tanzania embarked on constructing a standard-gauge railway from Dar es Salaam to Mwanza, via Isaka, to combat climate change and boost the economy. However, construction is accompanied by clearing of reserved forests (Pugu-Kazimzumbwi, Ruvu South) and has an effect on biodiversity. This book provides baseline information and detailed analyses of land-use and land-cover changes, soil and biodiversity as a function of developmental activities. Findings have indicated that areas are under pressure from various illegal anthropogenic activities. This chapter provides mitigation measures that will reduce the effect on biodiversity and foster food security, poverty reduction and development. The information provided sets a path towards monitoring and evaluating such a big project and suggests implementation of mitigation measures to safeguard biodiversity.

1.1 Background Information

Infrastructure services, including power, transportation, telecommunications, provision of water, and sanitation, are fundamental to national economic growth and development. If well managed, these infrastructures play an important role in stimulating the national economies of many countries (Solinen and Palmqvist, 2023; Zhang and Cheng, 2023). The recent experience of infrastructure investment in the People's Republic of China suggests an intertwined relationship between investment, urbanization and economic growth (World Bank, 2004). In one mechanism, urbanization generates demand for infrastructure investment for effective rural–urban linkages, which then drive economic growth via various channels including reducing transaction costs and raising productivity. Infrastructure development is also a tool which can promote urbanization through facilitating economic agglomeration towards hub cities.

Railways are a transportation infrastructure known to contribute to national economies worldwide with effective management techniques including systematic timetabling (Herranz-Loncán, 2011; Solinen and Palmqvist, 2023). One of the core advantages of a railway as the key national infrastructure is to drive down transport costs, increase rural–urban linkages, and improve access to markets. However, studies suggest that chronic inefficiency undermines this contribution, and this emphasizes the importance of preparation of an integrated development plan of the central railway corridor (Bogart, 2010). For efficiency, one therefore needs to understand the importance of integrating policy-makers in ensuring efficient infrastructure investment

*Corresponding author: masao.catherine@udsm.ac.tz; ndeutz@yahoo.com

DOI: 10.1079/9781800626942.0001

policies (Bogart, 2010; Solinen and Palmqvist, 2023). The adequacy of infrastructure that can contribute to diversifying production, expanding trade, coping with population growth, reducing poverty or improving environmental conditions helps determine a country's success by accommodating economic and urban growth (Calderón and Servén, 2004; World Bank, 2004). Infrastructure corridors have been used as a tool to stimulate multiplier growth of economies including human settlements in both rural and urban areas and as a stimulus to trade and commerce development. Policy-makers and planners have used infrastructure systems to attract both public and private investment for socio-economic development. However, the link between infrastructure and inclusive sustainable growth, including communities' livelihoods within the infrastructure corridors, remains a challenge for many developing countries.

To promote railway transportation, African countries are progressively making strides to develop new railway lines, and defunct ones are being renovated through international partnerships (Bouraima *et al.*, 2023). The Government of the United Republic of Tanzania (GoT), through the Tanzania Railways Corporation (TRC), embarked on a major railway revitalization programme in the country by rehabilitating and constructing new railway links. The construction of a standard-gauge railway (SGR) from Dar es Salaam to Mwanza, via Isaka (1219 km) is in progress. This partly follows the African Union (AU) and East African Community (EAC) decision in 2006 that all new railway development projects on the continent be implemented to a standard-gauge specification. These efforts are partly a response to the Convention on Climate Change and the recent Paris Agreement for parties to reduce the forecasted rise in global temperatures in the 21st century. Railways are among the activities that can contribute to the global effort of curbing emission of greenhouse gases and thus help in achieving the set goals by 2050.

The TRC intends to lay a separate SGR line alongside the existing metre-gauge railway from Dar es Salaam to Mwanza, via Isaka. The implementation of the project is being undertaken segment-wise. The first and second phases of the project start from Dar es Salaam traversing through the Coast, Morogoro and Dodoma regions to Makutupora Station in Manyoni district, Singida region, whereas the third and fourth phases are from Dodoma to Isaka, Kahama (Fig. 1.1). The railway passes through the regional headquarters of Dar es Salaam, Morogoro, Dodoma and Mwanza, which harbour the major train stations. There are also minor stations along the way including Dar es Salaam, Pugu, Soga, Ruvu, Ngerengere, Morogoro, Mkata, Kilosa, Kidete, Gulwe, Igunda, Dodoma, Bahi and Makutopora. The government has engaged Yapi Merkezi Company of Turkey for phases one to four, which are being reported in this book. The project is being implemented under a design-and-build contract mode.

The SGR Dar es Salaam–Makutupora alignment passes through, and in some places close to, coastal forests, eastern arc mountain forests and Miombo woodlands (Fig. 1.2) of Tanzania. These forests are globally important for their biodiversity and locally important in supporting the livelihoods of adjacent communities. Extensive studies compiled in this book have indicated that the areas are under high pressure from illegal logging and charcoal extraction. This threatens both the livelihoods of the communities depending on the forests and the biological values contained therein. Moreover, the SGR route passes close to, or through, two forest reserves (Pugu Kazimzumbwi Nature Forest and Ruvu South) and unreserved forests on village land (Fig. 1.2). There are also several larger protected areas in the vicinity of the SGR, which influence the biodiversity affected by the route. These are the Wami Mbiki Wildlife Management Area and Mikumi National Park. The establishment of the SGR is associated with the risks that may negatively impact the environment and biodiversity along the route. Therefore, understanding the biodiversity status and associated project risks along the SGR alignment is necessary to ensure that there is no net loss of biodiversity in accordance with International Finance Cooperation (IFC) Performance Standard 6 (PS6).

Existence of important biodiversity areas along the alignment call for strategic management of resources to ensure the long-term survival of species and continued ecosystem services for surrounding communities. It is also important to underscore that sustainable management of biodiversity is vital for

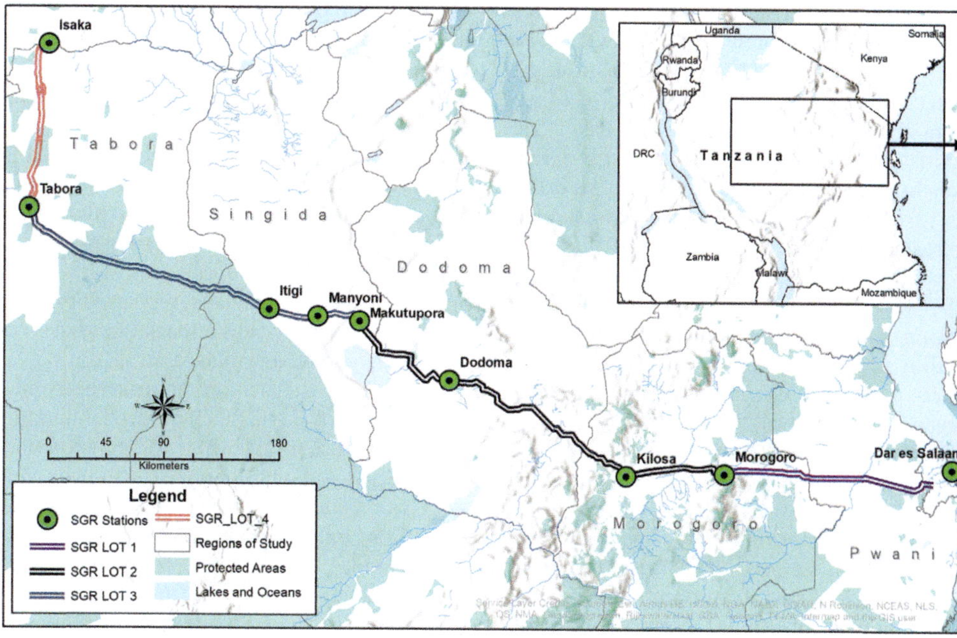

Fig. 1.1. Map of Tanzania showing the SGR line project area from Dar es Salaam to Isaka, Kahama.

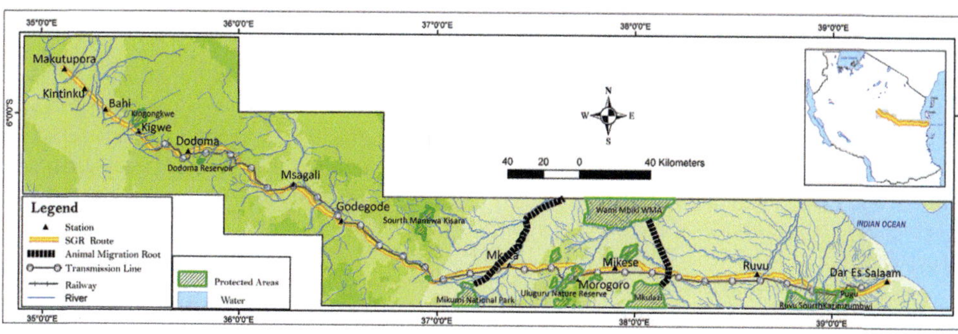

Fig. 1.2. Map showing important critical habitats (protected areas/forest reserves and wetlands) and wildlife corridors that are crossed or directly influenced by the SGR infrastructure.

economic growth and human well-being. When well managed, the biodiversity can provide the foundation for sustained inclusive growth, food security and poverty reduction. Moreover, the world's ecosystems regulate the air, water and soil on which we all depend, and form a unique and cost-effective buffer against extreme weather events and climate change (World Bank, 2019).

Existing SGR infrastructures elsewhere in Africa already sound an alarm, indicating the need for monitoring and evaluation studies that will inform the responsible authorities and the public on the progress of the natural ecosystems, post-construction, of such mega-projects. For example, the monitoring of SGR in Kenya has indicated post-construction environmental issues because of breaking up large areas of habitat into smaller, more isolated patches, which may not be able to support long-term natural processes (Okita-Ouma *et al.*, 2021; Koskei *et al.*, 2022). According to updates provided by the scientists, it appears that despite the initial processes of environmental and social assessment (ESIA) prior to project initiation, there was minimum adherence to the mitigation measures that ultimately resulted in widespread environmental degradation, which was not expected post-construction. Such feedback from the very initial SGR lines in East Africa should not be taken lightly but rather serve as a benchmark for better management of resources as we venture into this world-class economic infrastructure.

TRC is implementing a Biodiversity Action Plan (BAP) by ensuring that the biodiversity within the SGR footprint areas is restored, within which one of the important activities is to undertake baseline studies on various areas around critical habitats along the SGR alignment. The TRC was also obligated to contribute to the review of forest management plans for the areas that will be affected by the SGR alignment through data- and knowledge-sharing. Such an activity is in good progress, although the Pugu Kazimzumbwi status has been upgraded to nature forest reserve since March 2020 (Government Notice (GN) number 190) with a total area of 8965 ha consisting of three forest reserves, namely Pugu forest (2410 ha), Kazimzumbwi forest (4860 ha) from Kisarawe district, and Vikindu forest (1690 ha) from Mkuranga. The status of Ruvu South has remained as a forest reserve under GN No. 81 of 1979 in a map of JB No. 2868 with an area of 30,633 ha made from two forest reserves, namely Banda Forest Reserve and Kola Forest Reserve, gazetted under GN Nos 158 and 159, respectively.

As part of BAP implementation, biodiversity studies were conducted to understand the baseline status of the areas within, and at the vicinity of, the SGR alignment Dar es Salaam to Makutupora to Isaka, Kahama. Part of the studies included surveys on wetlands status, botanical studies, arthropods, galagos, birds and large mammals. The findings of the studies are reported in various reports that are communicated to respective authorities and are used as back-ups to implementation of project mitigation measures as per the ESIA studies. The same data have been compiled to develop this first book as a permanent consolidated reference that may be used by other similar projects elsewhere.

The book focuses on the biodiversity status and conservation of various taxa along the SGR alignment between Dar es Salaam and Makutupora, Dodoma, with special emphasis in the Pugu nature forest reserve and the Ruvu South forest reserve, Tanzania. Specifically, it provides baseline information about key habitats and the influence of infrastructure development on the biodiversity in these reserves. It is chronologically structured along a theoretical sequence, following an evolutionary hypothesis beginning with plants, followed by invertebrates, with emphasis on arthropods, and, lastly, status and diversity of amphibians, reptiles, birds and mammals. The latter has a special emphasis on bats, galagos and large mammals.

Although a lot of information has been published, this book is still much needed for making informed decisions following the construction of the SGR. It is envisioned that the book will serve as a benchmark for potential book series on railway infrastructure development in East Africa and Africa at large. Inadequate links between infrastructure and inclusive sustainable growth, including communities' livelihoods within the infrastructure corridors, remains a challenge to many developing countries. In most cases, infrastructure development is conducted with little or no inclusion of biodiversity considerations.

The book targets national and international scientists, students studying conservation biology, natural-resource managers, decision- and policy makers, as well as ecologists and conservation managers.

1.2 Chapter Organization

The book is organized into 12 chapters, namely: (1) Book Overview; (2) Effects of Standard-gauge Railway Development on Biodiversity and Forest Conservation: Case Study of Pugu Nature Forest Reserve and Ruvu South Forest Reserve in Tanzania; (3) Wetland Status within the Standard-gauge Railway Footprint from Dar es Salaam to Makutupora; (4) Floristic Structure, Composition, Diversity and Anthropogenic Pressure: The Implications of the Management of Biodiversity along the SGR in Pugu and Ruvu South Reserves, Tanzania; (5) Arthropod Assemblages of Pugu-Kazimzumbwi Nature Forest and Ruvu South Forest Reserves, Tanzania; (6) Macroinvertebrate and Fish

Assemblages as Indicators of Wetland Modification within the SGR Footprint; (7) Herpetofauna within the Standard-gauge Railway Footprint from Dar es Salaam to Makutupora; (8) The Avifauna of Pugu Hills and Ruvu South Coastal Forests amidst Railway Development; (9) Ecological Status of Galagos in Pugu Hills and Ruvu South Forest Reserves; (10) Bats of Pugu-Kazimzumbwi and Ruvu South Forests; (11) Wildlife Conservation Implications of Large Mammals along the Standard-gauge Railway from Dar es Salaam to Isaka, Kahama, Tanzania; and (12) Prospects on Biodiversity along the Standard-gauge Railway from Dar es Salaam to Makutupora, Dodoma, Tanzania.

It is important to note that chapters have been independently organized in various sections and subsections following the existing differences in the biodiversity components. A snapshot of some key findings has been provided below.

Chapter 2 provides the assessment that was undertaken by the Tanzania Forest Agency (TFS) by determining the implication of the SGR on the forest cover. It has been observed that approximately 156.54 ha of forest cover, representing 3% of the total forest area, has been degraded. Construction activities necessitated the removal of different vegetation types of about 8227.64 m^3, resulting in potential emissions of about 35,095 tons of CO_2 equivalent. Close monitoring and restoration and/or compensation of degraded land is recommended to recover some of the forests.

Chapter 3 explains the importance of wetlands in harbouring global biodiversity species of conservation value. Sixteen wetlands were identified along the SGR alignment Dar es Salaam to Makutupora, among which 16 were studied to understand their biodiversity value and water-quality parameters. The results show that most of the water-quality parameters were within acceptable ranges, but some sites showed elevated levels of turbidity, dissolved oxygen (DO), biological oxygen demand (BOD5), and total coliforms, indicating human impact. Soil disturbance was observed, potentially limiting plant growth, and disturbances along the SGR line affected species diversity. As expected, the wetlands are rich in terms of biodiversity content where 173 waterbirds were recorded, mainly hamerkop and pied kingfisher; 17 amphibian and 11 reptile species; 730 macroinvertebrates; and 1139 fishes. By the time of the survey, initial project mobilization had been conducted; thus, some of the effects of the SGR project were evident.

Chapter 4 provides a comprehensive list of plant species along the alignment where species of high conservation value (including IUCN red-listed species as well as endemic species) are provided with the status thereof. Specifically, seven plant community types in the two forest reserves; species richness for trees and shrubs ranging from 47 to 97 per ha, and density of 1782–3535 stems per ha, have been reported. The plant species diversity (Shannon-Weiner Diversity Index) in both forests was high, ranging from 3.4 to 4.1, and plant populations were expanding. Plant species of conservation concern identified were nine in PKNFR and six in RSFR. The chapter concludes that, due to the observed distribution pattern of threatened plant species within the reserves and coastal vegetation of Tanzania, the SGR project activities will have minimal negative impacts. It is, however, recommended that project mitigation measures be observed to ensure reduced effects.

Studies on galagos (Chapter 9) indicate existence of three species of galago in the areas. Rondo galago had a relatively higher density compared to other species. The overall population density for *G. rondoensis* in Pugu Hills Forest was 108.9 individuals per km^2 where *G. zanzibaricus* had 34.1 animals per km^2 and *O. garnettii* had 39.2 animals per km^2. The reason for the unexpected higher density of the *G. rondoensis* can be explained by the shy behaviour of the other two species as well as existing predators around the areas. Key recommendations emerging from this study include the need for habitat restoration in the galago areas.

Chapter 5 provides an account of arthropod assemblages of Pugu- Kazimzumbwi Nature Forest and Ruvu South Forest Reserves, Tanzania. The survey for antrhopods was conducted during wet and dry seasons where a reasonably high abundance and diversity of arthropods were sampled in both reserves probably because of similarities in the habitat types. The study concludes that the abundance of arthropods decreased with increasing distance from the railway line indicating that more disturbed areas had less encounters of the insects.

Previous studies had also reported a relationship between habitat disturbance and reduction in anthropod abundance stating more that insects that were habitat specialists are more negatively affected than others. Factors that have been attributed to declined abundance and diversity of anthropods include anthropogenic activities such as agriculture expansion, infrastructural development, livestock expansion, bush fires, timber and charcoal logging, mining, commercial firewood collection and as well as climate change. Both Pugu and Ruvu forests are known to have been facing disturbances at varying degrees and the SGR is regarded as a new stressor factor that was found to increase the disturbances to the anthropod habitats. Important recommendations emerging from the study includes the need for restoration activities that will be coupled with regular monitoring of the status of the anthropod populations in the areas.

Chapter 6 provides a detailed assessment of macroinvertebrate and fish assemblages to establishing whether these could be good indicators of levels for wetland disturbances. The findings indicated that, the abundance and diversity of both macroinvertebrates and fish in all sites that were sampled could explain the level of disturbance of a particular site suggesting their potentials to serve as reliable indicators for assessing human impacts on wetlands.

Chapter 7 provides and account of the herpetofauna within the SGR alignment from Dar es Salaam to Makutupora, Singida, Tanzania. Both day and night surveys were conducted to increase chances of encounters because of the natural behaviours of the species within this group. The data from the field survey was enriched through detailed literature review. Overall, fifty-one (51) amphibian and 107 reptile species were found along the SGR footprint from Dar es Salaam to Dodoma Region. Of conservation interest is one amphibian species (Ruvu spiny reed frog - *Hyperolius ruvuensis* which is already listed as Critically Endangered (CR) species listed under the International Union for Conservation of Nature (IUCN) and three reptile species that are listed as follows: one species is considered as Critically Endangered (CR), two species are listed as Vulnerable, one species is Near Threatened, three species are Data Deficient and six species are not evaluated. It is also worth noting that, out of the 51 amphibian species identified along the SGR route from Dar es Salaam to Makutupora, 27 species (52.7%) were distributed throughout the entire corridor. Twenty species were confined to the lowland regions between Dar es Salaam and Kilosa, while only four species were observed in areas above 500 meters in elevation. It has been pointed out clearly that, since the SGR crosses important wetlands that are home to amphibians and reptiles of conservation value, herpetofauna passes in form of tunnels, underpasses and drainage culverts to prevent blockage of the movements of animals between the two sides of the railway have been recommended as necessary mitigation measure.

Chapter 8 is an assessment of Avifauna of Pugu and Ruvu South Coastal Forests amidst Railway Development. The Birdlife international had previously reported two globally threatened bird species: Spotted Ground-thrush *Geokichla guttata* and Sokoke Pipit *Anthus sokokensis* which (amongst other species) have been given a special emphasis in the avifauna studies along the SGR alignment thus to ensure their conservation. Mist netting and transect using fixed-width point counts were applied as suitable methods for collecting the understory and skulking bird species while literature reviews for both published and grey material was also used to add data and information beyond the SGR alignment but within Pugu and Ruvu Forests. The findings have indicated similarity in species richness in the wet and dry season although there is a notable difference in diversity and composition between seasons. The wet season had the highest encounter of all counts in terms of diversity and abundance. This compilation has indicated that, at least 10 bird species that are not forest dependent but threatened with extinction are found in Pugu and Ruvu South forests thus necessitating the need to strengthen conservation efforts of these forests. The threatened species include Southern Ground-hornbill, Bateleur, Great Snipe, Hooded Vulture, White-backed Vulture, Rüppell's Vulture, Martial Eagle, Tawny Eagle, and Steppe Eagle. Amongst important recommnedations under this chapter is the need for forest enrichment planting to restore the degraded or modified habitats along SGR alignment using native plants species. The native plant species are expected to better support forest dependent bird species which will not do better in exotic dominated habitats.

Studies on galagos (Chapter 9) indicate existence of three species of galago in the areas. Rondo galago had a relatively higher density compared to other species. The overall population density for *G. rondoensis* in Pugu Hills Forest was 108.9 individuals per km^2 where *G. zanzibaricus* had 34.1 animals per km^2 and *O. garnettii* had 39.2 animals per km^2. The reason for the unexpected higher density of the *G. rondoensis* can be explained by the shy behaviour of the other two species as well as existing predators around the areas. Key recommendations emerging from this study include the need for habitat restoration in the galago areas.

Chapter 10 provides information on the bats recorded in the SGR alignment. Bats make up about 20% of the mammal species in the world, forming the second most speciose mammalian order providing important ecosystem services. A considerable number of species of bats were recorded in PKNFR and RSFR. The Natal long-fingered bat (*Miniopterus natalensis*) was recorded for the first time as the migratory species in PKNFR. The forests along the SGR are of considerable importance in conservation of bats and continuing to conserve them is recommended. The existence of wildlife corridors along the alignment is another important aspect that has been detailed in Chapter 11 and needs to be well monitored to ensure that the already reported impacts of Kenya SGR will not affect elephants and other vulnerable wildlife species in Tanzania. According to monitoring reports by Kenya Wildlife Service (KWS), it took only three months for the elephants to stop crossing the areas they used to because of the existence of the SGR alignment (Okita-Ouma *et al.*, 2021). Twenty-six species of large mammals were recorded in the areas; some are vulnerable and their habitat is already fragmented. The existence of the SGR alignment leads to a cumulative effect of fragmentation, thus proper mitigation options are key to such situations. Some of the proposed mitigation in the wildlife movement corridors include the construction of viaducts, especially in phase three of the project, because already the Kenya SGR model for use of underpasses had been recommended for phases 1 and 2, and, in a way, have already been implemented.

1.3 Rationale of the Book

It was earlier reported that transportation is an essential component of economic growth. Railways are regarded as a more cost- and energy-efficient way of transportation than motor vehicles and airplanes (Borda-de-Água *et al.*, 2017). Worldwide efforts are towards making railway transportation the backbone of sustainable transportation, although their construction is accompanied by negative effects on the biodiversity. Despite the existing risks associated with railway infrastructure worldwide, only a few books exist that have investigated biodiversity along such alignment. Among these are *Railway Ecology* (Borda-de-Água *et al.*, 2017) and *State of the Apes: Infrastructure Development and Ape Conservation* (Rainer *et al.*, 2018). Both books are focused on ecology and biodiversity aspects along railway infrastructure. *Railway Ecology* builds on the diverse European railway network, with work from North America, South America and Asia. *State of the Apes* presents original research and analysis, topical case studies and emerging tools and methods to inform debate, practice and policy with the aim of preventing and mitigating the harmful impacts of infrastructure projects on biodiversity.

In contrast, this book provides detailed analyses on biodiversity status, ranging from plants and various animal taxa (both invertebrates and vertebrates). In addition, it provides baseline information for the African context, which is an important element lacking for many construction development projects. Thus, the book provides important baseline data for benchmarking other infrastructure developments in sub-Saharan Africa. The book opens the way for other book series on Africa SGR projects that are rapidly developing. It is anticipated that it will be useful in igniting interest in railway transportation efficiency, both in terms of economic gains and environmental protection. The baseline data provided here lead the way towards monitoring of such large projects to see how the proposed mitigation measures are being implemented during construction and operation phases, whilst also noting some key lessons learned for similar projects being implemented elsewhere.

References

Bogart, D. (2010) A global perspective on railway inefficiency and the rise of state ownership, 1880–1912. *Explorations in Economic History* 47(2), 158–178.

Borda-de-Água, L., Barrientos, R., Beja, P. and Pereira, H.M. (2017) *Railway Ecology*. Springer, Cham, Switzerland.

Bouraima, M.B., Alimo, P., Agyeman, S., Sumo, P.D., Lartey-Young, G, *et al.* (2023) Africa's railway renaissance and sustainability: Current knowledge, challenges, and prospects. *Journal of Transport Geography* 106, 103487. DOI: 10.1016/j.jtrangeo.2022.103487.

Calderón, C. and Servén, L. (2004) The effects of infrastructure development on growth and income distribution. *Policy Research Working Papers*. DOI: 10.1596/1813-9450-3400.

Herranz-Loncán, A. (2011) The contribution of railways to economic growth in Latin America before 1914: A growth accounting approach. University of Barcelona. Available at: www.ub.edu/histeco/pdf/herranz-DT01.pdf (accessed 15 April 2024).

Koskei, M., Kolowski, J., Wittemyer, G., Lala, F., Douglas-Hamilton, I, *et al.* (2022) The role of environmental, structural and anthropogenic variables on underpass use by African savanna elephants (Loxodonta africana) in the Tsavo Conservation Area. *Global Ecology and Conservation* 38.

Okita-Ouma, B., Koskei, M., Tiller, L., Lala, F., King, L, *et al.* (2021) Effectiveness of wildlife underpasses and culverts in connecting elephant habitats: A case study of new railway through Kenya's Tsavo National Parks. *African Journal of Ecology* 59, 624–640. DOI: 10.1111/aje.12873.

Rainer, H., White, A. and Lanjouw, A. (eds) (2018) *State of the Apes: Infrastructure Development and Ape Conservation*. Cambridge University Press, Cambridge, UK.

Solinen, E. and Palmqvist, C.-W. (2023) Development of new railway timetabling rules for increased robustness. *Transport Policy* 133, 198–208. DOI: 10.1016/j.tranpol.2023.02.003.

World Bank (2004) *World Development Indicators*. World Bank, Washington, DC.

World Bank (2019) *World Development Report 2019: The Changing Nature of Work*. World Bank, Washington, DC. DOI: 10.1596/978-1-4648-1328-3.

Zhang, Y. and Cheng, L. (2023) The role of transport infrastructure in economic growth: Empirical evidence in the UK. *Transport Policy* 133, 223–233. DOI: 10.1016/j.tranpol.2023.01.017.

2 Effects of Standard-gauge Railway Development on Biodiversity and Forest Conservation: Case Study of Pugu Nature Forest Reserve and Ruvu South Forest Reserve in Tanzania

Elikana John*, Zainabu S. Bungwa, Juma R. Mwangi and Dos Santos Silayo
Tanzania Forest Services (TFS) Agency, Itega, Dodoma, Tanzania

Abstract

The standard-gauge railway line is being built to connect the port of Dar es Salaam with interior regions as well as neighbouring nations. The railway line goes through some protected areas, including Pugu and Ruvu forests. Assessment of the project's impact on forest conditions was conducted using inventory surveys and remotely sensed data. Results showed that there have been substantial impacts on the forest, resulting in loss of forest cover, runoff disturbances, habitat loss and ecosystem fragmentation. It was revealed that approximately 156.54 ha, representing 3% of total forest cover, has been lost or degraded. This was due to construction activities necessitating the removal of different vegetation types of about 8227.64 m^3, resulting in potential emissions of about 35,095 tons of CO_2 equivalent. However, EIA and ESIA have been observed to minimize the impact. We recommend close monitoring of restoration programmes in the respective forests.

Keywords: standard-gauge railway, forest reserves, ecosystem fragmentation, CO_2 emissions, mitigation strategies

2.1 Introduction

In light of substantial developments in trade and economic interactions between the countries, transportation infrastructure has grown in sub-Saharan Africa, which accounts for up to 90% of new infrastructure, including railways (Alamgir *et al.*, 2017; Nyumba *et al.*, 2021). It is a vital component of any built environment since it improves the speedy movement of products and services, hence improving the production processes in any economy (Shabana *et al.*, 2007). Therefore, railway technology has evolved considerably and reduced transportation costs and increased safety not only for passengers, railway workers and freight but also for the human populations living within the railway corridors (Frost, 2016).

Tanzania is no exception, with significant investment being made to improve its infrastructure network. The standard-gauge railway (SGR) line is being built by the government and will link the ports of Dar es Salaam on the Indian Ocean and Mwanza on the northern

*Corresponding author: elikana.maige@tfs.go.tz; elikanajohn@gmail.com

DOI: 10.1079/9781800626942.0002

Tanzanian shore of Lake Victoria with the neighbouring nations of Rwanda, Burundi and the Democratic Republic of the Congo (DRC). The large-scale construction of the SGR project has five sections, namely: Lot 1: Dar es Salaam–Morogoro; Lot 2: Morogoro–Makutupora; Lot 3: Makutupora–Tabora; Lot 4: Tabora–Isaka; and Lot 5: Isaka–Mwanza, with a total length of 1219 km (TRC, 2018). The construction of the SGR railway is anticipated to boost the government's ambitions for implementing the industrial policy and propel Tanzania's economy and meeting the country's development vision 2025. Furthermore, this railway will spark an economic revolution in the entire East Africa and Central African region, particularly for land-locked countries (TRC, 2018).

Experience shows that railway construction, as with any other project that goes through ecosystems, may have negative impacts on those ecosystems and on biodiversity conservation (Lucas *et al.*, 2017; Milewicz *et al.*, 2023). For example, a study conducted in Kenya post construction of the SGR showed that despite the anticipated advantages of the project and the need to meet the increasing demand for environmental accountability, there was a growing recognition of its impacts on the natural environment, especially in remote and fragile ecosystems (Ambani and Mulaku, 2021; Lape *et al.*, 2023). The impacts that have been reported are: interruption of wildlife migration and movement; alteration of species' behaviour; the bisection of basins and watersheds (Munyao *et al.*, 2020), and physical alteration of ecosystem structure and composition (Nyumba *et al.*, 2021). Likewise, the ecological interactions between predators and prey could be changed. Sang *et al.* (2022) reported that railway line construction could lead to increased water, soil and air pollution, and grassland fires. Therefore, achieving sustainable development requires balancing infrastructure development and biodiversity protection goals (Lape *et al.*, 2023).

History shows that railway infrastructure development in Tanzania dates back to the 1880s when the German administration in the colonial era constructed the Usambara railway line from Tanga port to the highlands of Amani forests in Muheza district, Tanga (Reichart-Burikukiye, 2013). Later, in 1905, a central line meter-gauge (1000 mm) railway (formerly known as the Tanganyika railway) was constructed running from Dar es Salaam to Kigoma with branches to Mwanza, Tanga, Kilimanjaro to mention a few (Limi *et al.*, 2017). Meanwhile, between 1968 and 1975, a 1860 km binational standard-gauge railway (1435 mm) linking the Southern Africa regional transport network to Eastern Africa's seaport of Dar es Salaam, offering both freight and passenger transportation services between and within Tanzania and Zambia, was constructed with the support of the Chinese government (Monson, 2008). The construction of these railway networks involved opening ways into different ecological landscapes. Some of the areas had different management status as either being general lands, forest reserves or game-reserved areas.

Over the past 12 years, the Tanzanian government through Tanzania Forest Services (TFS) Agency has continued to strengthen management of selected forest ecosystems with critical conservation importance for biodiversity and ecological functions. The decisions and actions have been to identify and upgrade biologically important forest reserves to become nature forest reserves (NFRs). The first nature forest reserve was Amani in Muheza Tanga, which forms part of the critical forest patches in the Usambara mountains and was declared in 1997. The network of NFRs has expanded to cover all the different forest types in the country (including the coastal forests, the northern volcanic, the southern highlands, Guineo-Congolian, Miombo and Miombo-Acacia forests) (Ract *et al.*, 2023). The current network of NFRs in Tanzania contains 23 reserves, covering an area of 965,755 ha. This is in line with the National Forest Policy Implementation Strategy (2021–2030), the National Environmental Policy of 2020 and the Environmental Master Plan of 2022, enhancing protected areas as being a cornerstone of global conservation efforts (Lopoukhine *et al.*, 2012).

The construction of SGRs as a strategic state project crosses different land-use categories including protected areas. It entails clearing of vegetation and the accompanying levelling operations that destroy the original topography and soil profile (Nyumba *et al.*, 2021). Paving a way for the railway line will entail the elimination, replacement or modification of the original natural vegetation. This impacts the protected areas that are equally important for protecting biodiversity, as well as for sequestering and storage of carbon, which is key in the fight against climate change. Even though the project has observed all environmental safeguard protocols there are still some social and ecological impacts on the protected areas, and biodiversity distributions and carbon dioxide emissions from land and vegetation clearance cannot be underestimated.

Therefore, there is a need to understand how SGR impacts the forests so as to enable managing authorities to plan for effective mitigation measures, and to contribute to planning and policy on future infrastructure projects. In so doing, the findings will aid Tanzania with its obligations to the international bodies on forest biodiversity conservation and climate change mitigation (John *et al.*, 2020) on the understanding that tropical forests store approximately 55% of the world's forest carbon (C) stocks, acting as C sinks through a positive balance of tree growth, recruitment and mortality (Pan *et al.*, 2011). Specifically, the study focused on addressing the following objectives: (i) assess and identify the magnitude of impacts caused by SGR development through the coastal forest reserves of Pugu Nature Forest Reserve and Ruvu South Forest Reserve; (ii) estimate quantities of CO_2 emitted due to land-use changes because of the new rail infrastructure; and (iii) develop and recommend mitigation measures and other conservation strategies to maintain the ecological balance of the forest reserves.

2.2 Material and Methods

2.2.1 Study area description

The study was conducted on a stretch of Lot 1, which spans from Dar es Salaam City to Morogoro Town, specifically on the two forest reserves: Ruvu South Forest Reserve and Pugu Nature Forest Reserve (hereafter, Ruvu SFR and Pugu NFR).

Ruvu south forest reserve

RSFR covers 30,633 ha. The forest lies between 6°53'S and 7°03'S and 38°46'E–39°02'E (Fig. 2.1A), 55 km south-west of Dar es Salaam, 20 km from Kibaha Town. The most northerly corner of Ruvu SFR borders the north-western end of Pugu NFR and is 2 km from the western edge of Kazimzumbwi. The forest spans Kibaha and Kisarawe districts in Coastal (Pwani) region.

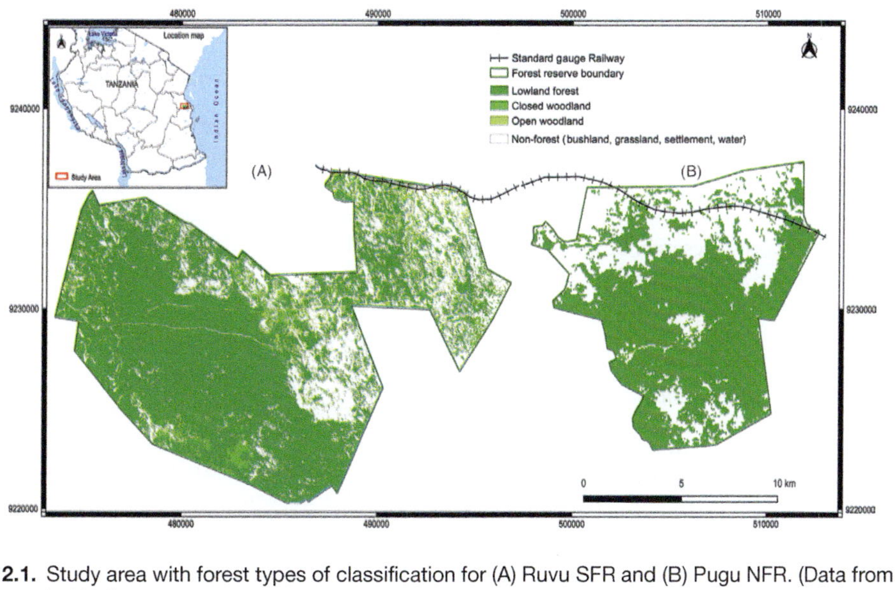

Fig. 2.1. Study area with forest types of classification for (A) Ruvu SFR and (B) Pugu NFR. (Data from John *et al.*, 2021)

Ruvu SFR supports Eastern Arc and coastal forest endemic vertebrates including four-toed elephant shrew, blotched genet, African civet, suni, bushy-tailed mongoose and giant pouched rat.

The forest is known as a coastal forest important bird area (Baker and Baker, 2002) hosting several rare and low-density forest bird species; and as a regional centre of endemism, with 33 endemic plant species (IUCN, 2012). Like other coastal forests, it is important for endemic and near-endemic elephant shrews such as *Rhynchocyon petersi*. However, little is known about the reserve's bats and galagos (BirdLife International, 2023).

Other mammals that had been recorded within the reserve include four diurnal primates, black-and-white colobus (*Colobus angolensis*) palliates, yellow baboons (*Papio cynocephalus*), Sykes monkeys (*Cercopithecus mitis*) and vervet monkeys (*C. aethiops pygerythrus*); the scrub hare (*Lepus saxatalis*); red-legged sun squirrel (*Heliosciurus rufobrachium*) and spotted hyena (*Crocucta Crocuta*). A total of 36 mammal species in 31 genera and 19 families have been recorded including the vulnerable little-collared fruit bat (*Myonycteris relicta*) and vulnerable lesser-poached rat (*Beamys hindei*) (Gwegime *et al.*, 2013).

Pugu nature forest reserve

The Pugu NFR is a coastal forest covering an area of 2410 ha located about 20 km south-west of Dar es Salaam, approximately between longitude 39°03'E and 39°07'E and latitude 6°52'S and 6°56'S (Fig. 2.1B). The reserve extends over the Pugu Hills area, adjacent to the Kazimzumbwi Forest Reserve. Despite the small size, the reserve is internationally recognized as a 'hotspot' of the diversity of endemism (Howell, 1981).

The forest is one of the last remaining vestiges of lowland coastal forest in Tanzania and contains a unique flora with 12 endemic species and one endemic genus, the East African coastal bird, the Sokoke pipit, and a number of bird species with unusual or restricted distributions, such as the Rondo dwarf galago. The reserve is a source of the Msimbazi river that flows into the Indian Ocean through Dar es Salaam City.

2.2.2 Data collection

Primary data

Biophysical data through field inventory was collected in February 2019 on both of the reserves. The targeted area of the SGR construction corridor was surveyed and mapped using hand-held GPS based on the two forest maps. The areas identified and impacted by SGR include the damping sites with their branch roads inside the reserves and the construction corridor (railway line and service roads).

Sampling design

A transect walk along the SGR corridor was established and circular sample plots of 15 m radius and 20 m from each end of the construction corridor on each side of the transect in both of the forest reserves. The parameters measured in the sample plots included diameter at breast height (dbh) for tree species with butt diameter greater than or equal to 20 cm, butt diameter for all tree species with butt diameter less than 20 cm, and height for all trees species measured for dbh. The distance between plots was 200 m and 400 m with the transect lengths of 4.93 km and 7.62 km transect in Pugu NFR and Ruvu South FR, respectively. This sampling design was based on the homogeneity of the vegetation in both forests.

Secondary data

Secondary data were gathered from existing management plans, previous studies on Biodiversity Action Plan of July 2019, and Environmental and Social Impact Assessment report of February 2019.

2.2.3 Biophysical data analysis

The volume estimation for the wood removed on SGR site, based on Mugasha *et al.* (2016) (Eqn 2.1), includes forest structure and stem density as number of stems per ha (N), basal area (G in m^2/ha), total volume (V in m^3/ha), species groups, and diameter.

$$V = 0.00076\ x\ Dbh^{2.3488} \qquad \text{(Eqn 2.1)}$$

where:
V = total tree volume (m^3/ha)
Dbh = diameter at breast height (cm)

2.2.4 Remotely sensed geospatial data

Remotely sensed geospatial data for the study area were acquired from John *et al.* (2021) and included the classified forest types. Both classified maps and forest vector boundary layers for both forest reserves were transferred to the geographic information system (GIS) for spatial analysis using QGIS, mainly estimating the size of the areas affected by SGR establishment.

2.2.5 CO_2 emissions from vegetation clearance

The construction of the SGR requires removal of existing vegetation. Therefore, the process may attribute the release of carbon stored in vegetation and soil due to land use and land-use changes (LULUC). This study considered the areas impacted to determine carbon emissions. The carbon emissions (tons C yr^{-1}) were estimated using the Njana *et al.* (2021) approach and the carbon density applied was obtained from Mauya *et al.* (2019) for different land covers in mainland Tanzania. The carbon densities were used as emission factors for forests that were transformed into non-forests. Therefore, the carbon emissions (tons C yr^{-1}) were calculated by multiplying the carbon density for each forest (tons C ha^{-1}) by the annual deforestation (ha yr^{-1}). The carbon densities were transformed into carbon dioxide equivalent (CO_2 e) by multiplying by 44/12, resulting in the expression of carbon emissions as CO_2 e (Njana *et al.*, 2021).

2.3 Results

2.3.1 Impacted area by SGR on Pugu NFR and Ruvu SFR

The direct footprint of the SGR is 2.95% and 0.28%, respectively, of the Pugu NFR and Ruvu SFR, representing a total area of 156.54 ha (Table 2.1) of the forest area affected during construction (Figs. 2.3 and 2.4). This reveals that approximately 3% of vegetation type as part of the protected forest area has been lost or degraded (Fig. 2.2A, B). Similarly, TRC (2019) indicated about 11% of the Pugu NFR is fragmented off the main block and extends for a combined distance of 12.6 km through or along the boundaries of the Pugu NFR and Ruvu SFR. Hence, the ecology of these areas cannot be restored, and the impact is deemed permanent.

2.3.2 Estimated volume of wood removed from Pugu NFR and Ruvu SFR

Results show that site clearance for construction of the said railway line involved removal of different vegetation types including trees of different species with a volume of 1635.56 m^3 on the area of 71.21 ha of Pugu NFR (Table 2.2); while a volume of 6592.08 m^3 was removed from Ruvu SFR (Table 2.3). Therefore, the removal of such vegetation amounting to 8227.64 m^3 would impact functionality and loss of habitat and ecosystem and reduce important remnants of the Northern Zanzibar-Inhambane Coastal Forest Mosaic eco-region unless mitigation measures are put in place. Meanwhile, the construction resulted in large surplus soil dumpsites that required proper management to avoid contamination and sedimentation of aquatic habitats (Fig. 2.3).

Table 2.1. Estimated area of the reserves impacted by the SGR footprint.

Reserve name	Area (ha)	Direct SGR footprint (ha)	Length (km)	% of the reserve affected
Pugu NFR	2410	71.21	4.93	2.95
Ruvu SFR	30,633	85.33	7.62	0.28
Total		156.54	12.55	3.23

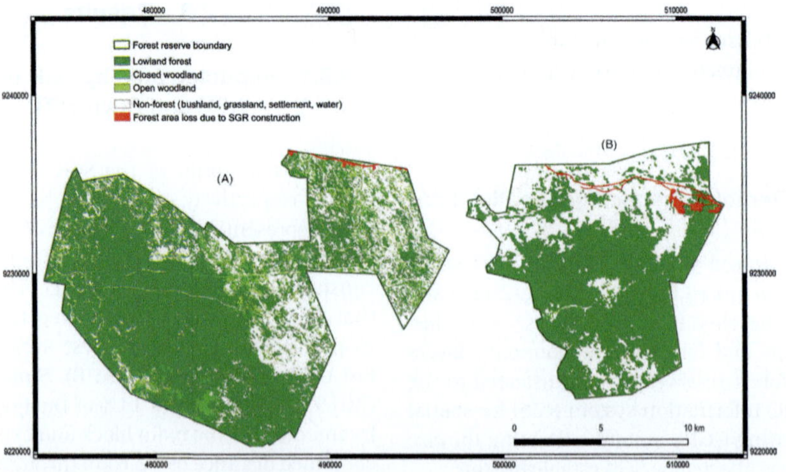

Fig. 2.2. Forest area affected in instituting the project through (A) Pugu NFR and (B) Ruvu SFR.

Table 2.2. The estimated volume of wood removed during construction in Pugu NFR.

Species name	Plant family	Volume (m^3/ha)	Disturbed area (ha)	Total volume (m^3)
Ficus lutea	Moraceae	7.02	71.21	499.81
Ficus stuhlmannii	Moraceae	4.02	71.21	286.42
Ficus sycomorus	Moraceae	8.57	71.21	609.99
Milicia excelsa	Moraceae	1.93	71.21	137.17
Senna siamea	Fabaceae	1.06	71.21	75.54
Trichilia emetica	Meliaceae	0.37	71.21	26.63
Total				1635.56

Table 2.3. Estimated volume of wood removed during construction in Ruvu SFR.

Species name	Plant family	Volume (m^3/ha)	Disturbed area (ha)	Total volume (m^3)
Afzelia quanzensis	Fabaceae	4.78	85.33	407.49
Bombax rhodognaphalon	Bombacaceae	48.5	85.33	4138.69
Cussonia arborea	Araliaceae	13.61	85.33	1161.57
Sclerocarya birrea	Anacardiaceae	5.13	85.33	438.01
Vitex keniensis	Lamiaceae	5.23	85.33	446.32
Total				6592.08

2.3.3 CO_2 emissions from vegetation clearance

The findings from Table 2.1 revealed the extent of the impacted areas from the two forest reserves. These areas were utilized to calculate the quantity of carbon emissions resulting from the conversion of forests into non-forested areas. The forest types designated as lowland forest in the two reserves (Fig. 2.2A,B) have a carbon density of 66.90t C/ha, while non-forest areas have a carbon density of 5.81t C/ha (Mauya *et al.*, 2019). Therefore, Table 2.4 indicates the total CO_2 emissions for the areas impacted by SGR are about 15,965 and 19,131 tons CO_2e, annually, from vegetation clearance for Pugu Nature Forest Reserve and

Fig. 2.3. Examples of area impacted by the project at (A–B) Pugu and (C–D) Ruvu South forest reserves.

Table 2.4. Estimated CO_2 emissions from vegetation clearance for the two reserves impacted by the SGR footprint.

Reserve name	Reserve area (ha)	Direct SGR footprint (ha)	Carbon density (t C/ha) (lowland to non-forest)	Factor	CO_2 emissions (tons CO_2e yr^{-1})
Pugu NFR	2410	71.21	61.09	3.6	15,965
Ruvu SFR	30,633	85.33	61.09	3.6	19,131
Total		156.54			35,096

Ruvu South Forest Reserve, respectively, making a total of 35,095 tons of CO_2e from vegetation removal due to SGR footprint.

2.4 Discussion

2.4.1 Ecosystem degradation, fragmentation and destruction

We assessed the impact of SGR construction on the coastal forests of the two-forest reserve of Pugu and Ruvu South. Our results indicate that the construction process led to the clearing of forest cover for about 2.95% and 0.28% of the Pugu NFR and Ruvu SFR, respectively, representing a total area of 156.54 (Table 2.1), resulting in habitat loss and fragmentation (Nyumba *et al.*, 2021). Therefore, fragmentation occurred for the two forest reserves (Fig. 2.2A, B) as large, continuous forests are divided resulting in smaller, often isolated, patches that may not be able to maintain viable populations in the long run (Fahrig, 2003). The results reveal that this can potentially disrupt the natural movement and migration patterns of various faunal species, leading to isolation of populations and potentially leading to genetic isolation and reduced biodiversity (e.g. Dorsey *et al.*, 2015). This can have significant consequences for both flora and fauna that rely on the forest ecosystem.

Soil compaction, excavation and soil movement from one area to another to establish the embankments were all part of the SGR construction. These actions might have disrupted and/or obstructed natural processes such as natural hydrology and animal migration pathways. Previous research has shown how road noise, vibrations, chemical pollution and human presence harm wildlife living near such heavy and busy infrastructure (Nyumba *et al.*, 2021). Similarly, creation of underpasses increases runoff, inevitably increasing the volumes and speed of the flow. This has caused flooding, soil erosion, channel modification and siltation of streams. Therefore, fragmentation may lead to isolated populations, making it challenging for species to find mates and maintain genetic diversity (Ony *et al.*, 2020).

2.4.2 Deforestation and habitat loss

The process of clearing removed trees with an estimated volume of 1635.56 m^3, and other vegetation, is contributing to deforestation. This has impacts on the two forest reserves' declining biodiversity, mostly caused by the removal of the forest cover of about 3%, which leads to habitat fragmentation and degradation (Fig. 2.2A, B). As a result of habitat loss, there is reduction of habitats available to support wildlife, which decreases numbers and increases the danger of population extinction (Wang *et al.*, 2021). The removed forest cover will impact all new species assemblages, functions, and the distribution of ecosystem services. A study by Requena Suarez *et al.* (2021) found that human-induced disturbances in some forest patches can benefit nearby forests in tree establishment through natural regeneration and tree growth; however, high rates of degradation and deforestation in African woodlands (McNicol *et al.*, 2018) limit this effect. As a result, new species connections, community assemblages and interactions emerge that did not previously exist.

2.4.3 Implications of carbon emissions from forest reserve

The estimated carbon emissions of about 35,096 tons CO_2e per year (Table 2.4) have substantial consequences for both local and global ecosystems, as well as for climate change. Hence, deforestation-induced carbon emissions exacerbate habitat degradation, posing a formidable obstacle to the flourishing of particular species. Likewise, it can disturb the natural flow of water, causing alterations in the distribution of rainfall and perhaps exacerbating the occurrence of droughts or floods, which can have adverse effects on the welfare of the local communities. Efforts to minimize these emissions include a blend of conservation, sustainable forest management, and restoration of the affected areas to alleviate the consequences of climate change. Therefore, carbon footprint analysis offers the necessary instruments for the sustainable development of infrastructure construction, especially in protected areas (Dimoula *et al.*, 2016).

2.4.4 Mitigation measures

The construction of the SGR is widely acknowledged to possess greater environmental friendliness compared to road vehicle transportation. However, it is imperative to recognize that the project still entails certain adverse effects, such as the clearance of vegetation, soil erosion, loss of wetlands, degradation, and emissions. Wang *et al.* (2021), in their observations on the same project, emphasized the need to put in place all mitigation measures as detailed in the ESIA reports and that any impacts should not be disregarded. Hence, the implementation of the necessary mitigation measures has the potential to minimize the indicated impacts observed in this study. The study found that the project is undertaking substantial and necessary measures in collaboration with relevant public and private institutions in addressing potential impacts, which include restoration and monitoring.

2.4.5 Restoration of the impacted areas

The restoration process is conducted on the affected area of about 3% in compliance with the Environmental Management Plan (Construction and Maintenance) of the impacted areas as stipulated in the Environmental and Social Impact Assessment report. The construction and maintenance activities of the SGR have resulted in several sites that necessitate restoration. These areas encompass significant locations such as borrow pits, stockpile sites, campsites and sidetracks. The majority of the impacted areas underwent the process of re-topsoiling and replanting with indigenous species to avoid or minimize invasive species; this was implemented as a component of the restoration strategy.

Windrows or rills of topsoil were replanted with grasses (Fig. 2.4) strategically arranged to create a temporary diversion bank, serving as an effective measure to minimize soil erosion. A rill was formed below the site using subsoil material, in conjunction with a sump or stacked rock containment structure, to restrict sediment and/or contaminant movement offsite. This aims at leaving a stable environment that is conducive to the establishment of landscapes characteristic to the area.

Similarly, rehabilitation of sidetracks requires removal of all temporary cross-drainage structures from watercourses, together with the removal and disposal of fill materials used for temporary approaches, crossings or embankments. All remaining areas have been reshaped to blend back in with pre-existing landforms. Table 2.5 presents a summary of the tree species that have been identified for the restoration efforts in both reserves, with the aim of preserving the original forest cover.

2.4.6 Monitoring

Regular monitoring of the condition of plants in the restored areas is strongly recommended to effectively monitor biodiversity in the affected areas. Typically, the features that are most frequently utilized for monitoring purposes are plant cover, plant density, frequency and biomass, which is a metric for measuring production (Borda-de-Água *et al.*, 2017). Hence, it is imperative to implement a monitoring plan that encompasses the assessment and changes on vegetation and understanding the behavioural

Fig. 2.4. Part of Pugu Nature Forest Reserve replanted with grasses (windrows) to minimize soil erosion after SGR line construction. (Google Earth image)

Table 2.5. Tree species identified for restoration of deforested forest at Pugu NFR and Ruvu SFR (P = Pugu NFR, R = Ruvu SFR)

Species	Plant family	Name of the reserve	Remark s
Millettia puguensis	Fabaceae	P, R	Endemic to Pugu and Ruvu
Baphia puguensis	Fabaceae	P, R	Endemic to Pugu and Ruvu
Whitfieldia orientalis	Acanthaceae	P, R	Coastal forest endemic
Pristimera graciflora	Celastraceae	P, R	Coastal forest endemic
Manilkara sansibarensis	Sapotaceae	P, R	Coastal forest endemic
Nesogordonia holtzii	Sterculiaceae	P, R	Coastal forest endemic
Rinorea welwitschii	Violaceae	P, R	Coastal forest endemic
Uvaria puguensis	Annonaceae	P, R	Endemic to Pugu and Ruvu
Pteleopsis myrtifolia	Combretaceae	P, R	Vulnerable
Melicia excelsa	Meliaceae	P, R	Vulnerable
Afzelia quanzesis	Fabaceae	P, R	Vulnerable
Soreindea madagascariensis	Anacardiaceae	P, R	Food for primates and birds
Strychnosis nocua	Loganiaceae	R	Food for primates and birds
Ficus sur	Moraceae	R	Food for primates and birds
Uapaca paludosa	Euphorbiaceae	R	Food for primates and birds

responses of fauna along the railway line. Monitoring is also crucial as the construction and operation of SGR may facilitate the spread of invasive plant species. Ascensão and Capinha (2017) found that invasive species can outcompete native vegetation, leading to changes in the composition of plant communities and potentially affecting the wildlife that depends on them.

2.5 Conclusion and Recommendations

2.5.1 Conclusion

The SGR's construction has resulted in a necessary slight reduction of forest cover of about 3% in both forest ecosystems with CO_2 emissions estimated at 35,096 tons of CO_2e yr^{-1}. The project also altered the landscapes leading to increased soil erosion, land degradation, flooding, sedimentation of water bodies, habitat destruction, and impeded wildlife movement in some areas. However, it was learned that implementation of the project clearly observed the EIA and ESIA. Therefore, nearly all necessary mitigation measures have been put in place with close monitoring by environmental experts.

2.5.2 Recommendations

It is recommended that similar studies be conducted in other SGR project sites (Lots) where activities of the project interfere with conservation areas in order to draw necessary lessons for interventions where necessary. Meanwhile, monitoring of mitigation measures that are currently underway for the two studied forests is important and, where necessary, specific studies on species and ecosystems' responses be designed as may be deemed necessary. It is further advised that other infrastructure development projects should undertake all necessary environmental and social assessments and implement the mitigation measures.

References

Alamgir, M., Campbell, M.J., Sloan, S., Goosem, M., Clements, G.R, *et al.* (2017) Economic, socio-political and environmental risks of road development in the tropics. *Current Biology* 27(20), R1130–R1140.

Ambani, M.M. and Mulaku, G.C. (2021) GIS assessment of environmental footprints of the Standard Gauge Railway (SGR) on Nairobi National Park, Kenya. *Journal of Environmental Protection* 12, 694–716.

Ascensão, F. and Capinha, C. (2017) Aliens on the move: Transportation networks and non-native species. In: Borda-de-Água, L., Barrientos, R., Beja, P. and Pereira, H.M. (eds) *Railway Ecology*. Springer Nature, pp. 65–80.

Baker, N.E. and Baker, E.M. (2002) *Important Bird Areas in Tanzania: A First Inventory*. Wildlife Conservation Society of Tanzania, DSM, Tanzania.

BirdLife International (2023) Important bird area factsheet: Kisarawe district coastal forests. Available at: http://datazone.birdlife.org/site/factsheet/kisarawe-district-coastal-forests-ibatanzania on 12/12/2023 (accessed 19 April 2024).

Borda-de-Água, L., Barrientos, R., Beja, P. and Pereira, H.M. (2017) *Railway Ecology*. Springer Nature.

Dimoula, V., Kehagia, F. and Tsakalidis, A. (2016) A holistic approach for estimating carbon emissions of road and rail transport systems. *Aerosol and Air Quality Research* 16(1), 61–68.

Dorsey, B., Olsson, M. and Rew, L.J. (2015) Ecological effects of railways on wildlife. In: Van der Ree, R., Smith, D.J. and Grilo, C. (eds) *Handbook of Road Ecology*. John Wiley & Sons, pp. 219–227.

Fahrig, L. (2003) Effects of habitat fragmentation on Biodiversity. *Annual Review of Ecology, Evolution, and Systematics* 34(1), 487–515.

Frost, M. (2016) Railway Management and Engineering. In: *Proceedings of the Institution of Civil Engineers-Transport*, 4th edn., Vol. 169. Thomas Telford, London.

Gwegime, J., Mwangoka, M., Mulungu, E., Perkin, A. and Nowak, K. (2013) The Biodiversity and forest condition of Ruvu South forest Reserve. *TFCG Technical Paper 37*.

Howell, K.M. (1981) Pugu Forest Reserve: biological values and development. *African Journal of Ecology* 19(1–2), 73–81.

IUCN (2012) *IUCN red list of threatened species*.

John, E., Bunting, P., Hardy, A., Roberts, O., Giliba, R, *et al.* (2020) Modelling the impact of climate change on Tanzanian forests. *Diversity and Distributions* 26(12), 1663–1686.

John, E., Bunting, P., Hardy, A., Silayo, D.A. and Masunga, E. (2021) A forest monitoring system for Tanzania. *Remote Sensing* 13(16), 3081.

Lape, O.A., Mwagha, S.M. and Siljander, M. (2023) Quantifying the environmental impact of Standard gauge railway (SGR) on land cover changes along the Nairobi-Kiambu corridor from 2016 to 2019. *Journal of Geography, Environment and Earth Science International* 27(4), 21–37.

Limi, A., Humphreys, R.M. and Mchomvu, Y.E. (2017) Modal choice between rail and road transportation: Evidence from Tanzania. *World Bank Policy Research Working Paper* 8174.

Lopoukhine, N., Crawhall, N., Dudley, N., Figgis, P., Karibuhoye, C, *et al.* (2012) Protected areas: Providing natural solutions to 21st century challenges. *Sapiens: Surveys and Perspectives Integrating Environment and Society* 5(2).

Lucas, P.S., de Carvalho, R.G. and Grilo, C. (2017) Railway disturbances on wildlife: Types, effects, and mitigation measures. In: Borda-de-Água, L., Barrientos, R., Beja, P. and Pereira, H.M. (eds) *Railway Ecology*. Springer Nature, pp. 81–99.

Mauya, E.W., Mugasha, W.A., Njana, M.A., Zahabu, E. and Malimbwi, R. (2019) Carbon stocks for different land cover types in Mainland Tanzania. *Carbon Balance and Management* 14, 1–12.

McNicol, I.M., Ryan, C.M. and Mitchard, E.T.A.A. (2018) Carbon losses from deforestation and widespread degradation offset by extensive growth in African woodlands. *Nature Communications* 9, 3045.

Milewicz, J., Mokrzan, D. and Szymański, G.M. (2023) Environmental impact evaluation as a key element in ensuring sustainable development of rail transport. *Sustainability* 15(18), 13754.

Monson, J. (2008) Liberating labour? constructing anti-hegemony on the TAZARA railway in Tanzania, 1965–76. *China Returns to Africa: A Rising Power and A Continent Embrace* 197–219.

Mugasha, W.A., Mwakalukwa, E.E., Luoga, E., Malimbwi, R.E., Zahabu, E, *et al.* (2016) Allometric models for estimating tree volume and aboveground Biomass in lowland forests of Tanzania. *International Journal of Forestry Research*.

Munyao, M., Siljander, M., Johansson, T., Makokha, G. and Pellikka, P. (2020) Assessment of human–elephant conflicts in multifunctional landscapes of Taita Taveta County, Kenya. *Global Ecology and Conservation* 24, e01382.

Njana, M.A., Mbilinyi, B. and Eliakimu, Z. (2021) The role of forests in the mitigation of global climate change: Emprical evidence from Tanzania. *Environmental Challenges* 4, 100170.

Nyumba, T.O., Sang, C.C., Olago, D.O., Marchant, R., Waruingi, L. *et al*. (2021) Assessing the ecological impacts of transportation infrastructure development: A reconnaissance study of the standard gauge railway in Kenya. *PLOS ONE* 16(1), e0246248.

Ony, M.A., Nowicki, M., Boggess, S.L., Klingeman, W.E., Zobel, J.M., Trigiano, R.N. and Hadziabdic, D. (2020) Habitat fragmentation influences genetic diversity and differentiation: Fine scale population structure of Cercis canadensis (eastern redbud). *Ecology and Evolution* 10(8), 3655–3670.

Pan, Y., Birdsey, R.A., Fang, J., Houghton, R, *et al*. (2011) A large and persistent carbon sink in the world's forests. *Science* 333, 988–993.

Ract, C.C., Burgess, N., Dinesen, L., Sumbi, P., Malugu, I, *et al*. (2023) Nature forest reserves in Tanzania and their importance for conservation. *bioRxiv*.

Reichart-Burikukiye, C. (2013) The railway in colonial East Africa. *Landscape, Environment and Technology in Colonial and Postcolonial Africa* 6, 62.

Requena Suarez, D., Rozendaal, D.M., De Sy, V., Gibbs, D.A., Harris, N.L, *et al*. (2021) Variation in aboveground Biomass in forests and woodlands in Tanzania along gradients in environmental conditions and human use. *Environmental Research Letters* 16(4), 044014.

Sang, C.C., Olago, D.O., Nyumba, T.O., Marchant, R. and Thorn, J.P. (2022) Assessing the underlying drivers of change over two decades of land use and land cover dynamics along the standard gauge railway corridor, Kenya. *Sustainability* 14(10), 6158.

Shabana, A.A., Zaazaa, K.E. and Sugiyama, H. (2007) *Railroad Vehicle Dynamics: A Computational Approach*. CRC Press, Boca Raton, Florida.

TRC (2018) Prime minister speech during his visit at Soga SGR Camp in Kibaha. Available at: https://www.trc.co.tz/publications/12 (accessed 15 June 2023).

TRC (2019) "*Environmental and Social Impact Assessment (ESIA) Report (Project no.0462176)*" Dar es Salaam–Makutopora, Tanzania, Tanzania Railways Corporation (TRC), Dar es Salaam: Standard Gauge Railway (SGR) Line Project.

Wang, J., Sekei, V.S., Ganiyu, S.A. and Makwetta, J.J. (2021) Research on the sustainability of the standard gauge railway construction project in Tanzania. *Sustainability* 13(9), 5271.

3 Wetlands Status within the SGR Footprint from Dar es Salaam to Makutupora, Singida

Edmond Alavaisha[1*], Heriel Moshi[2], Wilirk Ngalson[2], Jasson John[2] and Joel Nobert[1]

[1]Institute of Resource Assessment, University of Dar es Salaam, Tanzania; [2]College of Natural and Applied Sciences, University of Dar es Salaam, Tanzania

Abstract

Tanzania's standard-gauge railway (SGR) construction project crosses several wetlands, some of which are of international importance. We studied 16 wetlands, in the Pwani and Morogoro regions, to evaluate their biodiversity status despite human modification. Water-quality parameters were within acceptable ranges, but some sites there had elevated levels of turbidity, dissolved oxygen (DO), biological oxygen demand (BOD5), and total coliforms, indicating human impact. Disturbances affected soils and species diversity along the SGR line. Endangered species like *Xylotheca tettensis* and species of conservation value like *Dalbergia melanoxylon* were recorded. Invasive species including *Pistia stratiotes*, *Lantana camara*, *Calotropis procera*, *Xanthium stumarium* and *Ricinus communis* were identified, requiring immediate control. We documented 15 waterbird species, mainly hamerkop and pied kingfisher, 17 amphibian and 11 reptile species, 730 macroinvertebrates, and 1139 individual fish. Although no fauna species are of conservation concern, ongoing monitoring and restoration are essential to mitigate construction impact on wetlands and biodiversity.

3.1 Background Information

Wetlands, covering about 1.5% of the Earth's surface area (Ramsar Convention on Wetlands, 2018), play a crucial role in global biodiversity despite their relatively small area. While most wetlands are natural, only few are man-made (Maltby, 2009). These ecosystems, characterized by their seasonal or permanent inundation with water, create hydric soil that supports plants and animals adapted to wet environments. In tropical, arid and semi-arid regions, wetlands are often seasonal, drying out at least once a year. Even in their dry phases, wetlands remain significant, serving purposes like agriculture and supporting biodiversity not solely dependent on water availability. For instance, about 40% of wetland biodiversity is linked to agricultural and crop production (Zedler and Kercher, 2005).

Wetlands provide abundant food sources that attract diverse flora and fauna. Many species rely on wetlands at various stages of their life cycles. Dead plant leaves and stems break down in the water to form small particles of organic material called 'detritus'. This enriched material feeds many small aquatic insects, shellfish and small fish that are food for larger predatory fish, reptiles, amphibians, birds and mammals. The functions

*Corresponding author: edmondalavaish@yahoo.com; alavaisha.edmond@udsm.ac.tz

DOI: 10.1079/9781800626942.0003

of a wetland and the values of these functions to humans depend on a complex set of relationships between the wetland and the other ecosystems in the watershed. Therefore, maintaining and protecting wetlands improves human welfare.

Landscape geomorphology influences how water moves and is available to biodiversity over or through the soil. Soils host the zone of biogeochemical activity where plants and animals interact with the hydrologic cycle and other elemental cycles. A typical soil contains both mineral and organic materials as well as the adjacent water-filled and air-filled pore space. The physical and chemical properties of a soil may influence the processes that lead to wetland formation and function.

Tanzania is endowed with wetlands; about 10% of the surface area is covered by wetlands, i.e. mainly flood plains and rivers, streams and lakes, that have socio-ecological importance beyond the border (Kamukala, 1993). Most wetlands are associated with endemic and endangered species; thus, they need some conservation and management effort. Development of infrastructures such as roads and railways has been one factor, after agriculture, that affects wetland biodiversity. In Tanzania, as an effort to minimize the impacts and loss of wetlands, the government established sites designated as wetlands of international importance (Materu *et al.*, 2018). These areas are of socio-ecological importance and have an impact beyond their border. For example, they play an essential part in local and global water cycles and are the crucial link between water, food and energy.

Wetlands have been among the most productive areas that provide numerous beneficial services for people and wildlife. Some of these services, or functions, include protecting and improving water quality, providing fish and wildlife habitats, storing floodwaters, and maintaining surface water flow during dry periods. There is increasing recognition of the value and importance of wetland ecosystems over the last century, which led to the creation of laws, regulations and plans to restore and protect wetlands around the world. In Tanzania, wetlands protection largely falls under the NEMC (National Environment Management Council), which requires permits for dredging and filling activities in wetlands and monitors water-quality standards. Initiatives such as the 'no net-loss policy' aim to limit further wetland and biodiversity loss, requiring wetland creation, restoration or mitigation to offset wetland losses due to human activity. All these efforts are designed to protect or conserve wetlands and the ecosystem services they provide. The construction covers various fringed and flooded wetlands, some being seasonal while others are permanent. Among important large wetlands along the SGR footprint are Ruvu, Mkondoa and Mkata rivers. The baseline studies in these wetland ecosystems are important for ensuring compliance with requirements for local and international standards including the Tanzania Environmental Management Act 2004 and the IFC Performance Standards. This chapter therefore aims to present the status of biodiversity along the project footprint to promote the sustainable management of biodiversity resources and to maintain the ecological integrity of the biodiversity. The expected outputs will be number, types and location of wetlands, hyrology, water quality, soils and biota (flora and fauna) within the project areas.

3.2 Materials and Methods

3.2.1 Description of study area

The Tanzanian Railway Corporation (TRC), under the Ministry of Works, Transport and Communications, is implementing the first two phases of an electrified standard-gauge railway (SGR). The SGR is planned between Dar es Salaam and Mwanza via Isaka (1219 km), largely following the alignment of an existing metre-gauge railway originally built in the early 1900s. The first two phases of the SGR cover 541 km in total, between Dar es Salaam and Makutupora, via Dodoma, the country's administrative capital (Fig. 3.1).

This study covered areas between Dar es Salaam and Makutupora (see Table 3.1 for the sampling sites). The survey was conducted between April and May 2021 during the rainy season. Even though sampling was conducted during the rainy season, most areas in Dar es Salaam and Morogoro had less rainy characteristics; Dodoma and Singida regions were dry. Various survey methodology for flora, fauna and wetland environmental variables was conducted along the SGR footprints (within 60 m),

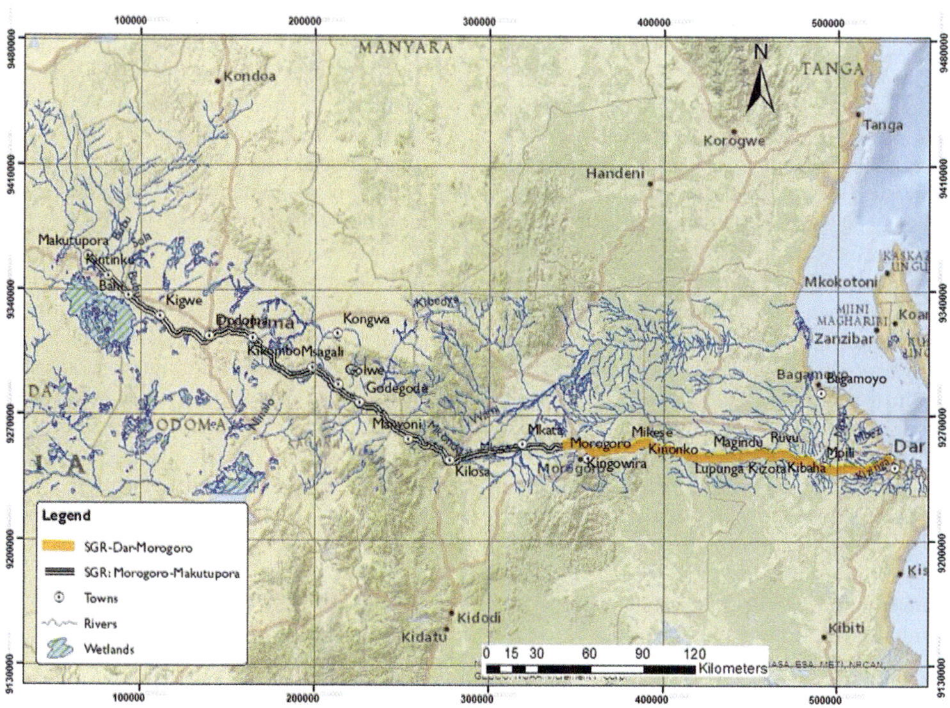

Fig. 3.1. Project location and railway line from Dar es Salaam to Makutupora.

Table 3.1. Study sites (sampling sites) along SGR line from Dar es Salaam to Makutupora via Dodoma.

Site	UTM zone	Easting	Northing	Description	Name
A	37S	522846	9241679	Flood wetland	Vingunguti
B	37S	509162	9240016	Fringe wetland	Msimbazi river
C	37S	495642	9239619	Fringe wetland	Mpiji river
D	37S	490465	9,240519	Fringe wet land	
E	37S	488751	9241307	Flood wet land	
F	37S	495601	9239469	Fringe wet land + stream	Kwamturuki swamp
G	37S	466928	9248698	Fringe wet land	
H	37S	460878	9247296	River	Ruvu river
I	37S	407936	9249221	Fringe	Ngerengere river 1
J	37S	356639	9,252422	River + swamp	Ngerengere river 2
K	37S	318884	9252701	Fringe	Mkata river
L	37S	272915	9250569	River	Mkondoa river
M	37S	246065	9264220	Fringe	Lumuma river
N	36S	804560	9313547	Fringe pond	Korongo idara ya maji
O	36S	793431	9316460	Flood wetland	Zuzu
P	36S	738351	9355673	Fringe	Bubu river

a total length of 541 km from Dar es Salaam to Makutupora, Dodoma region. The initial phase of SGR extends to the middle of the country cutting through a large part of most of the vegetation types occurring in Tanzania except montane forests. The existing vegetation, however, is highly disturbed due to anthropogenic pressure with major disturbances being agriculture and ongoing construction projects.

The rainfall is seasonal with a long rainy season between March and June and short rains between October and December (Rocky and Mligo, 2012) in the segment of Dar es Salaam to Morogoro. The rainfall distribution and vegetation cover varies with altitude, and mean annual temperature is 28°C, and the variation throughout the year is relatively low. The Dodoma segment of SGR, on the other hand, receives little rainfall throughout the year, with the driest months being June through August, whereas January, February and September are the months that receive most rainfall.

3.2.2 Data-collection approaches

In this chapter we used Ramsar criteria to define wetland as a distinct ecosystem that is flooded by water, either permanently or seasonally, where oxygen-free processes prevail. The primary factor that distinguishes wetlands from other landforms or water bodies is the characteristic vegetation of plants adapted to live in aquatic conditions. Rivers were also included despite availability of aquatic flora and hydraulic soil. The wetland status included types and location, hydrology, environmental flow, soils, water quality and macroinvertebrates, fish, flora and fauna (amphibians, reptiles and birds).

3.2.3 Wetland types, hydrology and water quality

Identifying and mapping all wetlands inside project footprint

The methodology used for the assignment comprised the following stages:

I. Catchment delineation from topographical data/DEM was performed for wetland sites and rivers within the project footprint. This involved using a digital elevation model (DEM) in GIS software to derive physical catchment characteristics such as slope, area and longest flow path. The DEM was also crucial for identifying topographic lowlands capable of retaining water through the flow-accumulation process in GIS.

II. Prior to fieldwork, GoogleEarth imagery and Landsat satellite images were reviewed for the project area. Remote-sensing techniques were used to classify and differentiate wetlands from other land covers. Areas with potential standing water, saturation-tolerant vegetation, or dark soils (indicating surface saturation) were identified using GIS and marked for field verification.

3.2.4 Fieldwork wetland mapping

Wetland determination data was prepared for capturing the attributes of the wetlands. Key information that was captured includes:

- location of the wetland (coordinates, watershed, including streams and tributaries in the vicinity of the wetland);
- status of the wetland (natural state, experienced minimal or minor disturbance or substantially disturbed);
- surrounding land use and habitat types (dominant species, water source, topography, adjacent vegetation types and land uses);
- vegetation characteristics;
- climate/hydrology of the sites; and
- type of the wetland (whether fringe or flood wetlands).

3.2.5 Hydrological data

Hydrological data were collected from the Wami-Ruvu Basin Water Board (WRBWB) and also from the Water Resources Engineering Department database. The data collected included flow data for all rivers crossing the SGR line and/or hydrologically connected to the wetlands and rainfall distribution in the vicinity of the wetlands. For the rivers which are not gauged, spot measurements of the flow were conducted. This entailed sampling, and analyzing the water quality in the field and laboratory using appropriate methods. Physico-chemical characterization of water samples from the wetland areas was conducted to assess the quality of water and to find out its pollution level and establish a method for monitoring.

Along the wetland (river/stream/swamp/flood plain) the transect was set to cover all micro-habitats and water samples were collected in replicates at mid-channel approximately 0.5 m depth in HCL acid-washed polythene bottles. The sampling also considered the morphology (depth and size) of the water body.

3.2.6 Water quality

Temperature, conductivity, pH, turbidity and dissolved oxygen (DO) were measured *in situ* using portable multi-parameter (HANNA, HI-9828, USA). At each point of *in situ* measurement, water samples were collected for the analysis of biological oxygen demand (BOD5), total coliform, nitrates-nitrogen (NO_3-N), ammonia-nitrogen (NH_3-N) and phosphates (PO_4). The analysis of water quality was done as per the standard methods described by APHA (1998) and Tanzania Bureau of Standards (2017).

3.2.7 Soil

Soil characteristics related to land-use change were assessed. Soil samples were collected using a soil auger from four random points within a plot, mixed to get a composite sample to avoid spatial heterogeneity. Additionally, one soil core per plot was taken for bulk density measurements. The composite samples were air-dried and sieved through 2 mm sieve and analyzed for soil texture, moisture, bulk density, pH and nutrients (TN, TP and SOC) in response to land-use change and habitat modification. SOC concentrations were analyzed by the Walkley-Black procedures (Klute *et al.*, 1986), whereby potassium dichromate ($K_2Cr_2O_2$) and concentrated sulphuric acid (H_2SO_4) were used, N concentrations were done following the Kjeldahl acid-digestion procedures (Klute *et al.*, 1986; Van Reeuwijk, 2002), and P determined through perchloric acid digestion and then measured calorimetry (Van Reeuwijk, 2002).

3.2.8 Flora

After the reconnaissance survey, a vegetation survey was conducted to classify the plant community types surrounding the wetlands affected by the SGR project. These types were categorized into broad groups like thickets, bushland, woodland and cultivations, based on their floristic composition and physiognomy. Sampling plots were then strategically distributed to capture a diverse range of communities, first by floristic types and then randomized. Locations typical of the sampled communities were identified based on expert judgement in line with Morris and Therivel (2001).

3.2.9 Identification of plant species of conservation significance

Species encountered in the field were initially identified by a botanist from the University of Dar es Salaam (UDSM). Those difficult to identify on site were collected, pressed, transported and compared with herbarium specimens at UDSM's Department of Botany, referencing checklists and flora like the Flora of Tropical East Africa (FTEA) and Flora Zambesiaca (FZ). Species names followed the nomenclature of Turrill and Milne-Redhead (1952). Threatened plant species were noted, and those within the SGR project area were assessed against the Convention on International Trade in Endangered Species of Wild Fauna and Flora (CITES), IUCN Red List, and LEAP Master-1996, for conservation status.

3.2.10 Invasive species

All invasive species within the sampled plant communities were identified using the available manuals, including *A Guide to Naturalized and Invasive Plants of Eastern Africa* (Witt and Luke, 2017) and a manual on management and control of invasive plant species in Tanzania.

Management and control measures for each species were obtained from the existing literature and best practices from other regions.

3.2.11 Fauna

Wetlands have a wide coverage of fauna that include vertebrates and invertebrates. The key fauna species of wetlands that were sampled during this mission included birds, amphibians, reptiles, macroinvertebrates and fish. Their conservation statuses were established based on IUCN Red List and other databases that are specific to fauna species.

3.2.12 Birds

A total count method was used to count all waterbirds at 16 identified locations/wetlands (A to P) along the SGR (within 60 m). The field team spent an average of 85 min at each wetland (maximum 163 min, minimum 26 min) counting birds and indices of breeding. However, the team spent more time for large wetlands and those difficult to navigate. All observations were conducted between the hours of 09:10 and 18:16. Bird identification was aided by binoculars and field guidebooks, following the naming conventions of BirdLife International (2021a).

3.2.13 Herpetofauna (amphibians and reptiles)

Several standard herpetofauna sampling techniques exist and have been successful in gathering baseline data on the diversity and abundance of herps. Due to differences in their lifestyles, no single method is sufficient for sampling all herpetofauna communities. Some species are arboreal (e.g. tree frogs, chameleons, geckoes), while others are terrestrial or aquatic, spending most or all of their lives in water (e.g. certain frogs, lizards, crocodiles). Sampling methods range from live-trapping, cover boards, quadrats, and transects to time-constrained or opportunistic searches (Grant *et al.*, 1992; Heyer *et al.*, 1994).

The herpetofauna in the project area were sampled during both day (0800–1800) and night (1900–0000). A time-constrained audio-visual survey method (Grant *et al.*, 1992; Heyer *et al.*, 1994; Doan, 2003) was chosen for its efficiency in detecting herps with minimal effort and its ability to provide comparable results over time. The selected wetlands within 60m along the SGR project site were surveyed by walking around while searching for herpetofauna in all available habitats, including vegetation, leaf litter, along wetlands and in rocks, by a pair of observers. The start and end times for each site were recorded. All encountered or heard herpetofauna species were identified, counted and recorded using standard field guides (Channing and Howell, 2006; Pickersgill, 2007; Harper *et al.*, 2010).

3.2.14 Macroinvertebrates and fish

Macroinvertebrates were collected along delineated water bodies within the project area using a modified kick-net method (using hand net mini-SASS with 1 mm mesh size) as described by Gerber and Gabriel (2002).The net was swept through vegetation, stony and gravel + sand + mud (GSM) biotope in opposing directions for 2 minutes to ensure equivalent sampling effort and to cover spatial variation. Obtained samples were sorted and preserved in 70% alcohol. Laboratory identification was performed up to the lowest taxonomic level possible using the macroinvertebrate guidebook by Gerber and Gabriel (2002).

A time-limited method of fishing using a net of 6 mm, based on FAME protocol was used to sample fish (Fig. 3.2). Most of the sampled fish were identified, counted and released *in situ*. Some fish that needed to be identified further were fixed in 70% alcohol and transferred to the laboratory. Identification of fish was done with the aid of *Field Guide to the Freshwater Fishes of Tanzania* (Eccles, 1992).

3.2.15 Assumptions and limitation of the study

This survey was conducted within a buffer of ~60 m along the SGR line from Dar es Salaam

Fig. 3.2. Sampling for fish species in Msimbazi river.

to Makutupora, assuming most of the impacts occurred within the boundary. However, some major impacts outside the boundary were pointed out. The assumption of boundary was important to align with the NEMC guideline but also to reduce the cost of the survey. Certain wetland habitats were excluded from this mapping because of the seasonal limitations. For instance, Dodoma was in the dry season, and most parts of Dar es Salaam were almost dry despite been the rainy season, while Dar es Salaam, Pwani and Morogoro regions receive bimodal rainfall and were experiencing the wet season. Consequently, in some areas it was challenging to clearly discern wetlands from seasonally and temporarily flooded areas. To overcome this challenge, surveyed sites were delineated as wetlands after confirming presence and richness of indicator species characteristic of the wetland. It is however important to note, time-constraint methods were used for fauna sampling to accommodate the short time of survey, possibly exclude some species that require more time to capture, including amphibians and reptiles.

3.3 Results and Discussion

A total of 16 wetlands of different statuses were identified during the survey from Dar es Salaam to Makutupora. Fig. A3.1 shows the SGR line from Dar es Salaam to Makutupora and the wetland areas it crosses. Characteristics of the wetlands with biodiversity potential along the line are as shown in Table A3.1. Fringe and flood wetlands were mainly located in Morogoro, e.g. along Mkondoa river and Kilosa. Other flood wetlands were observed around Zuzu and Kigwe in Dodoma. All wetlands were modified and had different levels of disturbances reflecting the intensity of human activities (e.g. construction) by either railway or service road constructions (Table A3.1). Despite modification and disturbances, wetland areas within the SGR footprint indicated potential habitat to support biodiversity. It was further noted that clearing of land, levelling and embarkment are likely to impact hydrology, soil and biodiversity integrity of the wetlands. Figure A3.1 and Tables A3.1-A3.9 available from https://www.cabidigitallibrary.org/doi/book/10.1079/9781800626942.0000.

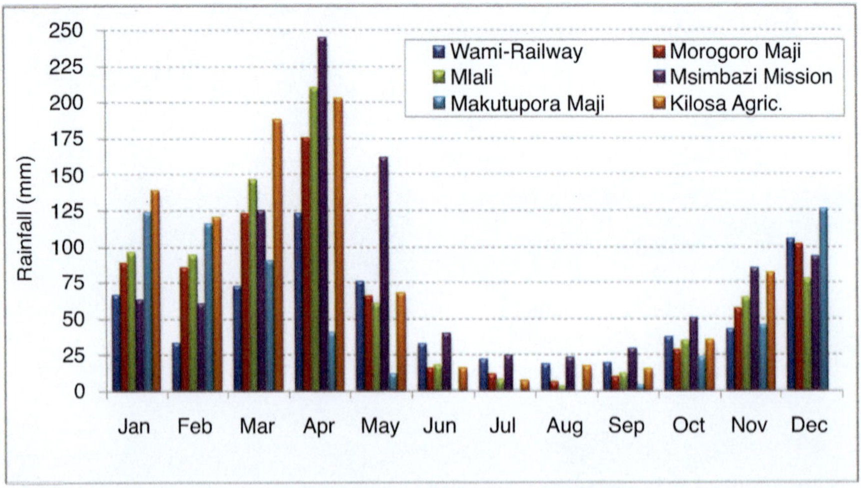

Fig. 3.3. Mean monthly rainfall for the selected stations in the project area.

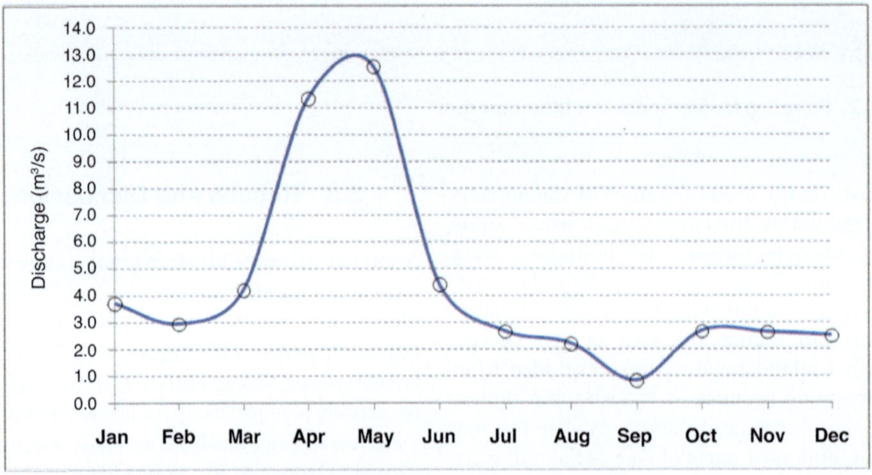

Fig. 3.4. Ngerengere river at Kingolwira – mean monthly flows (2010–2019).

Historical analysis of seasonal rainfall patterns in the project area indicates that the primary rainy season is March–May (MAM), the dry season is June–September (JJAS), short rains are in October–December (OND) and the intermediate season from January–February (JF) (Fig. 3.3). Figures 3.4–3.6 show the seasonal variability of flow for the gauged rivers at different SGR crossings.

3.3.1 Hydrological characteristics of the wetlands along the SGR route

The characterization involved identifying the location (coordinates) for GIS mapping and also the status and type of wetland, whether hydrologically connected to rivers, i.e. fringe wetlands, or flood-type (hydrologically disconnected from the river). Most of the wetlands

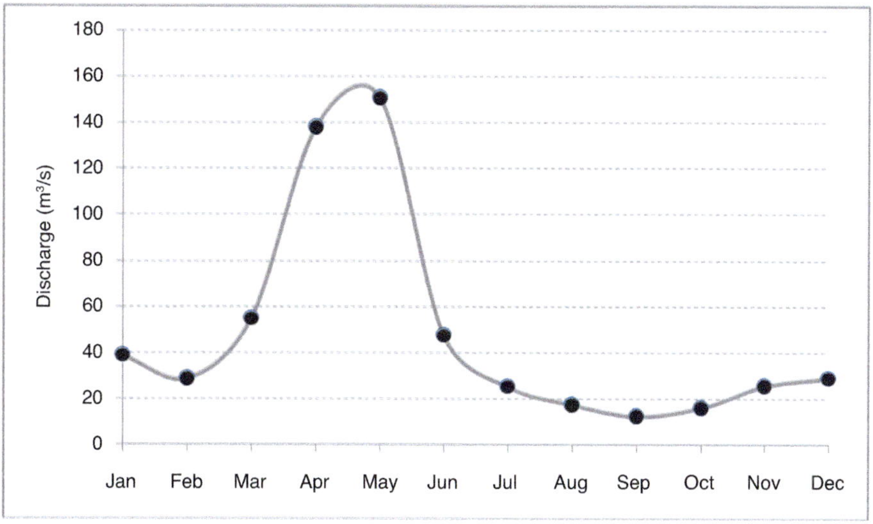

Fig. 3.5. Ruvu river at Morogoro bridge – mean monthly flows (2005–2019).

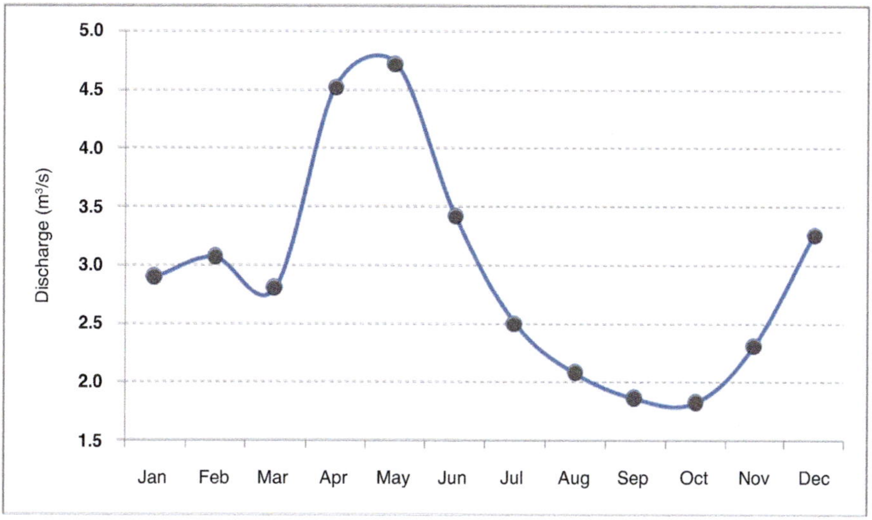

Fig. 3.6. Mkondoa river at Mbarahwe – mean monthly flows.

along the route were found to be fringe wetlands (hydrologically connected to rivers). The details of the wetlands in terms of location, status and type are presented in Table A3.2.

3.3.2 Water quality along the SGR line

The water quality varied among mapped wetland along the SGR line (Table 3.2). In general, most

Table 3.2. Water-quality parameters for the wetlands along the SGR route.

	Coordinates							
Site	Easting	Northing	Name of place	Parameters	U/S	D/S	Standards (TBS)	Remarks
A	522846	9241679	Vingunguti	pH	6.51	6.52	6.5–8.5	Satisfactory
				EC [µs/cm]	1354	1351	< 2000	Satisfactory
				Temp [°C]	27.00	26.93	20–35	Satisfactory
				DO [mg/l]	2.9	2.9		Satisfactory
				Turbidity [NTU]	135	129	5–25	Unsatisfactory
				Nitrate Nitrogen (NO_3-N) [mg/l]		4.8	10–75	Satisfactory
				Ammonia Nitrogen (NH_3-N) [mg/l]		1.90	< 1	Satisfactory
				Phosphate (PO_4) [mg/l]		0.14	< 2.3	Satisfactory
				BOD5[mg/l]		20	< 10	Unsatisfactory
				Total coliform [No./100mls]	5	2	< 10	Satisfactory
B	509162	9240016	Msimbazi	pH	6.9	6.87		Satisfactory
				EC [µs/cm]	1834	1901		Satisfactory
				Temp [°C]]	25.21	26.0		Satisfactory
				DO [mg/l]	2.4	2.8		Satisfactory
				Turbidity [NTU]	72	78		Unsatisfactory
				Nitrate Nitrogen (NO_3-N) [mg/l]		5		Satisfactory
				Ammonia Nitrogen (NH_3-N) [mg/l]		0.23		Satisfactory
				Phosphate (PO_4) [mg/l]		0.43		Satisfactory
				BOD5[mg/l]		10		Unsatisfactory
				Total coliform [No./100mls]	4	2		Satisfactory

Continued

Table 3.2. Continued

Site	Coordinates		Name of place	Parameters	U/S	D/S	Standards (TBS)	Remarks
	Easting	Northing						
C	495642	9239752	Mpiji	pH	6.78	6.74		Satisfactory
				EC [µs/cm]	485	504		Satisfactory
				Temp [°C]	26.7	27.0		Satisfactory
				DO [Mg/l]	1.4	1.4		Satisfactory
				Turbidity [NTU]	70.3	82.1		Unsatisfactory
				Nitrate Nitrogen (NO_3-N) [mg/l]		11.9		Satisfactory
				Ammonia Nitrogen (NH3-N) [mg/l]		0.77		Satisfactory
				Phosphate (PO_4) [mg/l]		0.01		Satisfactory
				BOD5[mg/l]		30		Unsatisfactory
				Total coliform [No./100mls]	12	8		Satisfactory
E	488751	9241307		pH	6.79	6.75		Satisfactory
				EC [µs/cm]	395	377		Satisfactory
				Temp [°C]	27.22	27.47		Satisfactory
				DO [Mg/l]	1.3	1.4		Satisfactory
				Turbidity [NTU]	82.1	87		Unsatisfactory
				Nitrate Nitrogen (NO_3-N) [mg/l]		2.3		Satisfactory
				Ammonia Nitrogen (NH_3-N) [mg/l]		0.97		Satisfactory
				Phosphate (PO_4) [Mg/l]		0.01		Satisfactory
				BOD5[mg/l]		30		Unsatisfactory
				Total coliform [No./100mls]	16	12		Unsatisfactory

Continued

Table 3.2. Continued

Site	Coordinates		Name of place	Parameters	U/S	D/S	Standards (TBS)	Remarks
	Easting	Northing						
G	466928	9248698		pH	7.04	7.53		Satisfactory
				EC [µs/cm]	2480	2445		Unsatisfactory
				Temp [°C]	33.32	32.64		Satisfactory
				DO [Mg/l]	1.2	1.2		Satisfactory
				Turbidity [NTU]	56	26.8		Unsatisfactory
				Nitrate Nitrogen (NO_3-N) [Mg/l]		3.2		Satisfactory
				Ammonia Nitrogen (NH_3-N) [Mg/l]		0.09		Satisfactory
				Phosphate (PO_4) [Mg/l]		0.01		Satisfactory
				BOD5[Mg/l]		10		Unsatisfactory
				Total coliform [No./100mls]	2	2		Satisfactory
H	460878	9247296	Ruvu	pH	8.21	8.31		Satisfactory
				EC [µs/cm]	508	503		Satisfactory
				Temp [°C]	28.43	28.46		Satisfactory
				DO [Mg/l]	1.4	1.4		Satisfactory
				Turbidity [NTU]	102	65		Unsatisfactory
				Nitrate Nitrogen (NO_3-N) [Mg/l]		25		Satisfactory
				Ammonia Nitrogen (NH_3-N) [Mg/l]		1.83		Satisfactory
				Phosphate (PO_4) [Mg/l]		2.18		Satisfactory
				BOD5[Mg/l]		20		Unsatisfactory
				Total coliform [No./100mls]	5	2		Satisfactory

Continued

Table 3.2. Continued

Site	Coordinates		Name of place	Parameters	U/S	D/S	Standards (TBS)	Remarks
	Easting	Northing						
I	407936	9247296	Ngerengere	pH	7.3	7.31		Satisfactory
				EC [µs/cm]	583	583		Satisfactory
				Temp [°C]	27.96	27.96		Satisfactory
				DO [Mg/l]	1.2	1.2		Satisfactory
				Turbidity [NTU]	223	220		Unsatisfactory
				Nitrate Nitrogen (NO_3-N) [mg/l]		11.1		Satisfactory
				Ammonia Nitrogen (NH_3-N) [mg/l]		1.38		Satisfactory
				Phosphate (PO_4) [mg/l]		1.21		Satisfactory
				BOD5[mg/l]		10		Unsatisfactory
				Total coliform [No./100mls]	13	9		Satisfactory
J	356639	9252422	Ngerengere	pH	7.66	7.64		Satisfactory
				EC [µs/cm]	494	493		Satisfactory
				Temp [°C]	25.06	25.19		Satisfactory
				DO [mg/l]	1.3	1.4		Satisfactory
				Turbidity [NTU]	48.7	50.2		Unsatisfactory
				Nitrate Nitrogen (NO_3-N) [mg/l]		1.9		Satisfactory
				Ammonia Nitrogen (NH3-N) [mg/l]		0.18		Satisfactory
				Phosphate (PO_4) [vg/l]		1.23		Satisfactory
				BOD5[mg/l]		30		Unsatisfactory
				Total coliform [No./100mls]	3	2		Satisfactory

Continued

Table 3.2. Continued

Site	Coordinates		Name of place	Parameters	U/S	D/S	Standards (TBS)	Remarks
	Easting	Northing						
K	318884	9252701	Mkata	PH	7.97	7.96		Satisfactory
				EC [μs/cm]	554	562		Satisfactory
				Temp [°C]	26.95	26.90		Satisfactory
				DO [mg/l]	3.1	3.2		Satisfactory
				Turbidity [NTU]	113	112		Unsatisfactory
				Nitrate Nitrogen (NO_3-N) [mg/l]		125		Unsatisfactory
				Ammonia Nitrogen (NH_3-N) [mg/l]		76		Satisfactory
				Phosphate (PO_4) [mg/l]		2.22		Satisfactory
				BOD5[mg/l]		20		Unsatisfactory
				Total coliform [No./100mls]	2	1		Satisfactory
L	272915	9250569	Mkondoa	pH	8.16	8.15		Satisfactory
				EC [μs/cm]	723	718		Satisfactory
				Temp [°C]	31.16	30.99		Satisfactory
				DO [mg/l]	1.3	1.3		Satisfactory
				Turbidity [NTU]	312	318		Unsatisfactory
				Nitrate Nitrogen (NO_3-N) [mg/l]		2.3		Satisfactory
				Ammonia Nitrogen (NH_3-N) [mg/l]		0.87		Satisfactory
				Phosphate (PO_4) [mg/l]		2.3		Satisfactory
				BOD5[mg/l]		20		Unsatisfactory
				Total coliform [No./100mls]	7	5		Satisfactory

Continued

Table 3.2. Continued

Site	Coordinates		Name of place	Parameters	U/S	D/S	Standards (TBS)	Remarks
	Easting	Northing						
M	246065	9264220	Lumuma	pH	8.35	8.35		Satisfactory
				EC [µs/cm]	572	570		Satisfactory
				Temp [°C]	28.69	28.70		Satisfactory
				DO [mg/l]	1.3	1.3		Satisfactory
				Turbidity [NTU]	394	401		Unsatisfactory
				Nitrate Nitrogen (NO_3-N) [mg/l]		15.1		Satisfactory
				Ammonia Nitrogen (NH_{3-}N) [mg/l]		0.26		Satisfactory
				Phosphate (PO_4) [mg/l]		0.01		Satisfactory
				BOD5[mg/l]		20		Unsatisfactory
				Total coliform [No./100mls]	17	13		Satisfactory
N	804560	9313547	Pond Water Dept/Mkonze	pH	7.62	7.63		Satisfactory
				EC [µs/cm]	1568	1577		Satisfactory
				Temp [°C]	23.06	23.03		Satisfactory
				DO [mg/l]	4.7	1.5		Satisfactory
				Turbidity [NTU]	2.0	1.8		Unsatisfactory
				Nitrate Nitrogen (NO_3-N) [mg/l]		0.2		Satisfactory
				Ammonia Nitrogen (NH_3-N) [mg/l]		0.06		Satisfactory
				Phosphate (PO_4) [mg/l]		0.07		Satisfactory
				BOD5[mg/l]		30		Unsatisfactory
				Total coliform [No./100mls]	8	3		Unsatisfactory

Continued

Table 3.2. Continued

Site	Coordinates		Name of place	Parameters	U/S	D/S	Standards (TBS)	Remarks
	Easting	Northing						
O	793431	9316460	Zuzu	pH	7.5	7.49		Satisfactory
				EC [µs/cm]	335	335		Satisfactory
				Temp [°C]	26.54	26.52		Satisfactory
				DO [mg/l]	2.3	2.3		Satisfactory
				Turbidity [NTU]	342	340		Satisfactory
				Nitrate Nitrogen (NO_3-N) [mg/l]		11.7		Satisfactory
				Ammonia Nitrogen (NH_3-N) [mg/l]		0.74		Satisfactory
				Phosphate (PO_4) [mg/l]		2.23		Satisfactory
				BOD5[mg/l]		20		Unsatisfactory
				Total coliform [No./100mls]	1	1		Satisfactory
P	738351	9355673	Bubu river	pH	8.54	8.55		Satisfactory
				EC [µs/cm]	5616	5609		Satisfactory
				Temp [°C]	27.72	27.75		Satisfactory
				DO [mg/l]	1.3	1.3		Satisfactory
				Turbidity [NTU]	109	117		Unsatisfactory
				Nitrate Nitrogen (NO_3-N) [mg/l]		13.3		Satisfactory
				Ammonia Nitrogen (NH_3-N) [mg/l]		1.03		Satisfactory
				Phosphate (PO_4) [mg/l]		0.62		Satisfactory
				BOD5[mg/l]		30		Unsatisfactory
				Total coliform [No./100mls]	17	15		Unsatisfactory

Note: U/S, upstream (water measured before crossing the SGR railway); D/S, downstream (water measured after crossing the SGR railway). All measurements were taken within the SGR footprint (within 60 m).

sampling sites were disturbed. However, both physical and chemical variables were within acceptable ranges of water quality for ecology as well as for aquatic environmental health and slightly above drinking-water standards (Tanzania Bureau of Standards (TBS), 2017; WHO, 2017). Human activities such as livestock keeping and railway construction were assumed to contribute to nutrient loading and increased water turbidity. It is difficult to disentangle all the human activities in the landscape to isolate the direct role of agriculture in water quality in multi-use regions like wetlands in Morogoro. Still, the effect of railway construction that is associated with high turbidity less dissolved oxygen can be defined as the major human-induced factor along the route of SGR.

3.3.3 Physical chemical characteristics of soils

The bulk density is the mass of dry soil per total volume, g/cm^3 (Table 3.3), and it is inversely related to the soil porous characteristics that indicate ability of soil to retain water. In general, a sandy soil indicated higher bulk density and sand a slightly lower bulk density. The pores within a sandy soil tend to be larger, while the pores in a clay soil are smaller with a more well-developed structure, lower bulk density, and larger pores.

Water relations of the soils differ regarding infiltration and percolation. For instance, all sites except site H (Ruvu) and sites O and P, in Dodoma and Singida, respectively, had clay

Table 3.3. Physical and chemical properties of soil related to land-use change and ecosystem functioning.

		Particle Size Distribution			Texture	Moisture	Bulk density	Soil Oorganic carbon	Total Nitrogen	Total Phosphorus
Site	pH	% clay	% silt	% sand	Class	%	g/cm^3	g/kg	g/kg	g/kg
A	5.5	24.12	1.64	74.24	Sandy clay loam	29.39	1.26	3.88	1.61	0.25
B	8.6	20.12	1.64	78.24	Sandy clay loam	13.32	1.67	3.30	0.84	0.16
C	8.0	16.12	1.64	82.24	Sandy loam	21.25	1.65	2.72	0.39	0.31
E	7.2	16.12	1.64	82.24	Sandy loam	21.25	1.73	2.52	0.60	0.49
H	7.6	44.12	13.64	42.24	Clay	22.22	1.51	9.12	0.74	0.25
I	7.8	16.12	3.64	80.24	Sandy loam	34.31	1.33	8.54	1.19	0.26
J	9.2	20.12	3.64	76.24	Sandy clay loam	42.24	1.27	7.76	1.26	0.15
K	8.8	44.12	7.64	48.24	Sandy clay	30.09	1.33	4.66	0.67	0.15
L	8.6	14.12	3.64	82.24	Sandy loam	17.71	1.81	39.58	0.63	0.45
M	8.2	38.12	7.64	54.24	Sandy clay	25.63	1.60	5.63	0.74	0.24
N	7.3	20.12	1.64	78.24	Sandy clay loam	20.47	1.46	8.34	0.74	0.24
O	5.0	74.48	10.00	15.52	Clay	18.10	1.70	9.70	1.30	0.35
P	7.2	40.12	15.64	44.24	Clay	25.08	1.46	5.43	0.77	0.21

soils with relatively small sand and silt contents. Thus, they tend to have poor infiltration and low percolation rates, and high water-storage capacity. Conversely, courser soils tend to have high rates of infiltration and percolation and good aeration, mainly because they have large pores that readily transmit water.

The bulk densities were above 1.3 for most sites (Table 3.3). Disturbed soils tend to have a high bulk density because of continuous human activities and removal of vegetation. The bulk density for topsoil ranges between 1.0 and 1.3 g/cm^3 (Kirkham, 2015); high bulk density creates small soil pores, which, despite holding water, may limit root growth.

Along with soil texture, the other prominent property that influences soil properties is pH. Soil acidity mainly influences the solubility of various elements in the soil, particularly some plant nutrients that influence productivity of wetlands. For instance, at low pH values (less than 5.8) the availability of nutrients, such as phosphorus and nitrogen, may be limited. Microbial activity is also reduced when soil acidity is high. At high pH values, above 7.5, the availability of phosphorus is also limited. Wetlands that are associated with water saturation (high moisture (see Table 3.3)) have increased organic matter content. The prolonged saturated and anaerobic conditions in wetland soils slow organic matter decomposition and lead to organic matter and nitrogen accumulation (Moyano *et al.*, 2013).

3.3.4 Flora sampling intensity and geographical location of sampled plots

The number of sampled plots for each project site during the analysis and their corresponding geographical location is summarized in Table 3.4. A total of 20 wetland sampling sites likely to be impacted by the SGR project were established, from which 36 plots were laid for collection of data on species composition. Based on the established plots, the total wetland area covered during the survey corresponds to an area of 1.44 ha, an area from which tree species were sampled. Shrub and herbaceous species, on the other hand, were both sampled from 108 nested plots in which shrubs contributed to 0.27 ha, whereas the herbaceous layer (grasses, herbs, forbs) was equivalent to 0.011 ha.

3.3.5 Status of the surveyed wetlands

During the survey, it was established that most wetlands were highly disturbed. The disturbance was mainly anthropogenic, mostly related to the ongoing SGR activities. Fig. 3.7A summarizes the status of various wetlands along the SGR alignment, 68% of which were highly disturbed. About 18% were moderately disturbed, whereas 15% were less disturbed. The less-disturbed wetland plots were 2, 11 and 13 while wetland plots 10, 12, 16, 17 and 18 were moderately disturbed and the remaining thus highly disturbed (Table A3.2).

The surrounding vegetation types (Table A3.2) showed that cultivations accounted for about 37% of the vegetation, characterized by a remnant of indigenous and exotic trees. The remaining 63% consisted of disturbed natural vegetation, divided into seven plant community fractions (Fig. 3.7B). These included bushlands with emergent trees, woodland, thickets, bushed grassland, riverine forest, wooded bushland, and bushed thickets. The woodlands were dominated by *Brachystegia*, *Combretum* and *Acacia* species, which covered large parts of Morogoro and Pwani regions while in Dodoma, *Acacia*, *Delonix alata* and *Adansonia* species were prevalent. Most bushland was dominated by *Combretuma obovata*, *Combretum mosambensis*, *Harrisonia* and *Dicrostachys* species. Coastal thickets were dominated by *Harrisonia*, *Dicrosthys*, *Capparis tomentosa* and *combretum* species. In addition, the riverine forest was characterized by *Ficus sur*, *Ficus cycomorus*, *Albizia*, *Acacia seyal*, *Syzygium cordatum* and *Kigelia africana*.

3.3.6 Physiognomic and floristic vegetation types characteristic of the wetlands

The major physiognomic category in the surveyed wetland plots was permanent swampy wetland. This was apportioned into eight physiognomic types (the number of each is shown): swampy wetland (18), swampy

Table 3.4. Geographical location of sampled plots for each wetland site.

Plot no.	UTM zone	Easting	Northing	Site	Altitude (m)
1	36M	779369	9320898		976
2A	36M	777146	9323317	O	943
2B	36M	777177	9323445	O	941
3	36M	774697	9325272		921
4A	36M	756284	9338508		834
4B	36M	756450	9338418		835
5	36M	748446	9345991		835
6A	36M	738507	9355375	P	845
6B	36M	738859	9355138	P	840
6C	36M	740896	9353354	P	844
7A	36M	819785	9316323	N	1122
7B	36M	819974	9316225	N	1120
7C	36M	820141	9316119	N	1115
8	37M	215475	9285510		782
9A	36M	251275	9260995		623
9B	36M	251375	9260932		620
10A	37M	246067	9264205	M	637
10B	37M	246101	9264310	M	675
11A	37M	276457	9244367	L	498
11B	37M	276466	9244499	L	502
12	37M	318870	9252653	K	403
13A	37M	356625	9252436	J	469
13B	37M	356742	9,252398	J	463
14	37M	359984	9,248975		470
15A	37M	405043	9250377		145
15B	37M	405112	9250463		191
16A	37M	407916	407916	I	172
16B	37M	407949	9249306	I	177
17A	37M	452238	9248080		68
17B	37M	452216	9248246		64
18A	37M	460752	9247245	H	28
18B	37M	460900	9247314	H	27
19A	37M	488679	9241244	E	164
19B	37M	488677	9241240	E	159
20A	37M	495350	9239676	C	165
20B	37M	495603	9239479	F	154

wooded grassland (4), bushed wetland (3), wetland cultivations (3), bushed wetland (3), bushed swampy grassland (2), riverine forest (2), bushed wetland with emergent trees (1). A few seasonal wetlands were also recorded. Seasonal wetlands were more evident where

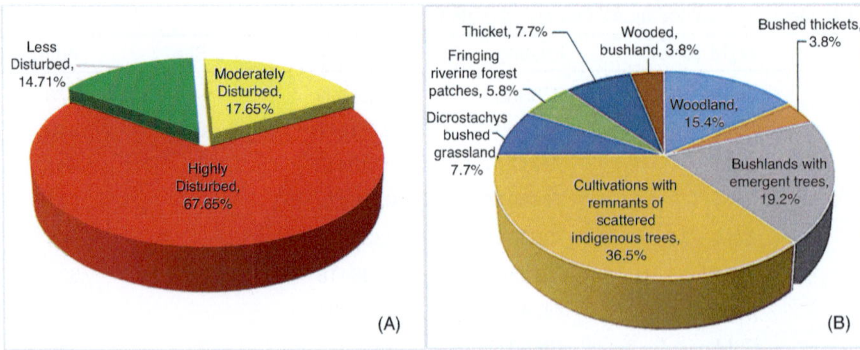

Fig. 3.7. (A) The status of disturbance in the surveyed wetlands; (B) the vegetation types surrounding the wetlands along the SGR alignment.

Fig. 3.8. Inundated wetlands formed as a result of uplifting on the river course across the SGR service road and railway alignment: (A) a gradually developing wetland; (B) flooded bushed woodlands; (C) flooded mango farmland; (D) dry *Tectona grandis* woodlots.

the construction of the SGR alignment and the associated service road caused uplifting on the facilities side. Such uplifting created embankments blocking the water flow in the rivers, which used to cross the SGR alignment.

In some areas, rivers were not completely blocked but their water flow was diverted through installed culverts that were unable to handle the volume, causing a backwash during the wet season. Sediments and debris, like fallen tree branches and logs, clogged or partially blocked the culverts, hindering the flow. These wetlands were dominated by terrestrial plant communities, with wetland indicator species like *Cyperus species*, *Typha capensis* and *Polygonum senegalense* slowly developing but still at low abundance and diversity compared to permanent wetlands. In these wetlands, tree species on the river banks and adjacent areas had withered, and most species, including trees and shrubs, had dried up due to prolonged waterlogging (Fig. 3.8). Tree species which were recorded to have died were *Dalbergia melanoxylon*, *Kigelia africana*, *Elaise guinense*, *Mangifera indica*, *Tectona grandis*, *Cocos nucifera*, *Combretum molle*, *Combretum collinum*, *Flueggia virosa*, *Acacia usambarense*, *Dicrostachys cinerea*, *Lonchocarpus bussei*, *Milletia dura*, *Catunaregum spinosa*, *Brachytegia bohemii*, *Brachystegia longifolia*, *Rytigynia* sp., *Bridelia carthatica*, *Piliostigima thoningii* and *Calatropis procera*. These artificially inundated wetlands were found more between the Ruvu and Ngerengere site and two wetlands between Dodoma and Makutupora.

A thorough analysis of the inundated wetlands formed revealed that plant communities affected in these inundated wetlands were well represented in the nearby vegetation communities. In addition, the area flooded as a result of the backwash in such wetlands was small and the flooding was seasonal. At 183 km, the partly clogged river water pushed back to the cultivations of rice, maize and banana farms, flooding a considerable part of the land which was under cultivation and leading to plant suffocation and death.

Based on floristic composition, vegetation of the surveyed wetland plots was diverse, having 16 vegetation types. The dominant floristic vegetation types and their plot numbers, in brackets, in descending order were *Cyperus* wetland (6), *Phragmites* wetland (6), *Hygrophila-Cyperus* wetland (3), *Typha-Commelina* wetland (3), *Phragmites-Pistia-Pennisetum* wetland (2) *Pluchea-Mimosa-Echnochloa* wetland (2), *Scleria-Hygrophila-Leersia* wetland (2), *Scripus-Tamarix-Typha* wetland (2), *Albizia-Ficus* riverine forest (1), *Cyperus-Leersia* wetland (1), *Cyperus-Phragmites* wetland (1), *Ipomoea-Hygrophila* wetland (1), *Oryza* cultivations (1), *Scripus-Tamarix* wetland (1), swampy *Brachystegia* woodland (1) and *Tamarix* wetland (1). As reported for the surrounding vegetation types, the wetland plant communities were highly disturbed. In some cases the entire wetland was already converted into rice cultivation and had lost all the characteristic features of a wetland including the indicator species and the characteristic spongy nature. Moreover, in some of the semi-dry areas near Makutupora, some farmers created artificial wetlands by draining water from the nearby rivers through constructed channels to irrigate paddy fields in the vicinity. These farmlands presented most features of a wetland but were dominated by crops, or weeds where they were abandoned.

3.3.7 General overview on the tree species richness

The total number of tree species identified in the present survey was 76 belonging to 28 families. The largest families were Caesalpinaceae and Mimosaceae, which had ten species each, followed by Papilionaceae and Combretaceae, in which five and four species were recorded, respectively. The predominant species in the woody layer were under genus *Albizia*, *Acacia* and *Cassia*. The families Euphorbiaceae (4) Tiliaceae and Verbenaceae were under-represented, each having three species, except in the former family where four species were recorded. Family Moraceae had three species while Bignoniaceae, Meliaceae, Palmae and Salvarodaceae had two species each. The remaining 15 families were represented by one tree species each (Table A3.3).

The observed number of tree species is low compared to the sampled areas as most wetlands were previously highly disturbed.

The number of species in the shruby layer was 99, which was almost double compared to tree species making the shruby layer, the second in dominance after the herbaceous layer. The identified species belong to 35 families, with the largest family and its species being: Euphorbiaceae (11), Rubiaceae (9), Mimosaceae (8), Malvaceae (7), Caesalpinaceae (6), Papilionaceae (6) and Asteraceae (5). Three families – namely Combretaceae, Tiliaceae and Verbenaceae – were under-represented, each having four species, while Apocynaceae had three species. A total of six families were represented by two species whereas 20 families were represented by one species (Table A3.4). The dominant species in the shrub layer were *Pluchea dioscorides*, *Cassia* and *Vernonia*, which signifies disturbance.

Regarding the herbaceous layer, in which 185 species were recorded, the predominant family was Poaceae, which was represented by 47 species. This was followed by families Papilionaceae, Cyperaceae, Asteraceae, Amaranthaceae and Labiatae, in which the number of species recorded were 18, 16, 15, 11 and 10, respectively. Two families, namely Malvaceae and Acanthaceae, had 9 and 6 each, respectively, whereas Euphorbiaceae and Solanaceae presented five species each. Convolvulaceae and Cucurbitaceae presented four species each, whereas Caesalpinaceae and Rubiaceae were represented by three species each. A total of 13 families had two species each whereas each of 19 families had one species (Table A3.5). The most common species in the herbaceous layer were recorded from *Cyperus*, *Leersia*, *Phragmites*, *Typha*, *Eragrostis*, *Ipomoea*, *Ludwigia*, *Hygrophila* and *Chloris* genus.

3.3.8 Plants species of conservation significance

The IUCN Red List stratifies species into nine categories based on criteria such as rate of decline, population size, area of geographic distribution, and degree of population and distribution fragmentation, for the purpose of conservation. In the IUCN Red List, 'threatened species' refers to three categories of species, namely Critically Endangered (CR), which refers to species which are at very high risk of extinction in the wild; Endangered (EN) – those at high risk of extinction in the wild; and Vulnerable species (VU) – those facing high risk of being endangered in the wild. The other designated categories of the IUCN Red List are species with no known individuals remaining (Extinct (EX)); species known only to survive in captivity, or as a naturalized population outside its historic range (Extinct in the Wild (EW)); species likely to become endangered in the near future (Near Threatened (NT)); and species at lowest risk, which do not qualify for a more at-risk category. Widespread and abundant taxa are included under the Least Concern (LC) category. The Data Deficient (DD) and Not Evaluated (NE) categories include species in which not enough data are available to make an assessment of their risk of extinction and those have not yet been evaluated against the criteria.

In the surveyed natural wetlands, no plant species of conservation concern was identified within the wet area. However, eight species of conservation concern were identified from the surrounding vegetation types. These species were *Baphia kirkii* (VU), around most wetlands at Pugu Ruvu; *Dalbergia vacciniifolia* (VU), in wetland plot 14; *Millettia impressa* (VU), in wetland plots 14 and 19B; *Millettia micans* (EN), in wetland 19A; *Khaya anthoceca* (VU), planted around wetland plot 14; *Crotalaria laburnoides* (EN) and *Pterocarpus tinctorius* (NT), along Mkondoa river in Kilosa; and *Milicia excelsa* (EN), around wetland 01 at Soga. Most species were identified in Pwani region, particularly in Pugu-Kazimzumbwi Nature Forest Reserve and Ruvu South Forest Reserve, both located in Pwani region. These species were also recorded in various areas in Morogoro region, particularly in the drier miombo woodlands around Ngerengere.

The basic principle of the Convention on International Trade in Endangered Species of Wild Fauna and Flora (CITES) is to control and monitor international trade on endangered and threatened species. The Convention establishes the international legal framework for co-operation between the producer and consumer and is essential for the conservation of wild species of plants and animals. As observed for the species of conservation concern, the surveyed wetlands had no plant species which fall under any of the CITES appendices. However, some of the CITES listed species were identified from the adjacent/surrounding vegetation types along the project footprint. These species were *Khaya anthoceca*, associated with wetland 14; and *Dalbergia melanoxylon* (Table A3.3), identified in inundated wetland plots 14, 15B, 17A and 19A. In addition, some of the tree species identified were *Milicia excelsa*, in and around wetland 01 and along Mkonda river in Kilosa, and *Pterocarpus tinctorius* in wetlands along Mkondoa river. These two species are highly exploited timber trees which are protected by the bylaws issued by the Forest Department of Tanzania

3.3.9 Endemic and invasive plant species

Endemic plant species are plants that are strictly confined to a particular region but not native to other areas. Based on the Flora of Tropical Africa (FTEA) and the List of East African Plants database (LEAP master), the endemic plant species encountered around the surveyed wetlands and their associated wetlands (in brackets) were: *Xylotheca tettensis* (plot 19A, Mkondoa wetland), a shruby species restricted in coastal forest of Dar es Salaam and Pwani regions; *Baphia kirkii* (19A), a tree species confined in coastal vegetation in Dar es Salaam, Tanga and Pwani regions; and *Dalbergia vaccinifolia* (14), restricted to the coastal vegetation.

A total of 19 invasive species were identified within the project area. Among the identified species, the most prominent and abundant species were *Pistia stratiotes, Lantana camara, Calotropis procera, Xanthium stumarium* and *Ricinus communis*. Others were *Datura stramonium, Leucaena leucocephala, Argemone mexicana, Ageratum conyzoides, Cassia abbreviate, Cassia spectabilis, Cassia obtusifolia, Cassia occidentalis, Hyptis suaveolens, Ipomoea cairica, Melia azedarach, Mimosa pigra, Parkinsonia aculeate, and Solanum campylacanthum*. Considering movements of people along the SGR and its associated service road, the chances of spreading seeds from these species to other sites are high and, environmentally, this is a concern which calls for immediate control measures given their invasive nature.

To effectively control these invasive species, mechanical control, such as slashing; hand pulling/uprooting prior to flowers or seed setting or at seedling stage should be employed. Also prescribed is burning, especially early at seedling stage or prior to flowers or seed setting. Additionally, chemical control, especially, the use of herbicides, i.e. foliar spray application, can be sought. Other measures include biological control, particularly the use of host-specific natural enemies (bacteria, fungi, viruses or parasitic or predatory organisms, such as insects), which feed on vital organs such as leaves and reproductive structures such as flowers and seeds. Where necessary, and for more effective control, integrated weed management in which two or more methods outlined above can be applied in combination.

3.3.10 Bird species and composition

A total of 173 individuals from 15 species of waterbirds, including African jacana (*Actophilornis africanus*), African openbill (*Anastomus lamelligerus*), African pygmy-kingfisher (*Ispidina picta*), black crake (*Zapornia flavirostra*), black-headed heron (*Ardea melanocephala*), cattle egret (*Bubulcus ibis*), dwarf bittern (*Ixobrychus sturmii*), grey heron (*Ardea cinerea*), hamerkop (*Scopus umbretta*), little egret (*Egretta garzetta*), long-tailed cormorant (*Microcarbo africanus*), pied kingfisher (*Cerylerudis* spp.), water thick-knee (*Burhinus vermiculatus*), white-faced whistling duck (*Dendrocygna viduata*) and yellow-billed stork (*Egretta garzetta*) were recorded from 16 locations/wetlands along the SGR (Table A3.6). However, two locations, K and M, were not productive. The most common bird was hamerkop, which was recorded in ten sampling locations out of 16, followed by the pied kingfisher, which was recorded in eight sampling locations. African openbill, African jacana, pied kingfisher and cattle egret were the most abundant waterbirds. Each of these birds had over 25 individuals recorded in the entire survey. None of the species recorded in this survey was of conservation concern. All are listed by IUCN as Least Concerne (BirdLife International, 2021a).

3.3.11 Breeding indices

A total of 70 nests of weavers and bishops were counted at nine wetlands. Numbers of nests were 5, 4, 10, 16, 2, 3, 12, 7 and 11 for site D, E, G, H, I, J, M, O and P, respectively, and one nest of hamerkop *Scopus umbretta* was recorded at site D. Although no attempts were made to check the activeness of the nests (to look for eggs and chicks), as accessibility was limited, birds were busy constructing nests, revealing that it was a breeding season for these species. The weavers and bishops that were sighted within the wetlands included African golden weaver *Ploceus subaureus*.

Although no species of conservation concern were recorded during the fieldwork, some of the wetlands such as Ruvu river flood plains have good habitats for both local and migratory waterbirds. Recently recognized as full species in the yellow weaver family, the Ruvu weaver *Ploceus holoxanthus* (Fjeldså *et al.*, 2020), is restricted to the Ruvu, Wami and Rufiji rivers in eastern Tanzania (Oschadleus *et al.*, 2021). Due to its restricted range and future threats for the wetlands, the Ruvu weaver is likely to be categorized as a threatened species.

The endangered (EN) migratory Madagascar pond heron *Ardeola idea* (BirdLife International, 2021b) has previously been recorded within the Ruvu Valley and Bagamoyo areas (Ndang'ang'a and Sande, 2008). It winters in African mainland from May to October and spends the rest of the period in its breeding season in Madagascar.

Although not commonly referred to as waterbirds, the breeding of the African golden weaver, golden-backed weaver and Zanzibar red bishop is usually associated with wetlands including swamps and seasonal or temporary wetlands. Unlike weavers and bishops, hamerkop constructs nests near permanent water bodies with availability of tall trees where the massive nest is positioned, and could act as a wetland indicator.

3.3.12 Amphibian and reptile species composition

During the study, 17 species of amphibians were recorded and 11 species of reptiles were recorded (Table 3.5). The reptiles' detection rate was low (Table A3.7). The most abundant species during the surveys were the tree frogs *Afrixalus fornasini* and *Hyperolius tuberilinguis*, the puddle frogs *Phrynobatrachus acridoides* and *P. natalensis* and the clawed frog *Xenopus muelleri* (see Table A3.7). Other species that are known to occur in some wetlands, including Ruvu river, but were not detected during the present study include the Nile crocodile (*Crocodylus niloticus*).

The species encountered during the surveys are widely distributed in sub-Saharan Africa. No species of conservation concern was found (Table 3.5). The area traversed by the SGR line is known to have other widely distributed species of amphibians and reptiles although they were not detected during the present study (see Channing and Howell, 2006; Spawls *et al.*, 2017). Reasons for failing to detect some of the species include limited time and the prevailing dry periods in some areas. Most species of amphibians and reptiles are active during the rainy season (Ngalason, 2010, 2016). Of all records made, neither the amphibian or reptile species were regarded as threatened, endangered, endemic or of conservation concern. The species obtained are either listed as Least Concern (LC) or as not assessed by the IUCN Red List (IUCN, 2021).

Amphibians are one of the groups mostly affected by infrastructure (Lucas *et al.*, 2017). Their dependence on aquatic habitats makes them vulnerable to the negative impacts of the structures. In contrast, some studies have shown reptiles to ignore railway disturbances (Lucas *et al.*, 2017). Railway infrastructure can provide habitats for herpetofauna, including rocky substrates, vegetation grown along the railway line, and wetlands formed by river inundations. Furthermore, the construction and operation of the infratructure can lead to increased movement of people and formation of new habitats for herpetofauna, including formation or expansion of wetlands, which can contribute to the spread of species (both native and invasive) that may eliminate or displace some species, reducing biodiversity and negatively affecting ecological communities.

3.3.13 Macroinvertebrates' and fishes' occurrences and distributions

Wetland macroinvertebrates and fishes suffer from multiple anthropogenic pressures. These are driven by increased human populations and the pressures resulting from increasing demand of wetland products and construction of infrastructure. Table 3.6 presents habitat features noted at each sampling site for macroinvertebrates and fishes. Most wetlands were fringe, some associated with swamps and ponds. Some sites had emerging

Table 3.5. Amphibian and reptile species recorded in wetlands during the surveys along SGR project area in Tanzania.

#	Generic, specific and common name	IUCN conservation status of the species
	Amphibians	
1	*Afrixalus brachynemis* (Short-legged spiny reed frog)	Least Concern (LC)
2	*Afrixalus fornasini* (Spiny reed frog)	Least Concern (LC)
3	*Arthroleptis stenodactylus* (Shovel-footed squaker)	Least Concern (LC)
4	*Hyperolius argus* (Argus reed frog)	Least Concern (LC)
5	*Hyperolius pusillus* (Translucent tree frog)	Least Concern (LC)
6	*Hyperolius tuberilinguis* (Yellow-green reed frog)	Least Concern (LC)
7	*Hyperolius viridiflavus* (Common reed frog)	Least Concern (LC)
8	*Leptopelis flavomaculatus* (Brown-backed tree frog)	Least Concern (LC)
9	*Phlyctimantis maculata* (Red-legged pan frog)	Least Concern (LC)
10	*Phrynobatrachus acridoides* (East African puddle frog)	Least Concern (LC)
11	*Phrynobatrachus natalensis* (Natal puddle frog)	Least Concern (LC)
12	*Phrynomantis bifasciatus* (Banded rubber frog)	Least Concern (LC)
13	*Ptychaderna anchietae* (Anchieta's ridged frog)	Least Concern (LC)
14	*Ptychadena mascareniensis* (Mascarene ridged frog)	Least Concern (LC)
15	*Pyxicephalus edulis* (Edible bullfrog)	Least Concern (LC)
16	*Sclerophrys gutturalis* (Common toad)	Least Concern (LC)
17	*Xenopus muelleri* (Müller's clawed frog)	Least Concern (LC)
	Reptiles	
1	*Agama lionotus* (Red-headed rock agama)	Not assessed
2	*Heliobolus spekii* (Speke's sand lizard)	Least Concern (LC)
3	*Hemidactylus mabouia* (Tropical house gecko)	Not assessed
4	*Hemidactylus platycephalus* (Tree gecko)	Not assessed
5	*Kinixys spekii* (Speke's hinged tortoise)	Not assessed
6	*Lygodactylus grotei* (Grote's dwarf gecko)	Least Concern (LC)
7	*Lygodactylus picturatus* (Yellow-headed dwarf gecko)	Least Concern (LC)
8	*Trachylepis striata* (Striped skink)	Not assessed
9	*Trachylepis varia* (Variable skink)	Not assessed
10	*Trachylepis maculilabris* (Speckle-lipped skink)	Not assessed
11	*Varanus niloticus* (Nile monitor)	Not assessed

wetland due to water blockage and damming activities during construction of SGR service roads and railways. Eight sites were in slow-water-moving areas dominated by small substrates and extensive aquatic macrophyte beds. These were indicative of most of the flooded areas, ponds and swamps. Seven sites were in moderate depth water. These sites were characteristic of the pelagic/profundal zone areas that extend to more than 3m in depth. At all these sites, major differences in habitat were associated with substrate composition and the presence of aquatic macrophytes.

3.3.14 Macroinvertebrates' abundances and taxa richness

A total of 730 macroinvertebrate individuals belonging to 10 orders and 22 families from 13 sampled wetlands were collected, identified

Table 3.6. Information on sampling sites along the SGR line.

Sampling site	Description	Name	Sampling	Depth	Stream habitat
A	Flood wetland	Vingunguti	M	Shallow	Slow
B	Fringe wetland	Msimbazi river	M, F	Moderate	Slow
C	Fringe wetland	Mpiji river	M, F	Moderate	Slow
D	Fringe wetland		NIL		
E	Flood wetland		M	Moderate	Slow
F	Fringe wetland + swamp	Kwamturuki river	F		
G	Fringe wetland		M	Shallow	Slow
H	River	Ruvu river	M, F	Moderate	Fast
I	Fringe	Ngerengere river 1	M	Moderate	Fast
J	River + swamp	Ngerengere river 2	M, F	Moderate	Fast
K	Fringe	Mkata river	NIL		
L	River	Mkondoa river	M, F	Moderate	Fast
M	Fringe	Lumuma river	M	Shallow	Fast
N	Fringe + pond	Korongo idara yamaji	M, F	Shallow	Slow
O	Flood wetland	Zuzu	M	Shallow	Slow
P	Fringe	Bubu river	M, F	Shallow	Slow

Sampling: "M", for macroinvertebrates; "F", = for fish, "NIL", for no sampling.

and counted. Among all identified macroinvertebrates common in Tanzanian rivers and wetlands (Gerber and Gabriel, 2002; Kaaya, 2014), the majority were from groups Odonata (20%), Coleoptera (19%), Ephemeroptera (13%), Trichoptera (8%), Diptera (4%) and Plecoptera (2%). The literature indicates that macroinvertebrates have a marked gradient in heterogeneity as a function of the gradient in water conditions (Elias *et al.*, 2014; Shimba and Jonah, 2016; Alavaisha *et al.*, 2019). Previous reports indicate that the taxa Coleoptera, Ephemeroptera and Trichoptera are abundant under moderate turbulent flow, associated with a high oxygen concentration and much particulate matter that makes a food source (Table A3.8 and Table 3.6).

3.3.15 Fish abundance and richness

A total of 1139 fish individuals from 23 different species were caught in wetland (Table A3.9). All fish species were common in the region (Eccles, 1992). *Tilapia niloticus* (34.24%) and *Astatotilapia* spp. (32.75%) had high abundance and were found in almost all wetlands (Table A3.9). There was also formation of new wetlands in low-lying land as a result of sand translocation and blockage of streams. For instance, in new emerging swamp (Fig. 3.9), despite development of riparian vegetation and reeds, mixing of water was very limited. The swamp was a potential fishing ground and a source of fresh water for communities.

3.4 Conclusion

The survey revealed about 16 important wetlands along the SGR footprint. The majority were slightly to moderately impacted by human activities, including contruction of the SGR and service road. Most wetlands were concentrated in the Morogoro segment, particularly in flooded areas along the Mkondoa and Ngerengere wetland system. Construction activities affected

Fig. 3.9. Emerging wetland following excavation of dry land at Kwamturuki.

both the physical and biological components of these wetlands. While water chemistry generally met satisfactory standards, high turbidity was recorded in some areas like Ruvu and Mkondo wetlands. The soils were disturbed with the likelihood of limiting growth of some plant species. Species distribution along the SGR line showed slight changes due to landscape alterations. The flora and fauna communities were indicative of wetland health and the biophysical environment. Although the survey identified plant species of conservation concern, such as threatened, endemic and CITES-listed species, such species were not restricted to the wetland habitats but were well distributed in the surrounding vegetation. Overall, the surveyed wetlands were disturbed, but the negative impact of the SGR project on the biological environment, which could lead to significant changes in wetland habitat and species composition, was moderate. Water diversion resulting from damming affected some tree species but also facilitated the formation of new wetlands. Similarly, fauna was affected, as indicated by low abundance and diversity, particularly at Zuzu and Msimbazi wetland. To ensure the sustainability of the project and safeguard natural wetlands, we propose minimizing land modification and conducting restoration efforts in heavily modified areas.

References

Alavaisha, E., Lyon, S.W. and Lindborg, R. (2019) Assessment of water quality across irrigation schemes: A case study of wetland agriculture impacts in Kilombero valley, Tanzania. *Water (Switzerland)* 11(4). DOI: 10.3390/w11040671.

APHA (1998) *Standard Methods for the Examination of Water and Wastewater*, 20th edn. American Public Health Association, Washington, DC.

BirdLife International (2021a) IUCN Red List for birds. Available at: http://www.birdlife.org (accessed 14 May 2021).

BirdLife International (2021b) Species factsheet: *Ardeola idae*. Available at: http://www.birdlife.org (accessed 14 May 2021).

Channing, A. and Howell, K.M. (2006) *Amphibians of East Africa*. Cornell University Press, Ithaca, New York.

Doan, T.M. (2003) Which methods are most effective for surveying rain forest herpetofauna? *Journal of Herpetology* 37(1), 72–81.

Eccles, D.H. (1992) *Field Guide to the Freshwater Fishes of Tanzania*. FAO species identification sheets for fishery purposes. Food and Agricultural Organization of the United Nations, Rome.

Elias, J.D., Ijumba, J.N., Mgaya, Y.D. and Mamboya, F.D. (2014) Study on freshwater macroinvertebrates of some Tanzanian rivers as a basis for developing biomonitoring index for assessing pollution in tropical African regions. *Journal of Ecosystems* 2014, 1–8. DOI: 10.1155/2014/985389.

Fjeldså, J., Christidis, L. and Ericson, P.G. (2020) *The Largest Avian Radiation: The Evolution of Perching Birds, or the Order Passeriformes*. Lynx Edicions, Cerdanyola del Vallès, Spain.

Gerber, A. and Gabriel, M.J.M. (2002) *Aquatic Invertebrates of South African Rivers Field Guide*. Department of Water Affairs and Forestry, South Africa. Available at: http://scholar.google.com/scholar?hl=en&btnG=Search&q=intitle:Aquatic+Invertebrates+of+South+African+Rivers+Field+Guide#0 (accessed 21 April 2024).

Grant, B., Tucker, A.D., Lovich, J.E., Mills, A.M., Dixon, P.M. *et al.* (1992) The use of coverboards and estimating patterns of reptile and amphibian biodiversity. In: Siegel, R. and Scott, N. (eds) *Wildlife 2001*. Elsevier Science Inc, pp. 379–403.

Harper, E.B., Measey, G.J., Patrick, D.A., Menegon, M. and Vonesh, J.R. (2010) *Field Guide to the Amphibians of the Eastern Arc Mountains and Coastal Forests of Tanzania and Kenya*. Camerapix Publishers International, Nairobi.

Heyer, W.R., Donnelly, M.A., McDiarmid, R.W., Hayek, L.-A.C. and Foster, M.S. (1994) *Measuring and Monitoring Biological Diversity: Standard Methods for Amphibians*. Smithsonian Books, Washington, DC.

IUCN (2021) *The IUCN Red List of Threatened Species*. Version 2021-1. International Union for Conservation of Nature, Cambridge, UK. Available at: https://www.iucnredlist.org (accessed 21 April 2024).

Kaaya, L. (2014) Biological assessment of tropical riverine systems using aquatic macroinvertebrates in Tanzania East Africa, PhD thesis, University of Cape Town. Available at: https://open.uct.ac.za/bitstream/item/8977/thesis_sci_2014_kaaya_lt.pdf?sequence=1 (accessed 21 April 2024).

Kamukala, G.L. (1993) An overview and scope of Tanzanian wetlands. Working paper. Available at: chrome extension://oemmndcbldboiebfnladdacbdfmadadm/https://aquadocs.org/bitstream/1834/521/1/Wetlands914.pdf (accessed 21 April 2024).

Kirkham, M.B. (2015) *Principles of Soil and Plant Water Relations*, 3rd edn. Elsevier/Academic Press. Available at: https://books.google.co.tz/books?hl=en&lr=&id=qa-zEAAAQBAJ&oi=fnd&pg=PP1&ots=zUWL9kHRj4&sig=xhi-JcKpoLl1mOOeZnSBdYzUjQ0&redir_esc=y#v=onepage&q&f=false (accessed 21 April 2024).

Klute, A., Danielson, R.E. and Sutherland, P.L. (1986) *Methods of Soil Analysis*, 2nd edn. American Society of Agronomy, Madison, Wisconsin.

Lucas, P., de Carvalho, R.G. and Grilo, C. (2017) Railway disturbances on wildlife: types, effects, and mitigation measures. In: Borda-de-Água, L., Barrientos, R., Beja, P. and Pereira, H.M. (eds) *Railway Ecology*. Springer Nature, pp. 81–99. DOI: 10.1007/978-3-319-57496-7.

Maltby, E. (2009) *Functional Assessment of Wetlands:Towards Evaluation of Ecosystem Services* CRC Press, Boca Raton, Florida.

Materu, S.F., Urban, B. and Heise, S. (2018) A critical review of policies and legislation protecting Tanzanian wetlands. *Ecosystem Health and Sustainability* 4(12), 310–320. DOI: 10.1080/20964129.2018.1549510.

Morris, P. and Therivel, R. (2001) *Methods of Environmental Impact Assessment*, 2nd edn. Spon Press, London. Available at: https://books.google.co.tz/books?hl=en&lr=&id=tRZnUBqloL8C&oi=fnd&pg=PR11&dq=Morris+and+Therivel+(2001).&ots=eW8kXKiVnJ&sig=svBZpjAPy5D5GXVreH6-1OC15fc&redir_esc=y#v=onepage&q=Morris and Therivel (2001).&f=false (accessed 21 April 2024).

Moyano, F.E., Manzoni, S. and Chenu, C. (2013) Responses of soil heterotrophic respiration to moisture availability: an exploration of processes and models. *Soil Biology and Biochemistry* 59, 72–85. DOI: 10.1016/j.soilbio.2013.01.002.

Ndang'ang'a, P.K. and Sande, E. (2008) *International Single Species Action Plan for the Madagascar Pond-heron (Ardeola idae)*, CMS Technical Series No. 20: AEWA Technical Series No. 39.

Ngalason, W. (2010) *Ecology of Anurans on the Lukwangule Plateau, Tanzania: Abundance, Ecology and Reproduction*. VDM Verlag Dr. Müller Aktiengesellschaft and Co. KG. Saarbrücken, Germany.

Ngalason, W. (2016) *Ecology and Altitudinal Distribution of Herpetofauna on Mount Kilimanjaro, Tanzania*. University of Dar es Salaam.

Oschadleus, H.D., Andersson, S. and Baker, N.E. (2021) History and biology of the reassigned ruvu weaver *Ploceus holoxanthus*. *Scopus: Journal of East African Ornithology* 41(1), 1–12.

Pickersgill, M. (2007) *Frog Search: Results of Expeditions to Southern and Eastern Africa*. Edition Chimaira, Frankfurt am Main, Germany.

Ramsar Convention on Wetlands (2018) *Global Wetland Outlook: State of the World's Wetlands and Their Services to People*. Available at: https://static1.squarespace.com/static/5b256c78e17ba335ea89fe1f/t/5b9ffd2e0e2e7277f629eb8f/1537211739585/RAMSAR+GWO_ENGLISH_WEB.pdf (accessed 21 April 2024).

Rocky, J. and Mligo, C. (2012) Regeneration pattern and size-class distribution of indigenous woody species in exotic plantation in Pugu forest reserve, Tanzania. *International Journal of Biodiversity and Conservation* 4(1). DOI: 10.5897/IJBC11.198.

Shimba, M. and Jonah, F. (2016) Macroinvertebrates as Bioindicators of water quality in the Mkondoa river, Tanzania, in an agricultural area. *African Journal of Aquatic Science* 41(4), 453–461. DOI: 10.2989/16085914.2016.1230536.

Tanzania Bureau of Standards (TBS) (2017) *National Environmental Standards Compendium*.

Turrill, W. and Milne-Redhead, E. (1952) *Flora of Tropical East Africa: Ranunculaceae*. University of Chicago Press, Chicago, Illinois. Available at: https://press.uchicago.edu/ucp/books/book/distributed/F/bo9855727.html (accessed 21 April 2024).

Van Reeuwijk, L.P. (2002) Procedures for soil analysis, 6th edn. International Soil Reference and Information Centre/Food and Agricultural Organization of the Nations. Available at: chrome-extension://oemmndcbldboiebfnladdacbdfmadadm/https://www.isric.org/sites/default/files/ISRIC_TechPap09.pdf (accessed 21 April 2024).

WHO (2017) *Guidelines for Drinking-water Quality: Fourth Edition Incorporating the First Addendum*. World Health Organization, Geneva, Licence. Available at: http://apps.who.int/iris/bitstream/handle/10665/254637/9789241549950-eng.pdf?sequence=1 (accessed 21 April 2024).

Witt, A. and Luke, Q. (2017) *A Guide to Naturalized and Invasive Plants of Eastern Africa*. CAB International, Wallingford, UK.

Zedler, J.B. and Kercher, S. (2005) Wetland resources: status, trends, ecosystem services, and restorability. *Annual Review of Environment and Resources* 30(1), 39–74. DOI: 10.1146/annurev.energy.30.050504.144248.

4 Floristic Structure, Composition, Diversity and Anthropogenic Pressure: Implications of the Management of Biodiversity along the SGR Segment in Pugu and Ruvu Forest Reserves, Tanzania

Heriel Moshi[1]*, Catherine A. Masao[2] and Henry Ndangalasi[1]
[1]College of Natural and Applied Sciences, University of Dar es Salaam, Tanzania;
[2]Institute of Resource Assessment, University of Dar es Salaam, Tanzania

Abstract

This chapter presents findings on floristic composition, structure and conservation status across a biodiversity hotspot, namely Pugu-Kazimzumbwi Nature Forest Reserve (PKNFR) and Ruvu South Forest Reserve (RSFR), dissected by SGR. The data were gathered systematically in plots established alongside SGR. Species of conservation importance were screened using flora manuals, CITES and IUCN database. Findings revealed seven expanding populations, richness (47–97), density (1782–3535) stems ha^{-1}, diversity (3.4–4.1) for trees and shrubs, respectively. In PKNFR, nine threatened species and seven endemics were identified, while in RSFR, threatened and endemics numbered six. In both reserves, CITES-listed species were three, invasives seven and anthropogenically disturbed species, including trees of conservation concern, 29. Therefore, the impacted forest segment hosts a significant number of conservation-sensitive plants and the combined effect of various disturbances on them cannot be underestimated. However, most identified conservation-sensitive species are widely distributed and unlikely to hinder project implementation.

4.1 Background

The Ministry of Works, Transport and Communications through the Tanzanian Railways Corporation (TRC) is implementing the first two phases of an electrified standard-gauge railway (SGR). This project, covering 1219 km, stretching from Dar es Salaam to Mwanza, is co-financed by the Tanzanian government and international finance. Part of the 541 km of the first phase extending from Dar es Salaam to Makutupora in Dodoma navigates through Pugu-Kazimzumbwi Nature Forest Reserve (PKNFR) and Ruvu South Forest Reserve (RSFR), which are important biodiversity hotspots.

PKNFR and RSFR are part of the East Africa coastal forests (Burgess and Clarke, 2000) and internationally recognized as part of the world biodiversity hotspot due to their salient restrictions in habitat and high species diversity and endemism (Myers *et al.*, 2000). Floristic species composition in coastal forests varies with topographic positions ranging from ridge tops

*Corresponding author: heriel2002@yahoo.com

DOI: 10.1079/9781800626942.0004

to along slopes and valley bottoms (Hawthorne, 1993). The study area contains a unique flora with several endemic species. Among the plant species of conservation concern include *Garcinia acutifolia, Fernandoa magnifica, Rothmannia macrosiphon, Baphia puguensis, Milletia puguensis* and *Uvaria decidua*. These species are listed under the International Union for Conservation of Nature (IUCN) Red List.

The threats to coastal forests biodiversity and habitats are exerted and exacerbated by anthropogenic activities especially shifting cultivation, wildfires, unsustainable logging especially the high valued timbers, uncontrolled settlement and charcoaling (WWF Tanzania Country Office, 2012). This has led to widespread forest degradation, fragmentation, habitat loss, increased number of threatened plant species in this forest, threats to plant survival, and changes in species composition and vegetation structure. The endemic species in this case are more impacted (Doggart, 2006; URT, 2015) and the forest reserves mostly threatened by degradation have been the PKNFR and RSFR, which are rich in natural resources.

PKNFR is part of the 34 global biodiversity hotspots (Mittermeier *et al.*, 2004; Rocky and Mligo, 2012). Besides its rich flora, PKNFR is rich in fauna species, particularly the endemic coastal birds and small mammals (Keith *et al.*, 1992). Due to degradation/fragmentation, its rich flora and fauna species, it has been suggested, include the small portions of remaining natural vegetation and habitats within PKNFR under a 'nature reserve' conservation instead of 'forest reserve' status, which are in danger of destruction in Tanzania. On the other hand, RSFR is one of the most important coastal forests in Tanzania, but in comparison to other coastal forest fragments it is relatively large. The forest also serves as an important resource for the surrounding communities and, more importantly, for water catchment. Its species composition is influenced by historical selective logging and unique soil and groundwater conditions (BirdLife International, 2012). RSFR also harbours several endemic plant species, and rare and low-density forest bird species (Baker and Baker, 2002; IUCN, 2023). The forest supports several vegetation types including disturbed dry riverine forest, thickets, woodland and scrub forest, large stands of grass-covered flood plain, and swamps in the centre of the reserve. Like PKNFR, RSFR has also been under constant anthropogenic pressure mainly from illegal logging and production of charcoal. The pattern of deforestation was exacerbated by land clearing by Sun Biofuel investments to the south of the reserve in 2008, which extended the illegal charcoal production network further to the interior of RSFR. It is worth noting that in recent years anthropogenic disturbance in both PKNFR and RSFR has shrunk due to conservation interventions by the management. However, the construction of the electrified SGR across PKNFR and RSFR in addition to the existing metre-gauge railway (MGR) is likely to have both direct and indirect impacts on the flora and biodiversity of the two reserves in general.

One way in which the SGR project is likely to impact the two forest reserves is through direct clearing of vegetation cover for construction activities in some segments. Such activities require cutting and filling, construction of bridges, underpasses and crossings for cattle and pedestrians, concrete batching plants, and support facilities such as stations and marshalling yards. These can largely impact soils and landscape as the area is completely uncovered of vegetation to remove the top and subsurface soils. Usually, excavated soil is required for earthworks such as construction of various infrastructure, e.g. permanent access and haulage roads for the main tracks and station tracks, platform stations and civil structures in parts of the forests. These earthworks and use of stripped soils, if not well managed, are likely to spread invasive plant species to the reserves and more so if invasive species are growing in the area where soil is excavated. Earthworks, if not properly managed, can also result in serious erosion leading to eyesores such as gullies. A meticulous understanding of the standing vegetation in areas where soil is excavated for such activities is therefore important to avoid spreading of alien invasive plant species from seeds and active soil seed banks from borrow pits to the forest reserves. It is on this basis that a detailed floristic survey on salient features of the standing vegetation in project sites such as borrow pits is important to minimize the chance of dispersing invasive species to the destination points, some of which are in PKNFR and RSFR.

Indirectly, dissection of the two reserves with the two railway systems (MGR and SGR) and associated road systems may increase the access to the interior of the reserves and exacerbate anthropogenic pressure on the standing vegetation. Co-ordinated efforts are thus needed to avoid further possible impacts and concomitant loss of flora and fauna of global conservation importance in these reserves. This is deemed vital in promoting sustainable management of biodiversity resources and sustaining ecological integrity of the protected areas likely to be impacted by the SGR project. In view of this, and as a beneficiary of international finance, the SGR project is implemented in line with International Finance Corporation (IFC)'s Performance Standards (PSs) (2012). As such, the Biodiversity Action Plan (BAP) developed for implementation during the design, construction and operation phases was thus considered both vital and inevitable. TRC, through its Environmental Department, thus commissioned botanical experts of the University of Dar es Salaam Department of Botany to provide expertise by conducting a meticulous baseline study to identify botanical features in areas navigated by the SGR alignment within PKNFR and RSFR.

This chapter presents an assessment of vegetation as part of a BAP that is envisaged to characterize the existing vegetation; identify the various plant communities and critical plant species habitats and their distinctive features; and document the vascular plants within the two reserves. The chapter further seeks to identify the typical elements of such flora and species of special conservation status, as recognized by the Tanzanian government, the IUCN and the Convention on International Trade on Endangered Species (CITES) of wild flora and fauna. Furthermore, investigation of the local plant communities and factors presently affecting their structure and composition are also presented. The results from the survey elucidate potential effects likely to emanate from the SGR project and its associated activities on the standing vegetation. As such, this chapter forms the basis for updating the existing forest management plans upon which project-related effects and adherence to implementation of BAP can be evaluated. It further provides mitigation, monitoring and rehabilitation programmes for project-related activities that affect the flora and fauna to ultimately safeguard the rich biodiversity resource in both PKNFR and RSFR. It is anticipated that through adoption of these practices the triangulation of benefits from ecosystem services, conservation needs and development priorities can be comprehended.

4.2 Methodology

4.2.1 Vegetation sampling

This study was conducted strictly to meet or exceed International Finance Corporation standards (IFC, 2012). It focused on the World Bank Environmental and Social Standard 6 (ESS6) and Performance Standard 6, which specifically applies to biodiversity conservation and sustainable management of living natural resources. This was necessary to clearly articulate survey aspects pertinent to protection and conservation of biodiversity, modified habitat, natural habitat, critical habitat, alien invasive plant species, and legally protected and internationally recognized areas.

The detailed survey was preceded by a reconnaissance survey to familiarize the team with the proposed forest reserves and decide on the best approach to conduct the detailed vegetation sampling and sampling intensity. Quantitative methods were employed in gathering data for use in enumerating plant species composition, abundance, dominance, co-dominance, richness, density and diversity. The plant community types in the projected area were classified, based on physiognomic features, into broad categories such as grassland, bushland, woodland, thicket and forest. Floristic naming of the identified plant community types was based on two or three dominant species selected using the relative magnitude of their estimated cover-abundance values. Transects were laid to cover large parts of the vegetation alongside the project footprint. Following the general classification of the plant community, sampling plots were stratified first by plant community types and randomized to ensure a full range of community diversity is sampled. Identification of locations that are typical of the communities sampled was made on best judgement (Morris and Therviel, 2001).

In the interests of capturing a broad sample of representative community types within the two selected forest reserves, a systematic plot-centred sampling method was adopted. On each transect, plots were distributed based on the observed vegetation heterogeneity. A Modified-Whittaker nested vegetation sampling method (Stohlgren *et al.*, 1995) was adopted. A detailed quantitative vegetation sampling was conducted in all plant community strata. To ensure proper representation of each sampled area within the reserves, plots were established within line transects in a section of not less than one kilometre across the SGR alignment. Special consideration was made to capture floristic variations along topographic positions as deemed necessary. Along the established transects, larger plots measuring 20 m × 20 m were used for sampling all trees with a diameter at breast height (DBH) equal to or greater than 10 cm. Within the larger plots, three nested plots of 5 m × 5 m were established to record all woody species with DBH less than 10 cm including seedlings, saplings, shrubs, lianas and vines. Three small plots measuring 1 m × 1 m were placed within the shrub plots to record all herbaceous species namely grasses, sedges and forbs. The Braun-Blanquet cover-abundance scale was used to estimate cover. The cover was expressed as a fraction or percentage of the reference area. In this case, a portable quadrat measuring 1 m^2 and divided by strings into 100 small squares each measuring 1 cm^2 was used. In addition, opportunistic sampling was conducted to capture species composition, in special micro-habitats not covered by the established transects and plots. A minimum sampling intensity of 0.02 was adopted to obtain a proper representation of the forest reserve dissected by the SGR. In all sites, plots were geo-referenced by geographic coordinates in the Universal Transverse Mercator (UTM) grid system using a hand-held global positioning system (GPS) unit.

4.2.2 Identification of plant species of conservation significance

All plant species were identified in the field to species level. Whenever this was not possible, the respective plant was given a code, specimens collected, pressed, and taken to the herbarium located at the Department of Botany, University of Dar es Salaam, for identification using relevant flora or matching with herbarium specimens. The nomenclature follows Turrill and Milne-Redhead (1952). Plant species of conservation concern in this area were identified based on the most recent taxonomic reports on coastal forest plants herbaria records, extensive fieldwork, Flora of Tropical East Africa online database and Flora Zambesiaca. Furthermore, the List of East African Plant species (LEAP Master-1996) and Plants of the World Online (POWO, 2023) were also consulted. The latest documents of CITES list appendices and the IUCN Red List of threatened plant species were used in screening plant species which fall under any of its respective appendices and categories. The IUCN threatened species in this context refers to Critically Endangered, Endangered and Vulnerable species categories. Plant species were further categorized based on threatened species: Endangered species (EN), referring to those at high risk of extinction in the wild; Critically Endangered species (CR), which are at extremely high risk of extinction in the wild; Vulnerable species (VU), those facing high risk of being endangered in the wild; Extinct species (EX) with no known individuals existing or known only to survive in captivity; Extinct in the Wild (EW), which are naturalized outside their historic range; and Near Threatened species (NT), referring to plant species likely to become endangered in the near future.

4.2.3 Anthropogenic disturbance

Assessment of the level of anthropogenic disturbance in the forest reserves was conducted by documenting all observations of disturbance and their intensities. The level of disturbance was assessed systematically along a strip of 10m wide on each side of the transect line for each transect line established. Incidences of disturbance were categorized, scored and recorded as (i) pole cutting; (ii) timber harvesting; (ii) traps; (iv) fire damage; (v) charcoal making; (vi) pit sawing; (vii) timber/planks/poles; (viii) animal trappings; (ix) cultivation; (x) grazing; and (xi) footpaths. The GPS coordinates were taken at the start and end of the disturbance transects. A digital camera was used to take photographs for reference.

4.2.4 Alien invasive plant species

The sampling for the alien invasive species employed both systematic and random sampling techniques. In this case, systematic sampling was achieved by recording all alien invasive species identified in established plots along the transects. Alien invasive species were recorded randomly at any point between or alongside plots. Ecological and human impact assessments of such plant species were also carried out.

4.3 Data Analysis

4.3.1 The percentage of dead trees due to disturbance

This was expressed as:

$$D = \frac{e}{b} * 100 \qquad \text{(Eqn 4.1)}$$

where e = total number of dead stems and b = total number of dead and live trees within the established plot/transect/site.

The causes of disturbance/mortality were compared in percentages in order to determine the most possible causes of mortality and their implication on the impact of anthropogenic activities on plant community structure and regeneration.

4.3.2 Density of trees, shrubs, herbs and saplings

Density (stems ha^{-1}) for each species was calculated by using the relationship:

$$Density = \frac{n}{A} \qquad \text{(Eqn 4.2)}$$

where n = total number of individual species i enumerated from the study site and A = total area of sampled plots.

Total study site tree density (k) was obtained by summing up the densities of each species and the proportion of study site tree density (G), i.e. relative density accounted for by species i was expressed as:

$$G = (h/k) * 100 \qquad \text{(Eqn 4.3)}$$

where h is the density of individual species i.

4.3.3 Stem diameter and population structure

Stem diameter values were obtained from the values read from the DBH tape measure for further characterization of the population structure and hence vertical stratification of the plant communities. DBH was classified in frequency intervals of 15 cm for all stems sampled in all transects from each plant community type to establish the plant community structure and forest stand structure in accordance with Parthasarathy and Karthikeyan, 1997 and Moshi (2007).

4.3.4 Species diversity, richness and abundance

Plant species diversity was calculated using the Shannon-Wiener diversity index (Shannon and Weaver, 1948), which is the preferred index as it combines two quantifiable measures, namely: species richness, which is the number of species in the community, and how the individuals are distributed (evenness or equitability), in the sample understudy. Species richness was obtained by enumerating the total number of species found in the sampling site, expressed as species per total area sampled.

4.4 Results and Discussion

4.4.1 Sampling intensity and geographical location of sampled plots

Table 4.1 summarizes the number of transect plots' locations and vegetation physiognomic types for sites surveyed in the present study. A total of four line transects were established, two

Table 4.1. Geographical location of sampled plots and vegetation types.

S/N	SITE	Plot	Eastings	Northings	Physiognomic vegetation type	Floristic vegetation type
1	PKNFR	T1P1	37M0512193	9239813	Bushland	*Harrisonia-Grewia* bushland
2	PKNFR	T1P2	37M0510735	9239885	Coastal thickets	*Pteliopsis-Markhamia* thickets
3	PKNFR	T1P3	37M0509792	9239653	Bushland with emergent trees	*Caesapinia-Grewia* bushland with *Milicia-Albizia* emergents
4	PKNFR	T1P4	37M0509184	9239917	Mixed coastal woodland	*Albizia-Milicia-Markhamia* woodland
5	PKNFR	T1P5	37M0509029	9239969	Riparian forest	*Baringtonia* riparian forest
6	PKNFR	T1P6	37M0509038	9240045	Bushland	*Hymenocardia-Landolphia* bushland
7	PKNFR	T1P7	37M0508391	9240373	Coastal thickets	*Flueggea-Harrisonia-Dichapetalum* bushland
8	PKNFR	T1P8	37M0508599	9240236	Bushland	*Pteliopsis-Dalbergia* bushlands
9	PKNFR	T1P9	37M0508240	9240412	Bushland with grassland patches	*Cassia-Uvaria-Flueggea* bushland with patches of *Panicum* grassland
10	PKNFR	T2P1	37M0508958	9240265	Bushed woodland	*Milletia-Albizia* bushed woodland
11	PKNFR	T2P2	37M0509183	9240247	Bushland with grassland patches	*Hymenocardia-Pteleopsis* bushland with patches of *Panicum-Hyparrhenia* grassland
12	PKNFR	T2P3	37M0509899	9239963	Bushland	*Leucaena-Flueggia* bushland
13	PKNFR	T2P4	37M0510887	9240133	Bushed woodland	*Cassia-Albizia* bushland
14	PKNFR	T2P5	37M0511974	9239902	Bushed woodland	*Milicia-Albizia* bushed woodland
15	PKNFR	T2P6	37M0512688	9239737	Coastal thickets	*Dalbergia-Uvaria* thickets
16	RSFR	T1P1	37M0492568	9240103	Bushed grassland	*Dicrostachys-Annona -Panicum-Hyparrhenia* bushed grassland
17	RSFR	T1P2	37M0492877	9239992	Bushed grassland	*Flueggea-Pterocarpus -Panicum-Cyperus* bushed grassland
18	RSFR	T1P3	37M0493484	9239972	Bushed grassland	*Hymenocardia – Panicum* bushed grassland
19	RSFR	T1P4	37M0494171	9239836	Bushland with grassland patches	*Hymenocardia-Markhamia-Millettia* bushland with patches of *Panicum-Rottboellia grassland*
20	RSFR	T1P5	37M0495085	9239618	Bushed woodland	*Brachystegia-Diplorynchus* bushed woodaland

Continued

Table 4.1. Continued

S/N	SITE	Plot	Eastings	Northings	Physiognomic vegetation type	Floristic vegetation type
21	RSFR	T1P6	37M0494672	9239634	Bushland with grassland patches	*Markhamia-Flueggia* bushland with patches of *Perotis-Heteropogon-Panicum* grassland
22	RSFR	T1P7	37M0495480	9239521	Bushed woodland	*Acacia-Spirostachys-Dombeya* bushed woodland
23	RSFR	T1P8	37M0495603	9239479	Wetland with emergent trees	*Cyperus-Echinochloa* wetland with Acacia emergents
24	RSFR	T2P1	37M0495350	9239676	Regenerating Bushed woodland	*Brachystegia-Diplorrynchus* woodland
25	RSFR	T2P2	37M0494864	9236834	Bushland	*Uvaria-Hymenocardia* bushland
26	RSFR	T2P3	37M0494359	9239918	Bushed woodland with grassland patches	*Markhamia-Millettia* bushed woodland with patches of *Panicum-Perotis-Cyperus* grassland
27	RSFR	T2P4	37M0493926	9239995	Bushed woodland with grassland patches	*Pterocarpus-Markhamia* bushed woodland with patches of *Perotis-Panicum-Digitaria* grassland
28	RSFR	T2P5	37M0493434	9240087	Bushland	*Hymenocardia* bushland
29	RSFR	T2P6	37M0495999	9239635	Seasonal Bushed wetland	*Cyperus-Pluchea* bushed wetland

in each of the two forest reserves (PKNFR and RSFR). They were arranged in such a way that one line covered the vegetation to the north and the second to the south of the SGR alignment. In both reserves, a total of 29 plots were established in which 15 were partitioned to PKNFR, with nine plots placed along transect 1 and six along transect 2. On the other hand, 14 plots were established in RSFR with eight plots along transect 1 and six in transect 2. The established transect covered a stretch of approximately 4.8 km in PKNFR and 3.04 km in RSFR. The sampled plots laid therein represent an area of 0.60 ha and 0.56 ha in PKNFR and RSFR, respectively, from which trees, shrubs, grasses and herbaceous species were sampled. The differences in the length of the four established transects across the two forests are a result of vegetation heterogeneity and variation in biophysical environment in those areas dissected by SGR project.

The vegetation in the two forests falls under the broad category of coastal mosaic vegetation type. However, at local scale, variations across this broad vegetation type were evident. Predominant vegetation physiognomic types in PKNFR were bushland, bushland with emergent trees, bushland with grassland patches, coastal thickets, mixed coastal woodland, bushed woodland and riparian forest. The dominant species in these communities were *Albizia glaberrima, Barringtonia racemosa, Caesalpinia bunduc, Senna siamea, Cassia spectabilis, Hymenaea verrucosa, Markhamia lutea, Dombeya shupangae, Dalbergia melanoxylon, Flueggea virosa, Harrisonia abyssinica, Hymenocardia ulmoides, Leucaena leucocephala, Grewia forbesii,* and *Pteleopsis myrtifolia*. The sub-dominant species on the other hand were *Milicia excelsa, Albizia versicolor, Uvaria kirkii, Dichapetalum mossambiscense, Landolphia kirkii,* and *Dalbergia obovata*.

On the other hand, part of RSFR dissected by the SGR project was characterized by bushed grassland, bushed wetland, bushed woodland with grassland patches, bushland with grassland patches, and inundated wetland with emergent trees. The dominant genera were *Acacia, Brachystegia, Cyperus, Dicrostachys, Flueggea, Hymenocardia, Markhamia, Pterocarpus* and *Uvaria*. Sub-dominant genera were *Spirostachys, Dombeya, Diplorhynchus Echinochloa, Pluchea, Annona* and *Pterocarpus*.

4.4.2 Species richness and density

The species richness for PKNFR and RSFR and their corresponding number of families are

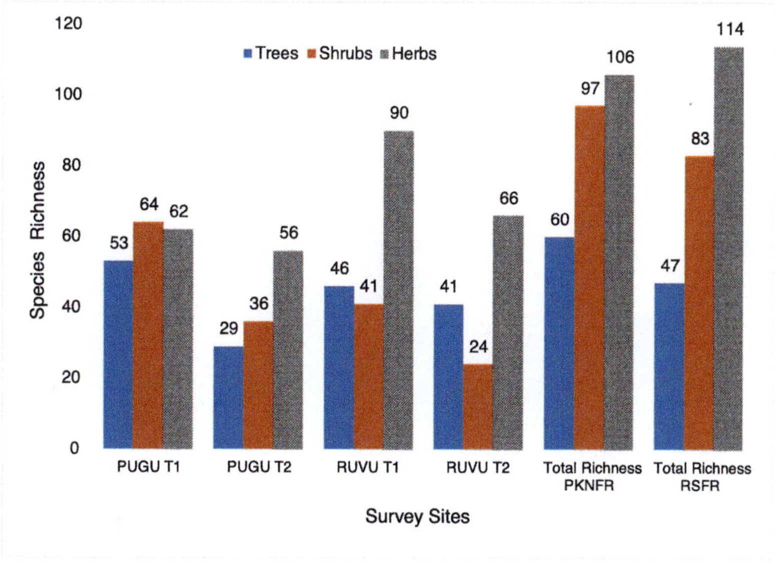

Fig. 4.1. Plant species richness for trees, shrubs, and herbs in PKNFR and RSFR.

summarized in Fig. 4.1. In PKNFR, the total number of families of trees was 26, while shrubs numbered 41 and herbs 34. The plant families that had a large number of species (numbers in brackets) in PKNFR were Caesalpinaceae (9) and Combretaceae (5) for trees; Apocynaceae (8), Rubiaceae (8), Papilionaceae (6), Annonaceae (5), Euphorbiaceae (5) and Mimosaceae (5) for shrubs; and Poaceae (27), Papilionaceae (12), Euphorbiaceae (7), Cyperaceae (6) and Acanthaceae (5) in the herbaceous layer. The recorded number families for the three vegetation layers above represented richness of 61 species, 98 shrubs and 106 herbaceous species (Appendix 4.2a–c).

In RSFR the species richness represented in brackets by each life form was trees (47), shrubs (84) and herbaceous species (114). The recorded tree species belong to 17 families, shrubs 37 families, and herbaceous species 33 families (Appendix 4.3a–c). The predominant families for tree species and their number of species in brackets for each was Mimosaceae (8), Caesalpinanaceae (6) and Papilionaceae (5). In the shrub layer it was Papilionaceae (8), Euphorbiaceae (7), Rubiaceae (7), Annonaceae (5) and Tiliaceae (5). The herbaceous layer was dominated by families Poaceae (29), Papilionaceae (13), Asteraceae (10), Cyperaceae (8), Acanthaceae (7) and Euphorbiaceae (5). The observed low species richness, particularly for the tree species, in both PKNFR and RSFR was largely related to their previous history, which suggests that large trees were harvested either for timber or for making charcoal.

The results showed variation in species density within and across sites. A similar trend in species richness was observed for the species density/ha in PKNFR and RSFR. Transect 2 had relatively low values in both reserves and more so in PKNFR. The overall species density in PKNFR for each life form was 3535 (trees), 1782 (shrubs) and 5542 (herbs), while in RSFR it was 2378 (trees), 2681 (shrubs) and 3149 (herbs)/ha.

The trees density (Fig. 4.2) in PKNFR in transect 1 was 1286 trees ha^{-1} and 1756 shrubsha^{-1} while in transect 2 it was low: 1092 and 925 shrubs ha^{-1}. The tree density in RSFR transect 1 was 1828ha^{-1} while shrubs were 1053ha^{-1}. In transect 2 the density went down to 1717

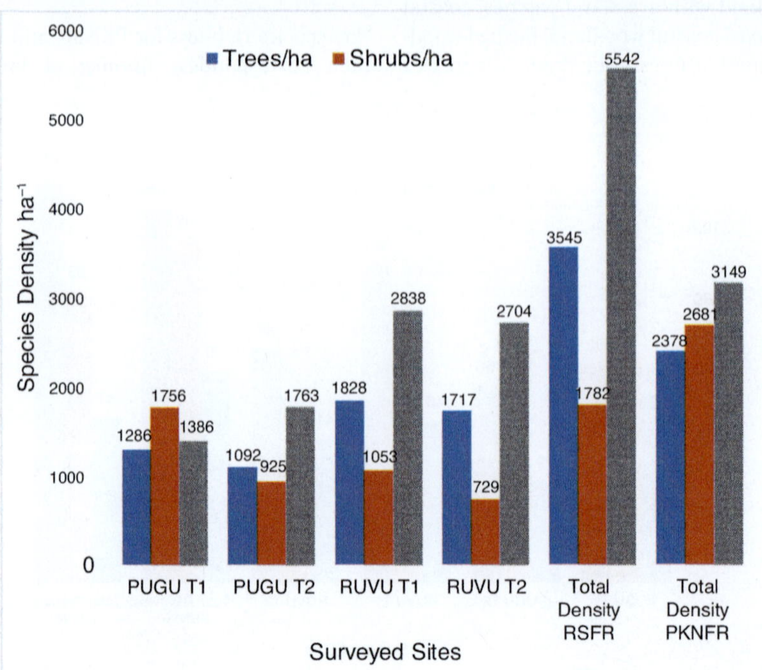

Fig. 4.2. Plant species density for trees, shrubs, and herbs in PKNFR and RSFR.

and 729 trees and shrubsha^{-1}, respectively. In both cases the shrubby species density decreased remarkably while for tree species the decrease was relatively marginal. It is worth mentioning that the observed discrepancies in species diversity did not directly relate to the current SGR project activities, but it was more related to the proximity of the reserves, particularly on transect 2 in both PKNFR and RSF, which was more on the settlements side. The low density in RSFR can thus be linked more with its proximity to the settlements, especially in parts dissected by the SGR, which was, to a lesser extent, in PKNFR. In RSFR, most shrubby species were mostly affected by charcoal making while for trees of DBH below five they were exploited more for poles.

4.4.3 Species diversity

The diversity of plant species in sampled transects across the two forest reserves is summarized in Fig. 4.3 and its inferences are explained below. The Shannon-Weiner diversity of the plant species within the studied project area ranged between 3.4 and 4.1. The tree species diversity did not vary remarkably between PKNFR and RSFR and across the transects. The overall species diversity in PKNFR was 3.54, 3.71 and 4.01 for trees, shrubs and herbs, respectively, while in RSFR it was 3.55, 3.43 and 4.11, respectively. It was interesting to note that although the variation in species richness and diversity in both sites was not remarkable, transect 2 in both PKNFR and RSFR presented comparatively low diversity values, which for trees went as low as 2.82 from 3.56 and for shrubs from 3.65 to 3.14 in PKNFR. As for the RSFR, the diversity of tree species in transect 1 was 3.37 but it went as low as 3.07 in transect 2. Similarly, the shrub species diversity recorded in transect 1 was 3.33 but in transect 2 it was 2.88. The diversity of plant species in the herbaceous layer did not vary remarkably across transects within the same site but it varied significantly across PKNFR and RSFR where it was higher in the latter than in the former. The findings showed

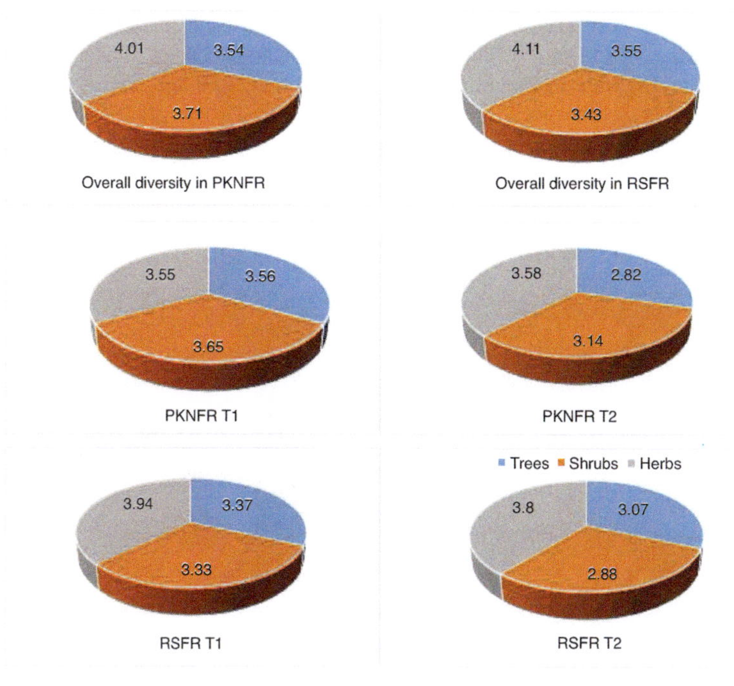

Fig. 4.3. Species diversity for trees, shrubs, and herbs in PKNFR and RSFR.

that transect 2 in both sites was relatively more on the disturbed site, which also happened to be more on the side bordering the settlements. This largely explains the observed variations in diversity between transects 1 and 2.

Theoretically, the Shannon-Weiner index ranges from 0–4.6 where a value near 0 indicates that every species in the sample is the same, and a value near 4.6 indicates that the diversity of species is higher, and individuals are evenly distributed between the plots. The Shannon and Wiener diversity indices mostly reported in the tropics range between 1.5 and 3.5 (Kent and Coker, 1992). Thus, the overall diversity of between 3.54 and 3.55 for trees and between 3.43 and 3.71 for shrubs recorded in both reserves is considered high, and it is more so for the herbaceous layer, which showed a diversity between 4.01 and 4.11. However, high species diversity between 4.01 and 4.11 in the herbaceous layer is an indication of a high degree of disturbance considering the nature of natural vegetation in both reserves, which is historically woodland/forest with a closed canopy cover. In such vegetation types, the woody species canopy cover is usually dense with low penetrant light to the ground cover. This usually renders species richness and diversity of the herbaceous layer low. The trees canopy was open mainly due to previous and ongoing anthropogenic disturbances. The open canopy has thus favoured the ground cover more among other light-loving species. The high anthropogenic pressure observed was also reflected in the varying diversity of the species in the herbaceous layer which was higher in RSFR transect 1 (3.94) and transect 2 (3.80). As for the PKFNR, it was 3.55 in transect 1 and 3.58 in transect 2.

4.4.4 Population structure

The study in forest population structure is the best way to understand regeneration and other dynamics which may either lead to increasing or decreasing stability of population structure. Population structure has also been used widely to examine the regeneration status of woody plants (Argaw *et al.*, 1999) as it is extremely dynamic and sensitive to changes in the level of regeneration (Cunningham, 2001; Moshi, 2007). The overall population structure of the project footprint summarized in Fig. 4.4 revealed that the majority of the tree stems were found in the lower DBH size classes, mostly between 5 cm and 19.9 cm and between 20 cm and 29.9 cm. The number of trees decreased progressively towards the higher DBH size classes giving rise to a characteristic 'inverted J-shaped' population distribution curve. This reveals presence of recent recruitment and regeneration in the lower size classes with a gradual decrease towards higher size classes and hence a stable and healthy population. Generally, the observed J-shaped

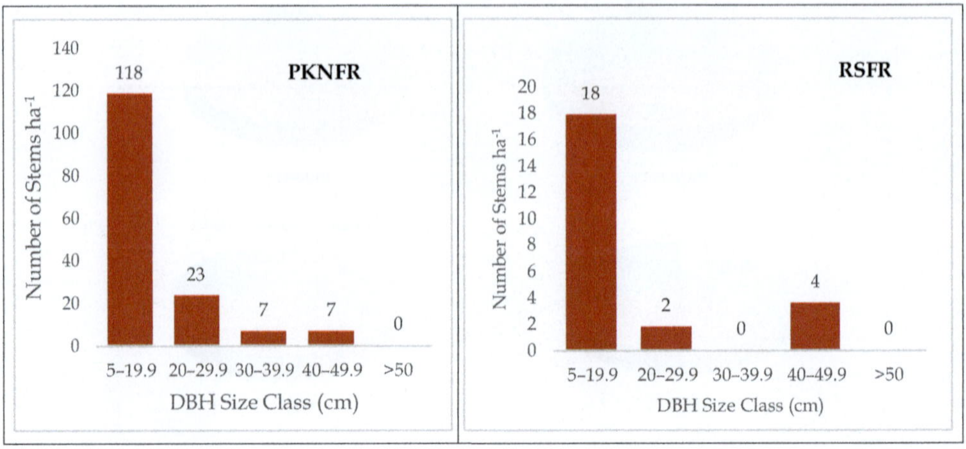

Fig. 4.4. DBH size class frequency distribution of trees identified in PKNFR and RSFR.

pattern of DBH size class distribution reveals that PKNFR and RSFR consist of an expanding population. Usually, the small-sized trees are significant reserve for replacing cut, large-sized and old trees (Kumlachew and Taye, 2003). The DBH size class between 5.0 cm and 19.9 cm accounted for about 76% of trees in all DBH size classes in PKNRF and 75% in RSFR, while the rest were distributed in the higher DBH size classes. Tree stems beyond DBH size class of 50 cm were not recorded in the present assessment in both PKNFR and RSFR. This is expected, given that concerted conservation efforts in PKNFR and RSFR started in recent years and the tree species are slowly regenerating from a serious previous anthropogenic pressure. It is worth noting that based on the obtained population structure, the surveyed sites in RSFR were relatively more disturbed. The DBH size class between 30 cm and 39.9 cm presented gaps which is an indication of selected pressure on tree species of such DBH size class, which in most cases are preferred for timber production. In PKNFR, species of high DBH size class above 30 cm were *Delonix regia*, *Afzelia quanzensis*, *Albizia glaberrima*, *Milicia excelsa* and *Ficus sur*, while in RSFR only *Bombax rhodognaphalon* was recorded. Stems of *Bombax rhodognaphalon* possibly remained standing as they are less favoured for timber.

4.4.5 Species of conservation importance and concerns

The IUCN red list species

The IUCN is the most comprehensive objective global tool for assessing conservation status of both plant and animal species. Knowledge on the conservation status of species is vital in catalyzing biodiversity conservation actions (IUCN, 2023). Table 4.2 and Fig. 4.5 summarize the findings from the survey conducted in both PKNFR and RSFR. During the survey, a total of nine plant species of conservation concern, as

Table 4.2. Threatened plant species recorded in Pugu-Kazimzumbwi Nature Forest Reserve.

S/N	Species name	Life form	IUCN status	Site	Transects/plots location
1	*Baphia kirkii*	Trees	VU	RSFR	T1P1, T1P7, T2P4
2	*Dalbergia melanoxylon*	Tree	NT	RSFR	T1P7, T1P5, T1P3, T1P2, T1P1, T2P5, T2P1
				PKNFR	T1P8, T1P4, T1P1, T2P4, T2P1, T2P2
3	*Dombeya acutangula*	Trees	CR	RSFR	T2P6, T2P5, T2P4
				PKNFR	T1P1, T2P6, T2P5, T2P2
4	*Milicia excelsa*	Tree	NT	PKNFR	T1P6, T1P3, T2P5
5	*Millettia micans*	Tree	EN	RSFR	T1P4, T1P6, T2P2, T2P3, T2P4
				PKNFR	T1P1
6	*Millettia usaramensis*	Tree	VU	PKNFR	T1P1, T1P4, T2P1
7	*Millettia impressa*	Shrub	VU	PKNFR	T1P2, T1P6, T1P7, T1P9, T2P5
8	*Uvaria kirkii*	Shrub	NT	RSFR	T1P7, T1P2, T1P1, T2P5
				PKNFR	T1P8, T1P6, T1P3, T1P2, T1P1
9	*Xylopia arenaria*	Shrub	VU	RSFR	T1P4
				PKNFR	T1P6, T1P2, T1P4, T1P5, T2P5

Fig. 4.5. Some of the threatened plant species identified in PKNFR and RSFR.

described in the various categories of the IUCN Red List of threatened species, were identified. Species of conservation concern were more in the PKNFR (8 species) than in RSFR (6 species), an observation which may be attributed to the level of anthropogenic disturbance which was more so in the RSFR than the PKNFR. The identified plant species fall under four out of the five major categories of the IUCN Red List of threatened species. In PKNFR, the threatened categories and the number of species in brackets were NT (4), VU (3), EN (1) and CR (1), whereas in RSFR the threatened species categories were VU (2), NT (2), EN (1) and CR (1).

In PKNFR and RSFR, a total of 15 and 13 plots, respectively, were found to host plant species of conservation concern as described in the various categories of IUCN Red List of threatened species. This implies that, on average, all plots in PKNFR had at least one plant species of conservation concern while in RSFR, 13 plots out of the 14 sampled plots recorded threatened species. Thus, while the identified threatened plant species were distributed in almost all plots, in some of the sampled plots the diversity and richness of threatened plant species were relatively high. For example, the number of plant species, in brackets, for the transects and plots sampled in PKNFR was: T1P1 (4), T1P8(3), T2P5 (4), T1P4 (3) T1P2 (3) and T1P6 (3), while in RSFR it was T1P1 (3), T2P5 (3), T1P7 (3), T2P4 (3) and T1P2 (2). Again, in this case, the PKNFR plots had relatively higher richness (4) of threatened species than in RSFR where a maximum of three species were observed.

Considering the small area surveyed for both PKNFR and RSFR and the large number of threatened plant species identified, the SGR alignment in the two forest reserves is thus crossing through critical habitats. This calls for effective conservation measures of the available plant resources. The impact of the SGR project on the plant species in most surveyed sites, however, was marginal given that a large part of it across the two reserves is aligned closer to and alongside the TRC old metre-gauge railway line, which translates to minimal clearance of vegetation cover in undisturbed areas. It is also worth mentioning that the anthropogenic disturbance which led to a remarkable species loss in these reserves has a long history as it emanated in the past. Currently, a large part of the vegetation is generally in the transitory succession phase. The fact that the identified species were generally widely distributed suggests their high abundance in both PKNFR and RSFR reserves. Most of the identified species were found in the moderately disturbed vegetation in a stretch of about 150m from the SGR towards the interior of the reserves. Only a few species were recorded below this average distance towards the SGR.

Endemic plant species

Endemic plant species refers to plants that are strictly native and confined to a particular region. Such species are of immense conservation concern because their loss in the native area will ultimately translate into extinction of such species. Based on FTEA, LEAP, POWO and *List of Targeted Plant Species, Subspecies and Varieties in the Eastern Arc Mountains and Coastal Forests of Tanzania and Kenya*, by Missouri Botanical Garden, a total of endemic plant species restricted to the coastal vegetation and two plant species only known to occur in Tanzania were identified. The names, habit and distribution of the identified endemic plant species are highlighted in Table 4.3.

The total number of plots in which coastal endemic species were identified in PKNFR and RSFR were seven and six, respectively, which implies that both forest reserves were rich in coastal endemic species. The number of endemic species identified in the established plot ranged between two and four. In this case, T1P1, T1P2, T1P3 and T1P6 in PKNFR recorded the highest number of species which was between two and three species. In RSFR, the most diverse plots were T1P1, T1P4, T1P7, T2P2, T2P4, which had two endemic plant species each. The remaining plots had one endemic plant species in both reserves. Sites with high concentration of endemic and threatened plant species constitute critical habitats; hence, they should be a focal point in the biodiversity monitoring and management plan.

CITES listed species

Tanzania became among the 183 CITES member states in 1978 and started the implementation of the convention in 1980 (CITES Secretariat, 2013). The primary role of the convention is to control and monitor international trade on endangered and threatened species. This is achieved through a well-established international legal framework for co-operation between the producer and consumer of the wild plant species. According to CITES Secretariat (2022), plant species are categorized into three appendices with Appendix I comprising all plant species at high risk of extinction. In this appendix trade in wild specimens without a special permit is prohibited. Appendix II includes species whose survival is not yet threatened but projected to be threatened in the absence of conservation efforts and/or regulations after a given period of time. International trade on such species is not allowed for both wild and artificially propagated or captive-breed specimens except through licensing. In Appendix III are plant species that need the co-operation of CITES member countries to prevent unsustainable or illegal exploitation at the request of a party under regulation of trade on the species. International trade on specimens of species listed therein is allowed only on presentation of the appropriate permits or certificates.

In the two surveyed forest reserves a total of four plant species, namely *Milicia excelsa*, *Dalbergia melanoxylon*, *Dalbergia obovata* and *Dalbergia arbutifolia*, which fall under Appendix II of the CITES were identified. The three *Dalbergia* species were more abundant in RSFR than in PKNFR while *Milicia excelsa* was identified in PKNFR only, in plots T1P6, T1P3, T2P5. In RSFR, the *Dalbergia melanoxylon* was identified in transect 1 in plots 1, 2, 3, 5, 7 and in transect 2 plot numbers 1 and 5; and *Dalbergia obovata* in transect 1 plot 7 and transect 2 plots 1 and 6. *Dalbergia arbutifolia* was recorded in transect 1 plot 7. In PKNFR, *Dalbergia melanoxylon* was distributed in transect 1 plots 1, 4, 8 and transect 2 plots 1, 2 and 4. *Dalbergia obovata* was recorded in transects 1 and 2, plots 3 and 6, respectively, while *Dalbergia arbutifolia* was observed in transects 1 and 2, plots 6 and 8.

National reserved trees

A total of 24 specified tree species are declared as national reserved trees throughout Tanzania's

Table 4.3. Endemic plant species recorded in PKNFR and RSFR.

S/N	Species name	Life form	Site	Transect and plot number	Physiognomic vegetation type
1	*Acacia zanzibarica*	Tree	RSFR	T1P5	Bushed woodland
				T1P8	Seasonal wetland with emergent trees
2	*Baphia kirkii*	Tree	RSFR	T1P1	Bushed grassland
				T1P7	Bushed woodland
				T2P4	Bushed woodland with grassland patches
3	*Combretum harrisii*	Tree	RSFR	T2P2	Bushland
			PKNFR	T1P6	Bushland
4	*Deinbollia borbonica*	Shrub	PKNFR	T1P1	Bushland
				T1P9	Bushland with patches of grassland
				T1P3	Bushland with emergent trees
5	*Millettia micans*	Tree	RSFR	T1P4 and T1P6	Bushland with grassland patches
				T2P2	Bushland
				T2P3, T2P4	Bushed woodland with grassland patches
			PKNFR	T1P1	Bushland
6	*Tetracera litoralis*	Shrub	PKNFR	T1P1	Bushland
7	*Xylopia arenaria*	Shrub	RSFR	T1P4	Bushland with grassland patches
				T1P6	Bushland
			PKNFR	T1P2	Coastal thickets
				T1P4	Mixed woodland
				T1P5	Riparian forest
				T2P5	Bushed woodland
8	*Uvaria kirkii*	Shrub	RSFR	T1P7	Bushed woodland
				T1P2	Bushed grassland
				T1P1	Bushed grassland
				T2P5	Bushland
			PKNFR	T1P8, T1P6	Bushland
				T1P3	Bushland with emergent trees
				T1P2	Coastal thickets
				T1P1	Bushland

mainland and national reserved tree species. Such trees are protected by the Forest Reserved Trees Order, (1995) ordinance cap. 389 section (17) 1 and forest act 2002 part viii, section 65, subsections 4a and 4b. In line with the issued order, reserved species, namely *Dalbergia melanoxylon* (mpingo), *Dalbergia obovata, Dalbergia arbutifolia, Milicia excelsa* (Mvule), *Afzelia quanzensis* (Mbabakofi) and *Pterocarpus angolensis* (Mninga), *Markhamia obtusifolia* (Mtalawanda), *Markhamia lutea* (Mtalawanda) and *Markhamia acuminate* (Mtalawanda) were identified from the two forest reserves during the survey. The actual location of all identified reserved species is specified in Table 4.4. More species were found in PKNFR (8) unlike in RSFR where six species were recorded.

The richness of plant species of conservation concern in various sites along the SGR alignment within PKNFR and RSFR

The highest number of species of conservation concern identified in PKNFR was seven in T1P1, four in each of T2P5 and T1P6, three in each of T1P3, T1P4 and T1P2, two species in each of T1P9, T2P1, T2P4 and T2P6 while the lowest was one recorded in T1P5. In RSFR, three species were recorded in each of plots T1P4, T1P8. T2P4, T2P5, T1P7 and T1P1, while T1P2 and T2P2 presented two species each. On the other hand, T1P5, T2P3, T2P6, T1P3 and T1P8 had one species each (Table 4.5).

Invasive plant species occurring within the study area

A total of seven alien invasive species were identified within the PKNFR and RSFR sites (Fig. 4.6). The identified species, transect and plot number (in brackets) were *Lantana camara* (PKNFR T1P1 and T1P9), *Calotropis procera* (PKNFR T1P9, RSFR T1P2), *Cassia spectabilis* (PKNFR T1P3), *Cassytha filiformis* (PKNFR T1P1), *Ricinus communis* (RFSR T1P1, T1P2, PKNFR TIP1, T2P3, T1P7, T1P9), *Xanthium strumarium* (PKNFR T1P1 and RSFR T1P2) and *Leucaena leucocephala*. The predominant invasive plant species within the reserves were *Xanthium strumarium, Calotropis procera* and *Lantana camara*. Although *Cassytha filiformis* was not widely spread it was still abundant in all plots where recorded. This species is both invasive and parasitic and has been reported to cause serious infestation in wild species and crop loss in cashew nut trees in Pwani, Lindi and Mtwara regions (Buriyo *et al.*, 2015). It usually spreads fast and outcompetes the host as it obtains both water and food from the vascular tissue of the host plants, which ultimately causes the death of the host.

To effectively control these invasive species, mechanical control such as slashing, uprooting prior to flowers or seed setting at seedling stage should be done. Under controlled environments, pre-arranged burning, especially prior to seed setting, is recommended to eliminate the mother plants while also discouraging seed dispersal. The chemical control method, especially the application of selective foliar herbicides, is necessary. Other measures include biological control, particularly the use of host-specific natural enemies (bacteria, fungi, viruses, or parasitic or predatory organisms, such as insects), which feed on vital organs such as leaves and reproductive structures such as flowers and seeds. To effectively control the alien invasive plant species, integrated weed management employing a combination of the methods outlined above can be adopted.

Anthropogenic disturbance in PKNFR and RSFR

Anthropogenic disturbances translating mostly into selective removal of mature tree species, burning or massive clearing of vegetation greatly affects the soil seedbanks, population structure, reproductive potential, spatial structure, and temporal dynamics of ecological systems. The PKNFR and RSFR have been subjected under severe anthropogenic pressure in the past and for quite a long time due to their proximity with the densely populated towns and city of Dar es Salaam. Although the two forests are currently protected by forest legislations, the demand for timber, poles, firewood and charcoal from the surrounding populations is still growing. This is because alternatives to these resources are still limited and, where available, the problem is still intensified by poverty, want of awareness of biodiversity conservation and its significance by the surrounding community, and insufficient

Table 4.4. Reserved trees and woody species identified in PKNFR and RSFR.

S/N	Species name	Site	Actual location	Associated physiognomic vegetation type
1	*Afzelia quanzensis*	RSFR	T1P2, T1P3	Bushed grassland
			T1P4, T1P6	Bushland with grassland patches
		PKNFR	T1P4	Mixed coastal woodland
			T2P1	Bushed woodland
2	*Dalbergia melanoxylon*	RSFR	T1P1, T1P2, T1P3	Bushed grassland
			T1P5, T1P7	Bushed woodland
			T2P1, T2P5	Bushland
		PKNFR	T1P8, T1P1	Bushland
			T1P4	Mixed coastal woodland
			T2P1, T2P4	Bushed woodland
			T2P2	Bushland with grassland patches
3	*Pterocarpus angolensis*	RSFR	T2P4 T1P5, T1P2	Bushed woodland with grassland patches
			37M0492733; UTM 9240066 between T1P1 & T1P2	Bushed grassland
4	*Dalbergia arbutifolia*	RSFR	T1P7	Bushed woodland
		PKNFR	T1P1, T2P8	Bushland
			T1P2, T2P6	Coastal thickets
5	*Dalbergia obovata*	PKNFR	T1P3	Bushland with emergent trees
		RSFR	T1P7	Bushed woodland
			T2P1	Bushed woodland
			T2P6	Seasonal bushed wetland
6	*Milicia excelsa*	PKNFR	T1P6	Bushland
			T1P3	Bushland with emergent trees
			T2P5	Bushed woodland
7	*Markhamia obtusifolia*	RSFR	T1P6, T1P4	Bushland with grassland patches
			T1P3, T1P2, T1P1	Bushland grassland
			T2P5, T2P2	Bushland
			T2P4, T2P3	Bushed woodland with grassland patches
		PKNFR	T1P9	Bushland with grassland patches
			T2P3	Bushland
8	*Markhamia lutea*	PKNFR	T1P3	Bushland with emergent trees
			T1P2	Coastal thicket
9	*Markhamia acuminata*	PKNFR	T1P9	Bushland with grassland patches
			T2P5	Bushed grassland

control by the forest authorities that reside far away from the forest reserve. Although recently anthropogenic pressure in the two forests has diminished to some extent due to protection, signs of anthropogenic pressure on the forest resources, structure, composition and stratification are still conspicuous and more so in RSFR. The patchy disturbance has resulted in formation of gaps and juvenile dominated populations, which, in the absence of adequate, timely and effective conservation practices, are likely to deepen, especially in RSFR.

Table 4.5. Richness of species of conservation concern per plot in RSFR and PKNFR.

RSFR plots	Species name	PKNFR plots	Species name
T1P4	*Xylopia arenaria, Baphia kirkii, Millettia micans*	T1P1	*Uvaria kirkii, Tetracera litoralis, Deinbollia borbonica, Dalbergia melanoxylon, Dombeya acutangula, Millettia micans, Millettia usaramensis*
T2P4	*Baphia kirkii, Dombeya acutangula, Millettia micans*	T2P5	*Xylopia arenaria, Millettia impressa, Milicia excelsa, Dombeya acutangula*
T2P5	*Dalbergia melanoxylon, Uvaria kirkii, Dombeya acutangula*	T1P6	*Combretum harrisii, Xylopia arenaria, Uvaria kirkii, Millettia impressa*
T1P7	*Baphia kirkii, Dalbergia melanoxylon, Uvaria kirkii*	T1P8	*Dalbergia melanoxylon, Uvaria kirkii*
T1P1	*Baphia kirkii, Dalbergia melanoxylon, Uvaria kirkii*	T1P3	*Deinbollia borbonica, Uvaria kirkii, Milicia excelsa*
T1P2	*Dalbergia melanoxylon, Uvaria kirkii*	T1P4	*Xylopia arenaria, Millettia usaramensis, Dalbergia melanoxylon*
T2P2	*Millettia micans, Combretum harrisii*	T1P2	*Xylopia arenaria, Uvaria kirkii, Millettia impressa*
T1P5	*Dalbergia melanoxylon*	T1P9	*Deinbollia borbonica, Millettia impressa*
T2P3	*Millettia micans*	T2P1	*Dalbergia melanoxylon, Millettia usaramensis*
T2P6	*Dombeya acutangula*	T2P4	*Dalbergia melanoxylon*
T1P3	*Dalbergia melanoxylon*	T2P6	*Dombeya acutangula*
T1P8	*Acacia zanzibarica*	T1P5	*Xylopia arenaria*

The results revealed a total of 29 tree species identified during the survey from different sampled plots within PKNFR and RSFR to have been impacted by at least one anthropogenic disturbance type. Various disturbances recorded within the reserves were illegal tree cutting for fuelwood and charcoal as well as selective tree logging for timber and poles (Fig. 4.7 and Appendix 4.1). As for the charcoal making, the most affected tree species based on the observed frequency of cut stems included one species of conservation concern, namely *Millettia micans*, an endangered and endemic plant species; and *Afzelia quanzensis*, which is a reserved tree species. In addition, several other tree species, namely *Grewia bicolor*, *Grewia conocarpa*, *Markhamia obtusifolia*, *Brachystegia bussei* and *Hymenocardia ulmoides*, were all harvested illegally for charcoal making. Charcoal making was evidenced by the presence of abandoned and fresh charcoal kilns and, in most cases, several cut trees of pole size. It was observed that, most trees cut for fuelwood and charcoal were resprouting, which indicates that if proper conservation and management practices are strictly enforced, the species can regenerate. Presence of kilns and/or their remains were observed more in RSFR than in PKNFR.

As for the disturbance caused by intentional fires, the findings revealed that the fires were not recent in most cases except around places where fresh and abandoned charcoal kilns were found. Based on personal communication with the forest officer, the most recent fire events occurred in the past 3 years but the effects on the vegetation cover were still evident and mostly they are set by hunters of small wild mammals. It was further observed that fires are more common during the dry season, some of which seem to have been caused deliberately. Dry-season fires were more rampant in RSFR than in PKFNR, possibly due to a more open nature of vegetation in RSFR. Species that were mostly affected included *Combretum harrissii*, a coastal endemic plant species, while others were *Brachystegia bussei*, *Xeroderris stuhlmannii*, *Pteleopsis myrtifolia*, *Flueggia virosa*, *Trema orientalis*, *Dichapetalum stulmannii*, *Harrisonia abyssinica* and *Hymenocardia ulmoides*. However, some of the tree species were actively regenerating from the partially burnt stumps. The regenerants from the stumps included *Combretum harrissii*, *Flueggia virosa*, *Diplorhynchus*

Fig. 4.6. Some of the invasive plant species identified in PKNFR and RSFR.

condylocarpon, *Markhamia obtusifolia*, *Pteleopsis myrtifolia* and *Hymenocardia ulmoides*, while other burnt, woody species had died completely. Usually, fire damage to vegetation has a higher magnitude than damage caused by pit sawing, fuelwood and pole cutting (Kielland-Lund, 1990). Depending on its frequency and intensity, fire has a tendency to kill or scorch trees and gradually reduce and modify tree forest cover. An increase in fire intensity may increase or maintain species diversity but it may also reduce diversity of woody species (Lyaruu, 1997). Thus, the control of fires is deemed vital as it tends to keep the vegetation open, eliminates certain species and selectively favours domination by fire-resistant species. The fire can effectively hold back recruitment of trees into mature-age classes. This can finally result in populations with uneven age distribution, and such populations are highly unstable.

Furthermore, during the survey it was established that, to a large extent, anthropogenic pressure on tree species for timber was particularly exerted on *Hymenaea verrucosa* and some of the threatened and reserved tree species, namely *Pterocarpus angolensis*, *Afzelia quanzensis*, *Milicia excelsa*. The evidence that these species were harvested for timber was drawn from the DBH size of the remaining stump, which, in most cases, was generally far above pole size and hence suitable for timber.

The exploitation of tree species for poles in the two forest reserves was the leading cause of disturbance recorded during the study. In most cases, poles and withies are the primary building materials and, to a lesser extent, the source of carving material in the surrounding communities. The seriously impacted species for pole cutting were *Millettia micans*, *Dalbergia melanoxylon*, *Pterocarpus angolensis*, *Afzelia quanzensis*,

Fig. 4.7. A representation of some of the anthropogenic disturbances reported in the two forest reserves.

Markamia obtusifolia, *Dombeya shupangae*, *Grewia bicolor*, *Albizia petersiana*, *Brachystegia bussei*, *Grewia conocarpa*, *Hymenocardia ulmoides*, *Diplorhynchus condylocarpon*, *Trema orientalis* and *Hymenaea verrucosa*. The first four tree species are species of conservation importance including endangered, endemic and reserved tree species.

Generally, the assessment of the major disturbance type in PKNFR and RSFR showed that extraction of poles was the leading type of anthropogenic disturbance as it accounted for approximately 44% of all anthropogenic disturbance types. This was followed by charcoal making (43%; Fig. 4.8). Illegal tree logging for timber and disturbance caused by intentional fires made little contribution as they accounted for 8% and 5% of the reported anthropogenic disturbances, respectively. The high incidences of pole cutting could possibly be due to less time required for felling the stems and the ease with which they can be moved out of the forests without being noticed by forest officers. On the contrary, a low level of disturbance emanating from illegal logging of trees for timber is an indication that, currently, this disturbance is not a major threat in the two reserves. However, the effect of selective logging of trees for timber in the past is still evidenced by open gaps.

It was further established that in the PKNFR, the leading anthropogenic disturbances were fire incidences, where 13 incidences corresponding to 62% of fire incidences, were recorded compared to RSFR, where eight incidences, equivalent to 38% of all incidences, were recorded. Additionally, a relatively large

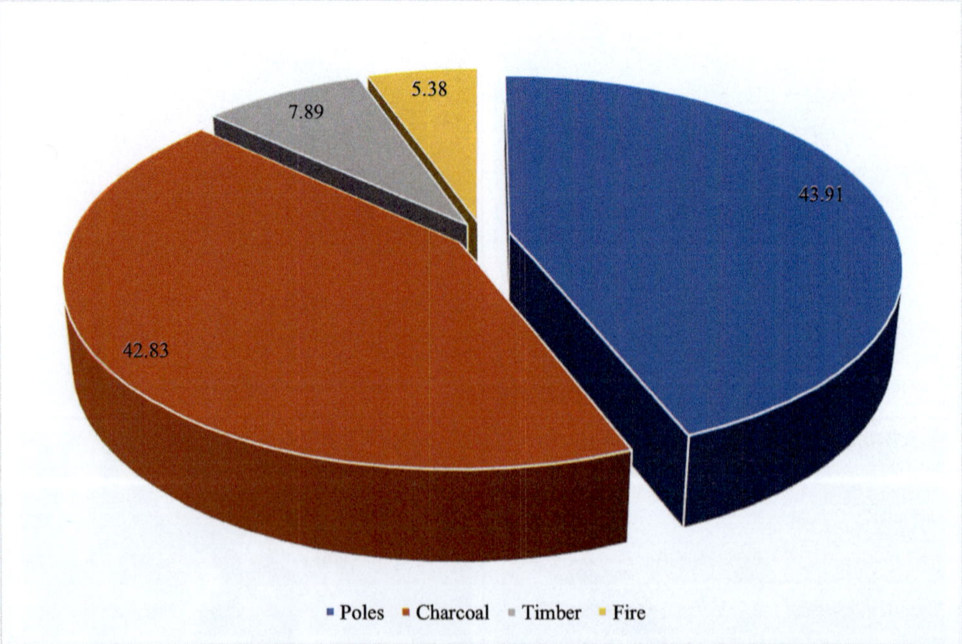

Fig. 4.8. Anthropogenic disturbance types and severity of their impacts in PKNFR and RSFR.

number of trees were exploited in PKNFR, in which 11 incidences, equivalent to 61% of recorded trees cut for timber, were reported in comparison to RSFR, where seven incidences (39%) were reported. In RSFR, incidences of trees extracted for charcoal making and poles were relatively higher compared to PKNFR. The numbers of trees cut for poles and charcoal in RSFR were 25 (64%) and 24 (67%), respectively, while in PKNFR it was 14 (36%) and 12 (33%), respectively (Fig. 4.9).

Grazing is another form of anthropogenic pressure observed in the study area. However, it was interesting to note that livestock grazing and/or its signs were not observed in PKNFR but in RSFR. Livestock grazing was indicated by presence of cow dung, herds of cattle and cattle trails, and foraged grasses. These indicators were seen in transect 2 plot 1 and in transect 1 plots 3, 5, 6, 7 and 8. Other signs of anthropogenic disturbance were evidenced by the presence of clear but informal footpaths/trails. This was observed in 12 plots of which two plots were found in PKNFR along transect 1 plots 3 and 9, while 10 trails were recorded in RSFR. Trails in RSFR were distributed in transect 1 plots 6, 2, 4, 5 and 7, as well as in transect 2 plots 2, 3, 4 and 5, and near the riverine vegetation. The presence of such footpaths indicates anthropogenic disturbance as these are the routes commonly used by the trespassers to access various timber and non-timber products from the forest. The fact that more trails were predominant in RSFR is an indication that anthropogenic pressure was higher in this reserve than in PKNFR.

The contribution of some of the anthropogenic disturbance types on the composition and structure of vegetation in both reserves was marginal. However, the loss of biodiversity emanating from the combined effect of various types of such disturbances cannot be underestimated in view of the ecological significance of both PKNFR and RSFR, which are the home of a significant number of endemic and threatened plant species. As such, effective conservation interventions are deemed necessary before the situation gets out of hand. Based on the observed anthropogenic disturbances, it is evident that the disturbed forest structure and composition of the standing vegetation are largely attributed to the previous and ongoing anthropogenic

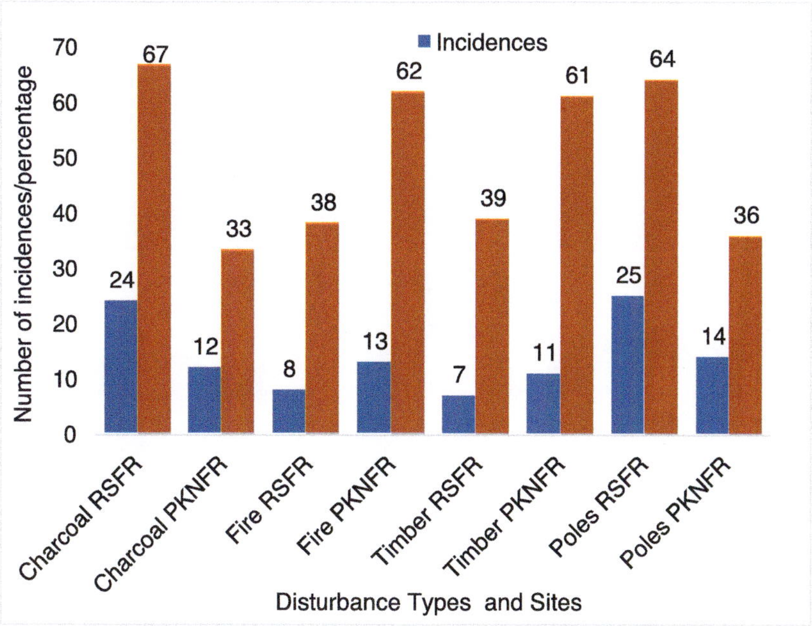

Fig. 4.9. A comparison of the severity of various anthropogenic disturbance impacts across PKNFR and RSFR.

disturbances. The SGR project and related activities are part of the latter.

Soil erosion

Erosion was observed in some parts of the forest reserve, particularly in PKNFR. This was evident in sites where soil was excavated, topsoil removed, and vegetation cover stripped off leaving bare soil. In addition, parts of roads used to access soil from the bore holes were eroded. In some areas, especially on the steep slopes, some parts were highly eroded, and some gullies were developing as a result of removed top vegetation cover and loosened soil, leading to a significant portion of soil washed off. Some of the areas severely affected by erosion had developed deep gullies measuring a width of approximately 2 m and depth of 3.5 m (Fig. 4.10). Sites that were severely affected by erosion in PKNFR were all along transect 1 on plot P1 near the Msimbazi river and SIDO area, where it covered plots P3, P4, P7 and P8. These sites should be rehabilitated and revegetated for effective control of soil erosion. Steep slopes should be reduced by constructing berms across the ramps and redirecting the surface runoff from the crest down the toes. Both tussock and stoloniferous grasses such as *Chloris gayana* and the use of eco-friendly soil-covering mesh should be employed.

4.5 Conclusion

The data for the findings presented in this report were collected towards the end of the wet season. This was important to identify the maximum number of plant species to enrich the species checklist with herbaceous species, which, in most cases, resprout during the wet season.

The SGR project has enormous environmental and socio-economic benefits at individual, national and regional level. Although the project is anticipated to improve people's livelihood in its coverage area, it also has potential to produce impact on the vegetation. The impacts from the SGR project on plant communities are likely to emanate from construction activities involving clearing of land for construction of various infrastructures. Such impacts will be more localized along the SGR right of way.

Fig. 4.10. Soil erosion associated with project activities in some parts of PKNFR.

However, the potential impacts identified can be alleviated and/or prevented if measures suggested under the Recommendations section are implemented adequately, aptly and correctly.

The findings of the present study revealed that both PKNFR and RSFR are under severe anthropogenic pressure. The major disturbances in the forest resources were predominantly illegal harvesting of tree species for timber, poles, fuelwood and charcoal making. Other anthropogenic activities reported were livestock grazing and intentional fires started by hunters of small mammals and intruders harvesting wild bee honey. The major driving pressure was primarily the proximity of the two reserves to the urban centres of Dar es Salaam City, and Kisarawe and Kibaha towns, which are characterized by high and growing populations dependent on forest resources coupled with poverty and poor forest management systems. In addition, anthropogenic disturbances were the cause of gully erosion, especially on steep slopes, and was exacerbated by excavation and vegetation removal.

Phytogeographically, vegetation in the PKNFR and RSFR is part of Zanzibar-Inhambane regional mosaic phytochorion. Within the PKNFR and RSFR, this vegetation type was characterized by several plant communities, namely bushland, bushland with emergent trees, bushland with grassland patches, coastal thickets, mixed coastal woodland, riparian forest and bushed woodland, which in most cases were regenerating.

The dominant woody species identified in both forest reserves were *Albizia glaberrima, Caesalpinia bunduc, Cassia siammea, Cassia spectabilis, Flueggea virosa, Harrisonia abyssinica, Pteleopsis myrtifolia* and *Hymenocardia ulmoides*. The subdominant woody species on the other hand were *Grewia forbesii, Albizia versicolor, Uvaria kirkii, Dichapetalum mossambiscense, Landolphia kirkii, Dalbergia obovata* and *Markhamia acuminata*.

Both PKNFR and RSFR are rich and diverse in terms of plant species. The observed Shannon-Weiner diversity of the plant species was remarkably high considering that diversity indices mostly reported in the tropics are in the range well below the values recorded in the two reserves.

Although the two forest reserves have been subjected to severe anthropogenic pressure in the past and, to some extent, to date, the overall plant species population structure in both forest reserves is expanding. In comparison to PKNFR, RSFR was relatively more affected by the ongoing anthropogenic pressure where, in most cases, trees were illegally logged for timber, poles and charcoal, as evidenced by several fresh and abandoned charcoal kilns in the reserve. The most logged tree species were *Pterocarpus angolensis, Baphia kirkii, Afzelia quanzensis, Milicia excelsa* and *Hymenaea verrucosa*. However, to a larger extent, the observed anthropogenic pressure could not be directly associated with SGR project activities except in some areas where land excavation and stripped topsoil resulted in gully erosion and removal of vegetation cover. Nevertheless, biodiversity loss emanating from the combined effect of various types of anthropogenic disturbance may be detrimental considering the ecological significance of both PKNFR and RSFR.

PKNFR and RSFR host a significant number of vulnerable, near-threatened, threatened, endangered, critically endangered, endemic, CITES-listed and national reserved plant species. As such, large parts of the reserve can be regarded as sensitive/critical habitat. However, based on the overall population structure obtained from both PKNFR and RSFR it is evident that most species, including the species of conservation concern, are regenerating and reported in other parts of the reserves and across coastal vegetation. The SGR project operations are thus less likely to induce significant negative impact to such habitats.

Several invasive species were identified within the two forest reserves in which *Lantana camara, Leucaena leucocephala* and *Xanthium strumarium* were dominant. These species should be controlled before they spread further into the reserves.

4.6 Recommendations

4.6.1 Mitigation measures against loss of species of conservation concern

The PKNFR and RSFR area impacted by the project hosts a significant number of diverse plant species of conservation concern. However, due to the distribution pattern of such plant species within the PKNFR, RSFR, coastal vegetation and other floristic regions of Tanzania, the project undertaking will have minimal negative impact on such species. As for the endemic and threatened plant species, an informed monitoring plan should be devised and implemented to ensure zero harm to the plant species. Environmental auditing of the species of conservation significance encountered in the project area should thus be done soon and on a regular basis to ascertain their status. Further surveys and literature review should be conducted to ascertain presence, regeneration potential, abundance and distribution of such species within the two forest reserves as the study was limited to vegetation cover alongside the SGR alignment.

Among the potential environmental impacts likely to occur when the project is fully commissioned are loss of vegetation, loss of species of conservation significance, loss of scenic beauty and overall loss of biodiversity. As mitigation actions, the proprietor should avoid unnecessary clearance of vegetation as much as possible, especially in the infrastructure's areas.

4.6.2 Rehabilitation/revegetation of cleared project footprint and erosion control

Areas within the PKNFR and RSFR stripped of their topsoil and vegetation cover or cut/excavated during the construction phase should be rehabilitated and revegetated using plant species indigenous to PKNFR and RSFR. To effectively curb soil erosion on steep slopes, it is proposed to rehabilitate such areas using indigenous tussock grasses and degradable mesh as has already been done in some steep sides of the constructed SGR alignment. Steep slopes should be reduced by constructing berms across the ramps and redirecting the runoff from the crest to reduce the fluvial effect of running water and denudation. Both tussock and stoloniferous grasses such as *Chloris gayana* and the use of soil covering mesh can be employed.

In cases where the native flora is disturbed, in the infrastructural areas, it should be replaced with native species. The checklist of tree species indigenous to PKNFR and RSFR are provided in the Appendices 4.2a–2c and 4.3a–3c, respectively, in the catalogue. In addition, the contractor is advised to close all access roads and diversions after construction of operational roads.

4.6.3 Monitoring and possible relocation of species of conservation concern

Clearance of trees and removal of species with high conservation values such as the endemic species, and CITES-listed and IUCN-listed threatened species, might be inevitable once the project takes off. As part of the monitoring plan, an inventory of trees likely to be cut and the number of trees to be planted as part of a compensatory tree-plantation programme should be in place well in advance. Monitoring of species of conservation concern and, if possible, relocation of such species to other suitable sites is recommended.

4.6.4 Monitoring of sensitive habitats

All critical habitats/sampling points identified to harbour threatened and endemic plant species should be monitored frequently. If possible, such sites should be set aside for *in situ* germplasm conservation of the host species. Effective monitoring of such species as well as introduction of other mitigation measures is paramount and may prove quite effective due to their confined dispersion.

4.6.5 Measures to alleviate anthropogenic pressure

To alleviate anthropogenic pressure from the surrounding population on ecosystems goods and services provided by the PKNFR and RSFR, especially building materials, fuelwood and medicinal plants, it is recommended to educate and encourage the neighbourhood to establish their own woodlots using the selected fast-growing woody species. This will ultimately lessen their dependency on forest products such as poles, fuelwood and charcoal from the forest reserves.

4.6.6 Management and control of alien invasive plant species

The project proprietor and forest reserve management should work closely to avoid introduction of alien invasive species in the forest reserve. All alien invasive plant species identified within the project area should be managed and controlled effectively. Among the recommended control and management interventions are removal and destroying of such plant species from the environment as quickly as possible. This can be accomplished through felling; slashing; bark stripping; ringbarking; hand pulling/uprooting prior to seed setting; chemical control, such as application of foliar herbicides; biological control, where necessary; or a combination of two or more of the above methods. In an environment where the population of alien invasive species have thrived in a wider area and proved difficult to eradicate, effective control measures should be instituted to curb further spread.

References

Forest Reserved Trees Order, section 17 (1) G.N. (1995) no 506 of 1995. United Republic of Tanzania.

Argaw, M., Teketay, D. and Olsson, M. (1999) Soil seed flora, germination and regeneration pattern of woody species in an acacia woodland of the rift valley in ethiopia. *Journal of Arid Environments* 43(4), 411–435.

Baker, N.E. and Baker, E.M. (2002) *Important Bird Areas in Tanzania: A First Inventory*. Wildlife Conservation Society of Tanzania, DSM, Tanzania.

BirdLife International (2012) Important bird areas factsheet: Kisarawe district coastal forests. Available at: http://www.birdlife.org on 07/12/2012 (accessed 26 April 2024).

Burgess, N.D. and Clarke, G.P. (2000) *The Coastal Forests of Eastern Africa*. IUCN Forest Programme, Gland, Switzerland and Cambridge, UK.

Buriyo, A.S., Kasuga, L., Moshi, H.N. and Nene, W.A. (2015) Ecological distribution and abundance of the parasitic weed *Cassytha filiformis* L. (Lauraceae) in major Cashew, *Anacardium occidentale* L. growing regions in Tanzania. *International Journal of Basic and Applied Sciences* 5(3), 109–116.

CITES Secretariat (2013) Convention on International Trade in Endangered Species of Wild Fauna and Flora. In: *Sixteenth meeting of the Conference of the Parties, Bangkok (Thailand), 3–14 March*, Consideration of Proposals for Amendment of Appendices I and II.

CITES Secretariat (2022) Convention on International Trade in Endangered Species of Wild Fauna and Flora. Available at: https://www.cites.org (accessed 26 April 2024).

Cunningham, A.B. (2001) *Applied Ethnobotany. People Wild Plant Use and Conservation*. Earthscan Publications Ltd, London.

Doggart, N. (2006) *Filling the Knowledge Gap: Methods Manual*. Tanzania Forest Conservation Group/ Museo Tridentino di Scienze Naturali, Dar es salaam, Tanzania.

Hawthorne, W.D. (1993) East African coastal forest botany. In: Lovett, J.C. and Wasser, S.K. (eds) *Biogeography and Ecology of the Rain Forests of Eastern Africa*. Cambridge University Press, Cambridge, UK, pp. 57–99.

IFC (2012) *International Finance Corporation: Performance Standards on Environmental and Social Sustainability*. World Bank Group.

IUCN (2023) The IUCN Red List of Threatened Species. Version 2023-1. Available at: https://www.iucnredlist.org (accessed 26 April 2024).

Keith, S., Urban, E.K. and Fry, C.H. (eds) (1992) *The Birds of Africa*, Vol. IV. Academic Press, London.

Kent, M. and Coker, P. (1992) *Vegetation Description and Analysis: A Practical Approach*. John Wiley and Sons, New York, pp. 167–169.

Kielland-Lund, J. (1990) Influence of grass fires on african landscape ecology. *Proceedings of a Joint Seminar/Workshop on Manage-Ment of Natural Resources in Tanzania under SUA/AUN Cooperation* 53, 46–54.

Kumlachew, Y. and Taye, B. (2003) The woody composition and structure of masha anderacha forest, south-western ethiopia. *Ethiopian Journal of Biological Science* 2(1), 31–48.

Lyaruu, H.V.M. (1997) *Plant community structure and land degradation in Mikumi National Park, Tanzania*. Institutionen för växtbiologi, Uppsala, pp. 1–35.

Mittermeier, R.A., Gil, P.R., Hoffman, M., Pilgrim, J., Brooks, T, *et al*. (2004) *Hotspots Revisited*. CEMEX, Mexico.

Morris, P. and Therviel, R. (eds) (2001) *Methods of Environmental Impact Assessment*, 2nd edn. Spon Press, London.

Moshi, H.N. (2007) Assessment of structure and soil seed bank diversity of the South Uluguru Mountains flora in Tanzania. MSc thesis, University of Dar es Salaam.

Myers, N., Mittermeier, R.A., Mittermeier, C.G., Da Fonseca, G.A. and Kent, J. (2000) Biodiversity hotspots for conservation priorities. *Nature* 403, 853–858.

Parthasarathy, N. and Karthikeyan, R. (1997) Plant biodiversity inventory and conservation of two tropical dry evergreen forests on the coromandel coast, south india. *Biodiversity and Conservation* 6, 1063–1083. Available at: https://doi.org/10.1023/A:1018328016810

POWO (2023) Plants of the World Online. Facilitated by the Royal Botanic Gardens, Kew, London. Available at: http://www.plantsoftheworldonline.org/ (accessed 9 May 2023).

Rocky, J. and Mligo, C. (2012) Regeneration pattern and size-class distribution of native tree species in exotic plantation in Pugu forest reserve, Tanzania. *International Journal of Biodiversity Conservation* 4(1), 1–14.

Shannon, C. and Weaver, W. (1948) The mathematical theory of communication. *Bell System Technical Journal* 27, 379–423, . Available at: http://dx.doi.org/10.1002/j.1538-7305.1948.tb00917.x

Stohlgren, T.J., Falker, M.B. and Schell, L.D. (1995) A modified-Whittaker nested vegetation sampling method. *Vegetatio* 17, 113–121.

Turrill, W. and Milne-Redhead, E. (eds) (1952) *Flora of Tropical East Africa*. Balkema, Rotterdam, The Netherlands.

URT (2015) The National Energy Policy. United Republic of Tanzania.

WWF Tanzania Country Office (2012) (N. D. Burgess, P. Harrison, P. Sumbi, J. Laizer, & A. Kijazi, Eds) Synthesis Baseline Report for Coastal Forests in Tamzania. WWF-Tanzania, Dar Es Salaam. Tanzania. Available at: http://coastalforests.tfcg.org/pubs/Coastal Forests GEF Baseline Synthesis 2012.pdf

Appendix 4.1. Anthropogenic disturbances recorded in PKNFR and RSFR and frequency of their occurrence.

Charcoal making (Species cut in around areas where charcoal kilns were recorded)			
S/N	Species name	Number of incidences	Site
1	*Grewia bicolor*	22	PT1P2
2	*Grewia bicolor*	22	RT1P4
3	*Grewia bicolor*	22	RT1P5
4	*Markamia obtusifolia*	21	PT1P2
5	*Markamia obtusifolia*	21	RT1P4
6	*Markamia obtusifolia*	21	RT1P5
7	*Brachystegia bussei*	15	RT2P1
8	*Grewia conocarpa*	13	RT2P2, RT1P4
9	*Grewia conocarpa*	9	PT1P2
10	*Hymenocardia ulmoides*	9	RT1P4, PT1P6
11	*Grewia conocarpa*	9	RT1P5
12	*Hymenocardia ulmoides*	8	PT1P2
13	*Hymenocardia ulmoides*	8	RT1P5
14	*Millettia micans*	7	RT1P4, RT2P2
15	*Millettia micans*	6	PT1P2
16	*Millettia micans*	6	RT1P5
17	*Afzelia quanzensis*	2	PT1P2
18	*Afzelia quanzensis*	2	RT1P4
19	*Afzelia quanzensis*	2	RT1P5
20	*Pteleopsis myritifolia*	1	PT1P2
21	*Xylopia arinaria*	1	PT1P2
22	*Pteleopsis myritifolia*	1	PT1P2
23	*Combretum harrissii,*	1	PT1P6
24	*Pteliopsis myritifolia*	1	PTP1
25	*Brachystegia microphylla*	1	RT1P2
26	*Pseudolachnostylis maproune-ifolia*	1	RT1P2
27	*Xylopia arenaria*	1	RT1P4
28	*Xylopia arenaria*	1	RT1P5
29	*Dombeya shupangae*	1	RT1P5
30	*Dombeya shupangae*	1	RT2P1
31	*Millettia micans*	1	RT2P2
32	*Combretum harrissii*	1	RT2P5
33	*Hymenocardia ulmoides*	1	RT2P5

Continued

Continued

Fire damage (burnt trees and ground vegetation)			
1	*Brachystegia bussei*	6	RT1P5
2	*Diplorhynchus condylocarpon*	4	RT1P5
3	*Xeroderris stuhlmannii*	2	RT1P5
4	*Combretum harrisii*	1	PT1P6
5	*Markhamia obtusifolia*	1	RT2P2
6	*Pteleopsis myrtifolia*	1	PT2P2
7	*Flueggia virosa*	Several populations in and extending outside the plot	RT2P1, PT1P9
8	*Hymenocardia ulmoides*	Several populations in and extending outside the plot	RT2P1, PT1P6, PT2P2
9	*Trema orientalis*	Several populations in and extending outside the plot	RT2P1
10	*Dichapetalum stulmannii*	Almost the entire plot	PT1P9
11	*Harrisonia abyssinica*	Almost the entire plot	PT1P9, PT1P9
12	*Combretum harrissii*	Almost the entire plot	PT1P5
13	*Hymenocardia ulmoides*	Almost the entire plot	PT1P5
14	*Burnt trees and grasses*	Almost the entire plot	PT1P6, RT2P5, PT1P8, PT1P8
Timber harvesting			
1	*Pterocarpus angolensis*	16	RT2P4, RT2P4, RT1P5
2	*Hymenocardia ulmoides*	13	RT2P2
3	*Afzelia quanzensis*	2	PT2P1, PT2P2
4	*Hymenaea verrucosa*	2	PT1P6, PT1P3
5	*Milicia excelsa*	2	PT1P6, PT1P3
6	*Albizia glaberrima*	1	PT1P3
7	*Albizia petersiana*	1	PT2P1
8	*Baphia kirkii*	1	RT2P4
9	*Millettia micans*	1	RT2P3
10	*Pteleopsis myrtifolia*	1	RT2P2
11	*Vitex schimperii*	1	PT1P1
12	*Hymenaea verrucosa*	1	PT1P6
13	*Milicia excelsa*	2	PT1P6
Pole cutting (plants or cut poles laying on the ground ready for transport)			
1	*Pterocarpus angolensis*	59	RT2P4, RT1P2, RT1P5
2	*Markamia obtusifolia*	35	RT1P4, RT2P1, RT2P2
3	*Dombeya shupangae*	22	RT2P1, RT1P6
4	*Grewia bicolor*	22	RT1P4
5	*Albizia petersiana*	20	PT1P8, PT2P1, PT2P1, PT2P1

Continued

Continued

6	*Brachystegia bussei*	10	RT1P2, RT1P5
7	*Grewia conocarpa*	9	RT1P4
8	*Hymenocardia ulmoides*	9	RT2P3, PT1P6
9	*Afzelia quanzensis*	9	RT1P3, RT1P4, RT2P1, PT2P2
10	*Hymenocardia ulmoides*	8	RT1P4
11	*Diplorhynchus condylocarpon*	7	RT2P1, RT1P5
12	*Trema orientalis*	7	RT2P2
13	*Hymenaea verrucosa*	7	RT1P3, PT1P6, RT2P1, PT1P3
14	*Millettia micans*	6	RT1P4
15	*Brackenridgea zanguebarica*	4	PT1P8
16	*Xeroderris stuhlmannii*	2	RT1P5
17	*Xylopia arenaria*	2	RT1P4, PT1P6
18	*Albizia glaberrima*	2	PT1P3
19	*Pteleopsis myritifolia*	2	PT1P2
20	*Dichapetalum stulmannii*	1	PT1P6
21	*Milicia excelsa*	1	PT1P3
22	*Spirostachys africana*	1	RT1P6

Appendix 4.2a. A checklist of tree species identified in PKNFR.

S/N	Family	Genus	Species	Author
1	Caesalpinaceae	*Afzelia*	*quanzensis*	Welw.
2	Mimosaceae	*Albizia*	*glaberrima*	(Schumach. & Thonn.) Benth.
3	Mimosaceae	*Albizia*	*petersiana*	(Bolle) Oliv.
4	Mimosaceae	*Albizia*	*versicolor*	Welw. ex Oliv.
5	Anacardiaceae	*Anacardium*	*occidentale*	L.
6	Moraceae	*Artocarpus*	*heterophyllus*	Lam.
7	Meliaceae	*Azadirachta*	*indica*	A.Juss.
8	Lecythidaceae	*Barringtonia*	*racemosa*	(L.) Spreng.
9	Caesalpinaceae	*Brachystegia*	*spiciformis*	Benth.
10	Euphorbiaceae	*Bridelia*	*cathartica*	G.Bertol.
11	Euphorbiaceae	*Bridelia*	*micrantha*	(Hochst.) Baill.
12	Connaraceae	*Byrsocarpus*	*orientalis*	(Baill.) Baker
13	Malvaceae	*Carpodiptera*	*africana*	Mast.
14	Caesalpinaceae	*Cassia*	*siamea*	(Lam.) Irwin et Barneby
15	Caesalpinaceae	*Cassia*	*spectabilis*	(DC.) H.S. Irwin & Barneby

Continued

Continued

S/N	Family	Genus	Species	Author
16	Caesalpinaceae	*Cassia*	*abbreviata*	Oliv.
17	Menispermaceae	*Cissampelos*	*mucronata*	A.Rich.
18	Palmae	*Cocos*	*nucifera*	L.
19	Combretaceae	*Combretum*	*pentagonum*	M.A.Lawson
20	Combretaceae	*Combretum*	*holstii*	Engl.
21	Combretaceae	*Combretum*	*harrisii*	Wickens
22	Combretaceae	*Combretum*	*molle*	G.Don
23	Burseraceae	*Commiphora*	*africana*	(A.Rich.) Engl.
24	Araliaceae	*Cussonia*	*arborea*	A.Rich.
25	Papilionaceae	*Dalbergia*	*melanoxylon*	Guill. & Perr.
26	Caesalpinaceae	*Delonix*	*regia*	(Bojer ex Hook.) Raf.
27	Ebenaceae	*Diospyros*	*verrucosa*	Hiern
28	Sterculiaceae	*Dombeya*	*acutangula*	Cav.
29	Palmae	*Elaeis*	*guineensis*	Jacq.
30	Palmae	*Elaeis*	*guineensis*	Jacq.
31	Moraceae	*Ficus*	*sur*	Forssk.
32	Moraceae	*Ficus*	*sycomorus*	L.
33	Tiliaceae	*Grewia*	*bicolor*	Juss.
34	Tiliaceae	*Grewia*	*microcarpa*	K.Schum.
35	Caesalpinaceae	*Hymenaea*	*verrucosa*	Gaertn.
36	Anacardiaceae	*Lannea*	*stuhlmannii*	Engl.
37	Mimosaceae	*Leucaena*	*leucocephala*	(Lam.) de Wit
38	Anacardiaceae	*Mangifera*	*indica*	L.
39	Euphorbiaceae	*Margaritaria*	*discoidea*	(Baill.) Webster
40	Bignoniaceae	*Markhamia*	*obtusifolia*	(Baker) Sprague
41	Bignoniaceae	*Markhamia*	*acuminata*	(Klotzsch) K.Schum.
42	Bignoniaceae	*Markhamia*	*lutea*	K. Schum.
43	Bignoniaceae	*Markhamia*	*obtusifolia*	(Baker) Sprague
44	Meliaceae	*Melia*	*azedarach*	L.
45	Moraceae	*Milicia*	*excelsa*	(Welw.) C.C.Berg
46	Papilionaceae	*Millettia*	*usaramensis*	Taub.
47	Papilionaceae	*Millettia*	*micans*	Taub.
48	Tiliaceae	*Muntingia*	*calabura*	L.
49	Combretaceae	*Pteleopsis*	*myrtifolia*	(M.A.Lawson) Engl. & Diels
50	Apocynaceae	*Rauvolfia*	*mombasiana*	Stapf
51	Caesalpinaceae	*Scorodophloeus*	*fischeri*	(Taub.) J.Léon
52	Anacardiaceae	*Sorindeia*	*madagascariensis*	DC.
53	Loganiaceae	*Strychnos*	*panganensis*	Gilg

Continued

Continued

S/N	Family	Genus	Species	Author
54	Euphorbiaceae	*Suregada*	*zanzibariensis*	Baill.
55	Myrtaceae	*Syzygium*	*cumini*	(L.) Skeels
56	Caesalpinaceae	*Tamarindus*	*indica*	L.
57	Combretaceae	*Terminalia*	*catappa*	L.
58	Ulmaceae	*Trema*	*orientalis*	(L.) Blume
59	Verbenaceae	*Vitex*	*schiliebenii*	Moldenke
60	Annonaceae	*Xylopia*	*arenaria*	Engl.
61	Rutaceae	*Zanthoxylum*	*chalybeum*	Engl.

Appendix 4.2b. A checklist of shrub species identified in PKNFR.

S/N	Family	Genus	Species	Author
1	Mimosaceae	*Acacia*	*brevispica*	Harms
2	Euphorbiaceae	*Alchornea*	*laxiflora*	(Benth.) Pax & K.Hoffm.
3	Sapindaceae	*Allophylus*	*africanus*	P.Beauv.
4	Annonaceae	*Annona*	*senegalensis*	Pers.
5	Asparagaceae	*Asparagus*	*africanus*	Lam.
6	Acanthaceae	*Barleria*	*taitensis*	S.Moore
7	Ochnaceae	*Brackenridgea*	*zanguebarica*	Oliv.
8	Caesalpinaceae	*Caesalpinia*	*bonduc*	(L.) Roxb.
9	Asclepiadaceae	*Calotropis*	*procera*	(Aiton) W.T.Aiton
10	Rubiaceae	*Canthium*	*stuhlmannii*	Bullock
11	Capparaceae	*Capparis*	*tomentosa*	Lam.
12	Apocynaceae	*Carissa*	*edulis*	(Forssk.) Vahl
13	Apocynaceae	*Carissa*	*sp.*	
14	Caesalpinaceae	*Cassia*	*abbreviata*	Oliv.
15	Caesalpinaceae	*Cassia*	*occidentalis*	L.
16	Rubiaceae	*Catunaregam*	*spinosa*	(Thunb.) Tirveng.
17	Rubiaceae	*Chassalia*	*umbraticola*	Vatke
18	Vitaceae	*Cissus*	*intergrifolia*	(Baker) Planch.
19	Verbenaceae	*Clerodendrum*	*myricoides*	(Hochst.) Vatke
20	Combretaceae	*Combretum*	*pentagonum*	M.A.Lawson
21	Boraginaceae	*Cordia*	*monoica*	Roxb.
22	Euphorbiaceae	*Croton*	*pseudopulchellus*	Pax
23	Papilionaceae	*Dalbergia*	*melanoxylon*	Guill. & Perr.
24	Papilionaceae	*Dalbergia*	*obovata*	E.Mey.
25	Papilionaceae	*Dalbergia*	*arbutifolia*	Baker

Continued

Continued

S/N	Family	Genus	Species	Author
26	Sapindaceae	*Deinbollia*	*borbonica*	Scheffler
27	Papilionaceae	*Desmodium*	*velutinum*	(Willd.) DC.
28	Dichapetalaceae	*Dichapetalum*	*mossambicense*	(Klotzsch) Engl.
29	Dichapetalaceae	*Dichapetalum*	*stulmannii*	Engl.
30	Mimosaceae	*Dichrostachys*	*cinerea*	(L.) Wight & Arn.
31	Dioscoreaceae	*Dioscorea*	*sansibarensis*	Pax
32	Boraginaceae	*Ehretia*	*amoena*	Klotzsch
33	Ebenaceae	*Euclea*	*natalensis*	A.DC.
34	Flagellariaceae	*Flagellaria*	*guineensis*	Schumach.
35	Euphorbiaceae	*Flueggea*	*virosa*	(Willd.) Voigt
36	Malvaceae	*Gossypium*	*barbadense*	L.
37	Tiliaceae	*Grewia*	*conocarpa*	K. Schum.
38	Tiliaceae	*Grewia*	*forbesii*	Harv. ex Mast.
39	Tiliaceae	*Grewia*	*platyclada*	K.Schum.
40	Sapindaceae	*Haplocoelum*	*africana*	F.G.Davies ined.
41	Simaroubaceae	*Harrisonia*	*abyssinica*	Oliv.
42	Guttiferae	*Harungana*	*madagascariensis*	Lam. ex Poir.
43	Apocynaceae	*Holarrhena*	*pubescens*	(Buch.-Ham.) G.Don
44	Linaceae	*Hugonia*	*castaneifolia*	Engl.
45	Phyllanthaceae	*Hymenocardia*	*ulmoides*	Oliv.
46	Rubiaceae	*Keetia*	*zanzibarica*	(Klotzsch) Bridson
47	Rubiaceae	*Lamprothamnus*	*zanguebaricus*	Hiern
48	Apocynaceae	*Landolphia*	*buchananii*	(Hallier f.) Stapf
49	Apocynaceae	*Landolphia*	*kirkii*	Dyer
50	Verbenaceae	*Lantana*	*camara*	L.
51	Capparaceae	*Maerua*	*angolensis*	DC.
52	Capparaceae	*Maerua*	*triphylla*	A.Rich.
53	Euphorbiaceae	*Manihot*	*esculenta*	Crantz
54	Sapotaceae	*Manilkara*	*sulcata*	(Engl.) Dubard
55	Papilionaceae	*Millettia*	*impressa*	Harms
56	Mimosaceae	*Mimosa*	*pigra*	L.
57	Rubiaceae	*Multidentia*	*dichrophylla*	(Mildbr.) Bridson
58	Ochnaceae	*Ochna*	*mossambicensis*	Klotzsch
59	Brackenregia	*Ochnaceae*	*Brackenridgea*	zanguebarica
60	Olacaceae	*Olax*	*dissitiflora*	Oliv.
61	Anacardiaceae	*Ozoroa*	*mucronata*	(Bernh.) R. Fern. & A. Fern.
62	Mimosaceae	*Parkia*	*filicoidea*	Welw. ex Oliv.
63	Rubiaceae	*Pentas*	*bussei*	K.Krause

Continued

Continued

S/N	Family	Genus	Species	Author
64	Asteraceae	*Pluchea*	*ovalis*	(Pers.) DC.
65	Compositae	*Pluchea*	*dioscorides*	(L.) DC.
66	Apocynaceae	*Plumeria*	*rubra*	L.
67	Rubiaceae	*Polysphaeria*	*parvifolia*	Hiern
68	Verbenaceae	*Premna*	*senensis*	Klotzsch
69	Myrtaceae	*Psidium*	*guajava*	L.
70	Guttiferae	*Psorospermum*	*febrifugum*	Spach
71	Vitaceae	*Rhoicissus*	*tridentata*	(L.f.) Wild & R.B.Drumm.
72	Anacardiaceae	*Rhus*	*natalensis*	Krauss
73	Euphorbiaceae	*Ricinus*	*communis*	L.
74	Apocynaceae	*Saba*	*comorensis*	(Bojer) Pichon
75	Celastraceae	*Salacia*	*madagascariensis*	(Lam.) DC.
76	Dracaenaceae	*Sansevieria*	*conspicua*	N.E.Br.
77	Passifloraceae	*Schlechterina*	*mitostemmatoides*	Harms
78	Scrophulariaceae	*Scoparia*	*dulcis*	L.
79	Mimosaceae	*Scorodophloeus*	*fischeri*	(Taub.) J.Léon
80	Apocynaceae	*Secamone*	*parvifolia*	(Oliv.) Bullock
81	Papilionaceae	*Sesbania*	*sesban*	(L.) Merr.
82	Sapotaceae	*Sideroxylon*	*inerme*	L.
83	Anacardiaceae	*Sorindeia*	*madagascariensis*	DC.
84	Loganiaceae	*Strychnos*	*panganensis*	Gilg
85	Loganiaceae	*Strychnos*	*madagascariensis*	Poir.
86	Dilleniaceae	*Tetracera*	*boiviniana*	Baill.
87	Dilleniaceae	*Tetracera*	*litoralis*	Gilg
88	Malvaceae	*Thespesia*	*danis*	Oliv.
89	Capparaceae	*Thylachium*	*africanum*	Lour.
90	Labiatae	*Tinnea*	*aethiopica*	Kotschy ex Hook.f.
91	Annonaceae	*Uvaria*	*acuminata*	Oliv.
92	Annonaceae	*Uvaria*	*kirkii*	Hook.f.
93	Annonaceae	*Uvaria*	*lucida*	Benth.
94	Asteraceae	*Vernonia*	*hildebrandtii*	Vatke
95	Annonaceae	*Xylopia*	*arenaria*	Engl.
96	Flacourtiaceae	*Xylotheca*	*tettensis*	(Klotzsch) Gilg

Appendix 4.2c. A checklist of herbaceous species identified in PKNFR.

S/N	Family	Genus	Species	Author
1	Malvaceae	*Abutilon*	*mauritianum*	(Jacq.) Sweet
2	Euphorbiaceae	*Acalypha*	*indica*	L.
3	Amaranthaceae	*Achyranthes*	*aspera*	L.
4	Amaranthaceae	*Aerva*	*lanata*	(L.) Schult.
5	Rubiaceae	*Agathisanthemum*	*bojeri*	Klotzsch
6	Compositae	*Ageratum*	*conyzoides*	L.
7	Bromeliaceae	*Ananas*	*comosus*	(L) Merr.
8	Poaceae	*Aristida*	*adoensis*	Hochst.
9	Acanthaceae	*Asystasia*	*gangetica*	(L.) T.Anderson
10	Asteraceae	*Bidens*	*pilosa*	L.
11	Acanthaceae	*Blepharis*	*maderaspatensis*	Heine ex Roth
12	Acanthaceae	*Blepharis*	*madagascariensis*	Benoist
13	Poaceae	*Brachiaria*	*brizantha*	(A.Rich.) Stapf
14	Cyperaceae	*Bulbostylis*	*buchananii*	C.B.Clarke
15	Caesalpinaceae	*Cassia*	*absus*	L.
16	Caesalpinaceae	*Cassia*	*mimosoides*	L.
17	Lauraceae	*Cassytha*	*filiformis*	L.
18	Poaceae	*Chloris*	*gayana*	Kunth
19	Menispermaceae	*Cissampelos*	*mucronata*	A.Rich.
20	Araceae	*Colocasia*	*esculenta*	(L.) Schott
21	Commelinaceae	*Commelina*	*africana*	L.
22	Commelinaceae	*Commelina*	*benghalensis*	Wall.
23	Tiliaceae	*Corchorus*	*trilocularis*	L.
24	Papilionaceae	*Crotalaria*	*goodiiformis*	Vatke
25	Papilionaceae	*Crotalaria*	*lanceolata*	E.Mey.
26	Amaranthaceae	*Cyathula*	*polycephala*	Baker
27	Cyperaceae	*Cyperus*	*exaltatus*	Retz.
28	Cyperaceae	*Cyperus*	*hemisphaericus*	Boeck.
29	Cyperaceae	*Cyperus*	*difformis*	L.
30	Vitaceae	*Cyphostemma*	*adenocaule*	(A.Rich.) Wild & R.B.Drumm.
31	Papilionaceae	*Desmodium*	*barbatum*	(L.) Benth.
32	Papilionaceae	*Desmodium*	*velutinum*	(Willd.) DC.
33	Poaceae	*Dichanthium*	*annulatum*	(Forssk.) Stapf
34	Dichapetalaceae	*Dichapetalum*	*arenarium*	Breteler
35	Gramineae	*Digitaria*	*abyssinica*	(A.Rich.) Stapf
36	Poaceae	*Digitaria*	*scalarum*	(Schweinf.) Chiov.
37	Dioscoreaceae	*Dioscorea*	*sansibarensis*	Pax

Continued

Continued

S/N	Family	Genus	Species	Author
38	Poaceae	*Echinochloa*	*haploclada*	(Stapf) Stapf
39	Poaceae	*Eleusine*	*indica*	(L.) Gaertn.
40	Asteraceae	*Emilia*	*coccinea*	(Sims) Sweet
41	Euphorbiaceae	*Euphorbia*	*heterophylla*	L.
42	Euphorbiaceae	*Euphorbia*	*hirta*	L.
43	Euphorbiaceae	*Euphorbia*	*prostrata*	Aiton
44	Flagellariaceae	*Flagellaria*	*guineensis*	Schumach.
45	Cyperaceae	*Fuirena*	*umbellata*	Rottb.
46	Colchicaceae	*Gloriosa*	*superba*	L.
47	Gramineae	*Harpachne*	*schimperi*	A.Rich.
48	Gramineae	*Heteropogon*	*contortus*	(L.) Roem. & Schult.
49	Malvaceae	*Hibiscus*	*cannabinus*	L.
50	Labiatae	*Hoslundia*	*opposita*	vahl
51	Gramineae	*Hyparrhenia*	*rufa*	(Nees) Stapf
52	Poaceae	*Hyparrhenia*	*filipendula*	(Hochst.) Stapf
53	Labiatae	*Hyptis*	*suaveolens*	Poit.
54	Papilionaceae	*Indigofera*	*volkensii*	Taub.
55	Papilionaceae	*Indigofera*	*arrecta*	A.Rich.
56	Convolvulaceae	*Ipomoea*	*sinensis*	(Desr.) Choisy
57	Compositae	*Launaea*	*cornuta*	(Hochst. ex Oliv. & Hiern) C.Jeffrey
58	Poaceae	*Leersia*	*hexandra*	Sw.
59	Onagraceae	*Ludwigia*	*stolonifera*	(Guill. and Perr)
60	Papilionaceae	*Macrotyloma*	*axillare*	(E.Mey.) Verdc.
61	Poaceae	*Melinis*	*repens*	(Willd.) Zizka
62	Malvaceae	*Melochia*	*corchorifolia*	L.
63	Mimosaceae	*Mimosa*	*pigra*	L.
64	Mimosaceae	*Mimosa*	*pudica*	L.
65	Papilionaceae	*Mucuna*	*sp.*	
66	Musaceae	*Musa*	*separia*	L.
67	Labiatae	*Ocimum*	*basilicum*	L.
68	Polygonaceae	*Oxygonum*	*sinuatum*	(Meisn.) Dammer
69	Gramineae	*Panicum*	*atrosanguineum*	A.Rich.
70	Gramineae	*Panicum*	*heterostachyum*	Hack.
71	Poaceae	*Panicum*	*coloratum*	L.
72	Poaceae	*Panicum*	*deflexum*	Schumach.
73	Poaceae	*Panicum*	*maximum*	Jacq.
74	Poaceae	*Panicum*	*trichocladum*	K.Schum.

Continued

Continued

S/N	Family	Genus	Species	Author
75	Mimosaceae	*Parkia*	*filicoidea*	Welw. ex Oliv.
76	Passifloraceae	*Passiflora*	*foetida*	L.
77	Poaceae	*Pennisetum*	*polystachion*	(L.) Schult.
78	Gramineae	*Perotis*	*hildebrandtii*	Mez
79	Euphorbiaceae	*Phyllanthus*	*amarus*	Schumach. & Thonn.
80	Euphorbiaceae	*Phyllanthus*	*maderaspatensis*	L.
81	Solanaceae	*Physalis*	*peruviana*	L.
82	Poaceae	*Pogonarthria*	*squarrosa*	(Roem. & Schult.) Pilg.
83	Polygalaceae	*Polygala*	*erioptera*	DC.
84	Papilionaceae	*Rhynchosia*	*minima*	(L.) DC.
85	Papilionaceae	*Rhynchosia*	*hirta*	(Andr.) Meikle & Verdc.
86	Papilionaceae	*Rhynchosia*	*resinosa*	(A.Rich.) Baker
87	Poaceae	*Rottboellia*	*cochinchinensis*	(Lour.) Clayton
88	Acanthaceae	*Ruellia*	*tuberosa*	L.
89	Passifloraceae	*Schlechterina*	*mitostemmatoides*	Harms
90	Cyperaceae	*Scleria*	*foliosa*	Hochst. ex A.Rich.
91	Malvaceae	*Sida*	*acuta*	Burm.f.
92	Poaceae	*Sorghum*	*bicolor*	(L.) Moench
93	Rubiaceae	*Spermacoce*	*senensis*	(Klotzsch) Hiern
94	Gramineae	*Sporobolus*	*pyramidalis*	P.Beauv.
95	Scrophulariaceae	*Striga*	*asiatica*	(L.) Kuntze
96	Papilionaceae	*Stylosanthes*	*fruticosa*	(Retz.) Alston
97	Taccaceae	*Tacca*	*leontopetaloides*	(L.) Kuntze
98	Poaceae	*Themeda*	*triandra*	Forssk.
99	Acanthaceae	*Thunbergia*	*alata*	Bojer ex Sims
100	Euphorbiaceae	*Tragia*	*furialis*	Bojer
101	Asteraceae	*Tridax*	*procumbens*	L.
102	Tiliaceae	*Triumfetta*	*rhomboidea*	Jacq.
103	Typhaceae	*Typha*	*capensis*	(Rohrb.) N.E.Br.
104	Asteraceae	*Vernonia*	*hildebrandtii*	Vatke
105	Dracaenaceae	*Sansevieria*	*conspicua*	N.E.Br.
106	Solanaceae	*Solanum*	*incanum*	L.

Appendix 4.3a. A checklist of tree species identified in RSFR.

S/N	Family	Genus	species	Author
1	Mimosaceae	*Acacia*	*brevispica*	Harms
2	Mimosaceae	*Acacia*	*mellifera*	(Vahl) Benth.
3	Mimosaceae	*Acacia*	*nigrescens*	Oliv.
4	Mimosaceae	*Acacia*	*robusta*	Burch
5	Mimosaceae	*Acacia*	*zanzibarica*	(S.Moore) Taub.
6	Caesalpinaceae	*Afzelia*	*quanzensis*	Welw.
7	Mimosaceae	*Albizia*	*amara*	(Roxb.) Boivin
8	Mimosaceae	*Albizia*	*petersiana*	(Bolle) Oliv.
9	Mimosaceae	*Albizia*	*versicolor*	Welw. ex Oliv.
10	Anacardiaceae	*Anacardium*	*occidentale*	L.
11	Papilionaceae	*Baphia*	*kirkii*	Baker
12	Bombacaceae	*Bombax*	*rhodognaphalon*	K.Schum.
13	Caesalpinaceae	*Brachystegia*	*bussei*	Harms
14	Caesalpinaceae	*Brachystegia*	*microphylla*	Harms
15	Euphorbiaceae	*Bridelia*	*cathartica*	G.Bertol.
16	Caesalpinaceae	*Cassia*	*singueana*	Delile
17	Combretaceae	*Combretum*	*holstii*	Engl.
18	Combretaceae	*Combretum*	*molle*	G.Don
19	Combretaceae	*Combretum*	*zeyheri*	Sond.
20	Burseraceae	*Commiphora*	*africana*	(A.Rich.) Engl.
21	Papilionaceae	*Dalbergia*	*melanoxylon*	Guill. & Perr.
22	Apocynaceae	*Diplorhynchus*	*condylocarpon*	(Müll.Arg.) Pichon[
23	Sterculiaceae	*Dombeya*	*acutangula*	Cav.
24	Sterculiaceae	*Dombeya*	*shupangae*	K.Schum.
25	Tiliaceae	*Grewia*	*bicolor*	Juss.
26	Tiliaceae	*Grewia*	*conocarpa*	K. Schum.
27	Caesalpinaceae	*Hymenaea*	*verrucosa*	Gaertn.
28	Hymenocardiaceae	*Hymenocardia*	*ulmoides*	Oliv.
29	Caesalpinaceae	*Julbernardia*	*globiflora*	(Benth.) Troupin
30	Anacardiaceae	*Lannea*	*schweinfurthii*	(Engl.) Engl.
31	Anacardiaceae	*Lannea*	*stuhlmannii*	Engl.
32	Euphorbiaceae	*Margaritaria*	*discoidea*	(Baill.) Webster
33	Bignoniaceae	*Markhamia*	*obtusifolia*	(Baker) Sprague
34	Papilionaceae	*Millettia*	*micans*	Taub.
35	Euphorbiaceae	*Pseudolachnostylis*	*maprouneifolia*	Pax
36	Combretaceae	*Pteleopsis*	*myrtifolia*	(M.A.Lawson) Engl. & Diels
37	Papilionaceae	*Pterocarpus*	*angolensis*	DC.

Continued

Continued

S/N	Family	Genus	species	Author
38	Malvaceae	*Rhodognaphalon*	*schumannianum*	A.Robyns
39	Anacardiaceae	*Sclerocarya*	*birrea*	(A.Rich.) Hochst.
40	Euphorbiaceae	*Spirostachys*	*africana*	Sond.
41	Ulmaceae	*Trema*	*orientalis*	(L.) Blume
42	Verbenaceae	*Vitex*	*doniana*	Sweet
43	Verbenaceae	*Vitex*	*ferruginea*	Schumach. & Thonn.
44	Verbenaceae	*Vitex*	*mombassae*	Vatke
45	Apocynaceae	*Voacanga*	*africana*	Stapf
46	Papilionaceae	*Xeroderris*	*stuhlmannii*	(Taub.) Mendonça & E.C.Sousa
47	Rutaceae	*Zanthoxylum*	*chalybeum*	Engl.

Appendix 4.3b. A checklist of shrub species identified in RSFR.

S/N	Family	Genus	species	Author
1	Mimosaceae	*Acacia*	*brevispica*	Harms
2	Euphorbiaceae	*Alchornea*	*laxiflora*	(Benth.) Pax & K.Hoffm.
3	Sapindaceae	*Allophylus*	*africanus*	P.Beauv.
4	Annonaceae	*Annona*	*senegalensis*	Pers.
5	Moraceae	*Antiaris*	*toxicaria*	Lesch.
6	Euphorbiaceae	*Antidesma*	*venosum*	Tul.
7	Asteraceae	*Aspilia*	*mossambicensis*	(Oliv.) Wild
8	Ochnaceae	*Brackenridgea*	*zanguebarica*	Oliv.
9	Asclepiadaceae	*Calotropis*	*procera*	(Aiton) W.T.Aiton
10	Rubiaceae	*Canthium*	*Klotzsch*	Klotzsch
11	Apocynaceae	*Carissa*	*spinarum*	L.
12	Malvaceae	*Carpodiptera*	*africana*	Mast.
13	Caesalpinaceae	*Cassia*	*abbreviata*	Oliv.
14	Rubiaceae	*Catunaregam*	*spinosa*	(Thunb.) Tirveng.
15	Rubiaceae	*Chassalia*	*umbraticola*	Vatke
16	Vitaceae	*Cissus*	*intergrifolia*	(Baker) Planch.
17	Verbenaceae	*Clerodendrum*	*myricoides*	(Hochst.) Vatke
18	Combretaceae	*Combretum*	*harrisii*	Wickens
19	Combretaceae	*Combretum*	*pentagonum*	M.A.Lawson
20	Boraginaceae	*Cordia*	*monoica*	Roxb.
21	Papilionaceae	*Crotalaria*	*goodiiformis*	Vatke
22	Araliaceae	*Cussonia*	*arborea*	A.Rich.
23	Papilionaceae	*Dalbergia*	*melanoxylon*	Guill. & Perr.
24	Papilionaceae	*Dalbergia*	*obovata*	E.Mey.

Continued

Continued

S/N	Family	Genus	species	Author
25	Dichapetalaceae	*Dichapetalum*	*mossambicense*	(Klotzsch) Engl.
26	Dichapetalaceae	*Dichapetalum*	*stulmannii*	Engl.
27	Mimosaceae	*Dichrostachys*	*cinerea*	(L.) Wight & Arn.
28	Ebenaceae	*Diospyros*	*usaramensis*	Gürke
29	Ebenaceae	*Diospyros*	*verrucosa*	Hiern
30	Apocynaceae	*Diplorhynchus*	*condylocarpon*	(Müll.Arg.) Pichon[
31	Sterculiaceae	*Dombeya*	*acutangula*	Cav.
32	Boraginaceae	*Ehretia*	*amoena*	Klotzsch
33	Ebenaceae	*Euclea*	*natalensis*	A.DC.
34	Flacourtiaceae	*Flacourtia*	*indica*	(Burm.f.) Merr.
35	Euphorbiaceae	*Flueggea*	*virosa*	(Willd.) Voigt
36	Guttiferae	*Garcinia*	*livingstonei*	T.Anderson
37	Tiliaceae	*Grewia*	*bicolor*	Juss.
38	Tiliaceae	*Grewia*	*conocarpa*	K. Schum.
39	Tiliaceae	*Grewia*	*forbesii*	Harv. ex Mast.
40	Tiliaceae	*Grewia*	*microcarpa*	K.Schum.
41	Tiliaceae	*Grewia*	*mollis*	Juss.
42	Sapindaceae	*Haplocoelum*	*inopleum*	Radlk.
43	Simaroubaceae	*Harrisonia*	*abyssinica*	Oliv.
44	Guttiferae	*Harungana*	*madagascariensis*	Lam. ex Poir.
45	Apocynaceae	*Holarrhena*	*pubescens*	(Buch.-Ham.) G.Don
46	Labiatae	*Hoslundia*	*opposita*	vahl
47	Linaceae	*Hugonia*	*castaneifolia*	Engl.
48	Phyllanthaceae	*Hymenocardia*	*ulmoides*	Oliv.
49	Papilionaceae	*Indigofera*	*rhynchocarpa*	Baker
50	Rubiaceae	*Lamprothamnus*	*zanguebaricus*	Hiern
51	Apocynaceae	*Landolphia*	*kirkii*	Dyer
52	Papilionaceae	*Lonchocarpus*	*bussei*	Harms
53	Euphorbiaceae	*Margaritaria*	*discoidea*	(Baill.) Webster
54	Celastraceae	*Maytenus*	*senegalensis*	(Lam.) Exell
55	Papilionaceae	*Millettia*	*micans*	Taub.
56	Rubiaceae	*Multidentia*	*crassa*	(Hiern) Bridson & Verdc.
57	Ochnaceae	*Ochna*	*mossambicensis*	Klotzsch
58	Olacaceae	*Olax*	*dissitiflora*	Oliv.
59	Anacardiaceae	*Ozoroa*	*mucronata*	(Bernh.) R. Fern. & A. Fern.
60	Rubiaceae	*Pentas*	*bussei*	K.Krause
61	Asteraceae	*Pluchea*	*dioscorides*	(L.) DC.

Continued

Continued

S/N	Family	Genus	species	Author
62	Rubiaceae	*Polysphaeria*	*parvifolia*	Hiern
63	Vitaceae	*Rhoicissus*	*tridentata*	(L.f.) Wild & R.B.Drumm.
64	Anacardiaceae	*Rhus*	*natalensis*	Krauss
65	Euphorbiaceae	*Ricinus*	*communis*	L.
66	Celastraceae	*Salacia*	*leptoclada*	Tul.
67	Celastraceae	*Salacia*	*madagascariensis*	(Lam.) DC.
68	Rhamnaceae	*Scutia*	*myrtina*	(Burm.f.) Kurz
69	Papilionaceae	*Sesbania*	*sesban*	(L.) Merr.
70	Euphorbiaceae	*Spirostachys*	*africana*	Sond.
71	Loganiaceae	*Strychnos*	*panganensis*	Gilg
72	Loganiaceae	*Strychnos*	*madagascariensis*	Poir.
73	Euphorbiaceae	*Suregada*	*zanzibariensis*	Baill.
74	Dilleniaceae	*Tetracera*	*boiviniana*	Baill.
75	Malvaceae	*Thespesia*	*danis*	Oliv.
76	Capparaceae	*Thylachium*	*africanum*	Lour.
77	Meliaceae	*Turraea*	*robusta*	Gürke
78	Annonaceae	*Uvaria*	*acuminata*	Oliv.
79	Annonaceae	*Uvaria*	*kirkii*	Hook.f.
80	Annonaceae	*Uvaria*	*lucida*	Benth.
81	Compositae	*Vernonia*	*hildebrandtii*	Vatke
82	Papilionaceae	*Xeroderris*	*stuhlmannii*	(Taub.) Mendonça & E.C.Sousa
83	Annonaceae	*Xylopia*	*arenaria*	Engl.
84	Flacourtiaceae	*Xylotheca*	*tettensis*	(Klotzsch) Gilg

Appendix 4.3c. A checklist of herbaceous species identified in RSFR.

S/N	Family	Genus	species	Author
1	Papilionacea	*Abrus*	*precatorius*	L.
2	Euphorbiaceae	*Acalypha*	*indica*	L.
3	Amaranthaceae	*Achyranthes*	*aspera*	L.
4	Rubiaceae	*Agathisanthemum*	*bojeri*	Klotzsch
5	Asteracea	*Ageratum*	*conyzoides*	L.
6	Poaceae	*Aristida*	*adoensis*	Hochst.
7	Poaceae	*Aristida*	*barbicollis*	Trin. & Rupr.
8	Asteracea	*Aspilia*	*mossambicensis*	(Oliv.) Wild

Continued

Continued

S/N	Family	Genus	species	Author
9	Acanthaceae	*Asystasia*	*gangetica*	(L.) T.Anderson
10	Acanthaceae	*Blepharis*	*affinis*	Lindau
11	Acanthaceae	*Blepharis*	*maderaspatensis*	Heine ex Roth
12	Poaceae	*Brachiaria*	*brizantha*	(A.Rich.) Stapf
13	Cyperaceae	*Bulbostylis*	*buchananii*	C.B.Clarke
14	Caesalpinaceae	*Cassia*	*mimosoides*	L.
15	Caesalpinaceae	*Cassia*	*absus*	L.
16	Lauraceae	*Cassytha*	*filiformis*	L.
17	Poaceae	*Chloris*	*gayana*	Kunth
18	Menispermaceae	*Cissampelos*	*mucronata*	A.Rich.
19	Vitaceae	*Cissus*	*quadrangularis*	L.
20	Vitaceae	*Cissus*	*rotundifolia*	(Forssk.) Vahl
21	Commelinaceae	*Commelina*	*africana*	L.
22	Commelinaceae	*Commelina*	*benghalensis*	Wall.
23	Tiliaceae	*Corchorus*	*trilocularis*	L.
24	Acanthaceae	*Crabbea*	*velutina*	S.Moore
25	Leguminosae	*Crotalaria*	*retusa*	L.
26	Papilionacea	*Crotalaria*	*goodiiformis*	Vatke
27	Papilionacea	*Crotalaria*	*lanceolata*	E.Mey.
28	Papilionaceae	*Crotalaria*	*laburnoides*	Klotzsch
29	Cucurbitaceae	*Cucurbita*	*maxima*	(Duchesne) Poir.
30	Poaceae	*Cynodon*	*dactylon*	(L.) Pers.
31	Cyperaceae	*Cyperus*	*difformis*	L.
32	Cyperaceae	*Cyperus*	*distans*	L.f.
33	Cyperaceae	*Cyperus*	*exaltatus*	Retz.
34	Cyperaceae	*Cyperus*	*hemisphaericus*	Boeck.
35	Vitaceae	*Cyphostemma*	*adenocaule*	(A.Rich.) Wild & R.B.Drumm.
36	Poaceae	*Dactyloctenium*	*aegyptium*	(L.) Willd.
37	Papilionacea	*Desmodium*	*barbatum*	(L.) Benth.
38	Poaceae	*Dichanthium*	*annulatum*	(Forssk.) Stapf
39	Dichapetalaceae	*Dichapetalum*	*arenarium*	Breteler
40	Poaceae	*Digitaria*	*abyssinica*	(A.Rich.) Stapf
41	Poaceae	*Digitaria*	*milanjiana*	(Rendle) Stapf
42	Poaceae	*Echinochloa*	*haploclada*	(Stapf) Stapf
43	Asteracea	*Emilia*	*coccinea*	(Sims) Sweet
44	Poaceae	*Eragrostis*	*aspera*	(Jacq.) Nees
45	Asteracea	*Ethulia*	*conyzoides*	L.f.
46	Euphorbiaceae	*Euphorbia*	*heterophylla*	L.

Continued

Continued

S/N	Family	Genus	species	Author
47	Euphorbiaceae	*Euphorbia*	*hirta*	L.
48	Flagellariaceae	*Flagellaria*	*guineensis*	Schumach.
49	Cyperaceae	*Fuirena*	*umbellata*	Rottb.
50	Colchicaceae	*Gloriosa*	*superba*	L.
51	Asteracea	*Gutenbergia*	*polycephala*	Oliv. & Hiern
52	Poaceae	*Harpachne*	*schimperi*	A.Rich.
53	Poaceae	*Heteropogon*	*contortus*	(L.) Roem. & Schult.
54	Malvaceae	*Hibiscus*	*cannabinus*	L.
55	Labiatae	*Hoslundia*	*opposita*	vahl
56	Acanthaceae	*Hygrophila*	*auriculata*	(Schumach.) Heine
57	Poaceae	*Hyparrhenia*	*filipendula*	(Hochst.) Stapf
58	Poaceae	*Hyparrhenia*	*rufa*	(Nees) Stapf
59	Labiatae	*Hyptis*	*suaveolens*	Poit.
60	Papilionacea	*Indigofera*	*arrecta*	A.Rich.
61	Papilionacea	*Indigofera*	*volkensii*	Taub.
62	Convolvulaceae	*Ipomoea*	*sinensis*	(Desr.) Choisy
63	Oleaceae	*Jasminum*	*fluminense*	Vell.
64	Cyperaceae	*Kyllinga*	*elata*	Steud.
65	Cyperaceae	*Kyllinga*	*erecta*	Schumach.
66	Asteracea	*Launaea*	*cornuta*	(Hochst. ex Oliv. & Hiern) C.Jeffrey
67	Poaceae	*Leersia*	*hexandra*	Sw.
68	Onagraceae	*Ludwigia*	*stolonifera*	(Guill. and Perr)
69	Onagraceae	*Ludwigia*	*stolonifera*	(Guill. and Perr)
70	Papilionacea	*Macrotyloma*	*axillare*	(E.Mey.) Verdc.
71	Commelinaceae	*Murdannia*	*simplex*	(Vahl) Brenan
72	Poaceae	*Oryza*	*punctata*	Steud.
73	Polygonaceae	*Oxygonum*	*sinuatum*	(Meisn.) Dammer
74	Amaranthaceae	*Pandiaka*	*rubro-lutea*	(Lopr.) C.C.Towns.
75	Poaceae	*Panicum*	*atrosanguineum*	A.Rich.
76	Poaceae	*Panicum*	*coloratum*	L.
77	Poaceae	*Panicum*	*heterostachyum*	Hack.
78	Poaceae	*Panicum*	*maximum*	Jacq.
79	Poaceae	*Panicum*	*trichocladum*	K.Schum.
80	Poaceae	*Pennisetum*	*polystachion*	(L.) Schult.
81	Poaceae	*Perotis*	*hildebrandtii*	Mez
82	Euphorbiaceae	*Phyllanthus*	*amarus*	Schumach. & Thonn.
83	Solanaceae	*Physalis*	*peruviana*	L.

Continued

Continued

S/N	Family	Genus	species	Author
84	Asteracea	*Pluchea*	*dioscorides*	(L.) DC.
85	Polygalaceae	*Polygala*	*erioptera*	DC.
86	Papilionacea	*Rhynchosia*	*resinosa*	(A.Rich.) Baker
87	Papilionaceae	*Rhynchosia*	*hirta*	(Andr.) Meikle & Verdc.
88	Poaceae	*Rottboellia*	*cochinchinensis*	(Lour.) Clayton
89	Acanthaceae	*Ruellia*	*tuberosa*	L.
90	Acanthaceae	*Ruellia*	*patula*	Jacq.
91	Apocynaceae	*Secamone*	*parvifolia*	(Oliv.) Bullock
92	Pedaliaceae	*Sesamum*	*angustifolium*	(Oliv.) Engl.
93	Papilionacea	*Sesbania*	*sesban*	(L.) Merr.
94	Malvaceae	*Sida*	*alba*	L.
95	Zingiberaceae	*Siphonochilus*	*aethiopicus*	(Schweinf.) B.L.Burtt
96	Solanaceae	*Solanum*	*incanum*	L.
97	Rubiaceae	*Spermacoce*	*chaetocephala*	DC.
98	Rubiaceae	*Spermacoce*	*senensis*	(Klotzsch) Hiern
99	Poaceae	*Sporobolus*	*pyramidalis*	P.Beauv.
100	Scrophulariaceae	*Striga*	*asiatica*	(L.) Kuntze
101	Araceae	*Stylochiton*	*puberulus*	N.E.Br.
102	Papilionacea	*Stylosanthes*	*fruticosa*	(Retz.) Alston
103	Papilionacea	*Tephrosia*	*villosa*	(L.) Pers.
104	Poaceae	*Themeda*	*triandra*	Forssk.
105	Euphorbiaceae	*Tragia*	*furialis*	Bojer
106	Poaceae	*Tragus*	*berteronianus*	Schult.
107	Tiliaceae	*Triumfetta*	*rhomboidea*	Jacq.
108	Typhaceae	*Typha*	*capensis*	(Rohrb.) N.E.Br.
109	Poaceae	*Urochloa*	*mosambicensis*	(Hack.) Dandy
110	Asteracea	*Vernonia*	*glabra*	(Steetz) Vatke
111	Asteracea	*Vernonia*	*hildebrandtii*	Vatke
112	Asteracea	*Vernonia*	*poskeana*	Vatke & Hildebr.
113	Malvaceae	*Waltheria*	*indica*	L.
114	Curcubitaceae	*Zehneria*	*scabra*	(L.f.) Sond.

5 Arthropod Assemblages of Pugu-Kazimzumbwi Nature Forest and Ruvu South Forest Reserves, Tanzania

Anitha Byabato Philbert* and Bruno Alberto Nyundo
Department of Zoology and Wildlife Conservation, University of Dar es Salaam

Abstract
Arthropods of Pugu-Kazimzumbwi Nature Forest Reserve (PKNFR) and Ruvu South Forest Reserve are poorly studied. A survey of arthropods was carried out in the two forests in connection with an environmental impact assessment for the SGR project. The aim was to provide baseline information on the abundance, distribution and diversity of arthropods along the SGR line so that any habitat alteration during the current and future constructions can be easily monitored. A total of 1876 and 1401 arthropods were collected from PKNFR and RSFR, respectively, during the wet season, whereas 2077 and 2760 individuals were collected, respectively, during the dry season. The abundance of arthropods increased towards the railway line along the gradient. In terms of diversity, five classes of arthropods (Arachnida, Insecta, Malacostraca, Diplopoda and Chilopoda) were identified. The study seeks to promote sustainable management of biodiversity resources through the adoption of practices that integrate conservation needs and development priorities.

5.1 Background Information

The diverse arthropod assemblage of the Tanzanian coastal forests remains largely unknown and its systematics is patchy. Studies on arthropods in the two forests (Pugu and Ruvu) have mainly focused on larger taxa such as Lepidopterans (Congdon and Collins, 1997; Howell *et al.*, 2013). Coleopterans and Orthopterans have been studied to some degree (Hemp, 2017) with other groups of arthropods such as millipedes receiving little attention (Enghoff, 2016). Apart from forests, arthropods are found in nearly any habitat ranging from small temporary pools, terrestrial, deserts, under trees, in the soil, in large lakes, small springs and large rivers (Townsend *et al.*, 1997). In the ecosystem, arthropods play an important role as pollinators, food source in various trophic levels, decomposers, scavengers, nutrient cyclers and recyclers, etc. In addition, arthropods can be used as indicator species; their presence or absence is used to indicate the condition of water quality and other environmental parameters. For example, the larvae of caddisflies, mayflies and stoneflies cannot survive in polluted water so a stream with these larvae is assumed to have good water quality (Espinosa *et al.*, 2020). Butterflies in particular, are used as indicators of environmental health because they are sensitive to disturbances such as habitat fragmentation, pollution and pesticide residues, changes in climate and other factors such as temperature, precipitation, humidity and light

*Corresponding author: panitha@cartafrica.org

 Biodiversity and Conservation along an East African Railway (eds C.A. Masao *et al.*)
DOI: 10.1079/9781800626942.0005

level (Fleishman and Murphy, 2009). Other arthropod indicator species include spiders, bees and ants which are normally used to indicate environmental degradation, whereas ground beetles can indicate disturbance or pollution in the soil (Lesch and Bouwman, 2018; Uehara-Prado *et al.*, 2009). When used as bio indicators, arthropods reflect abiotic and biotic components of the ecosystem, represent environmental change and/or correlate the diversity patterns of biodiversity species in a particular habitat, community or biome (Andersen *et al.*, 2004).

Tropical forests are among the richest terrestrial ecosystems where the majority of both flora and fauna species are found and where over 50% of the global biodiversity is distributed (Collen *et al.*, 2008). However, maintenance of arthropod diversity has become an issue of international concern; with biodiversity decline continuing unabated, tropical ecosystems are particularly affected by this loss (Chowdhury *et al.*, 2017). The drastic decline of arthropods and other biodiversity is attributed to habitat destruction and fragmentation caused by developmental activities and poor conservation practices. Habitat loss is a very common result of anthropogenic activities such as agriculture, logging and construction. Habitat alteration may affect species richness, resulting either in decreases or, more rarely, in increases in the diversity of specific insect groups (McKinney, 2008). The effect of human activities on the global environment has been a mass-destruction event on a geological timescale as well as causing widespread variations in the distribution of organisms (Gallou *et al.*, 2017). Human activities can affect, both positively and negatively, the distribution and richness of arthropod species.

Forests are important natural resources on the Earth; apart from being major reserves of biodiversity and environmental services, they regulate climate, stabilize soils, sequester carbon, clean air, and are also sources of water (catchment area) and nutrient recyclers (Ramirez *et al.*, 2003). The two forests (Pugu Hills and Ruvu South) harbour rich and diverse fauna, including many forest-dependent and endemic species of arthropods. A total of 200 species of insects have been recorded from these forests including some East African coastal forest endemics such as the azure hairstreak (*Hemiolaus* sp.), the gold-banded forester (*Euphaedra neophron*), the forest queen (*Euxanthe wakefieldi*) and the spotted sylph (*Astictopterus stellata*) (Howell *et al.*, 2013). In addition, checklists of selected groups of arthropods have been produced from the two forests, including 59 orthopteran species (Hemp, 2017) of which 12 are probably endemic to the area, and another list has been produced for millipedes, reporting eight and seven species from Pugu and Ruvu forest reserves, respectively.

The ongoing SGR construction project crosses some important biodiversity hotspots in Tanzania including Pugu Hills and Ruvu South Forest reserves, wildlife corridors, and other areas with varied conservation status. The two forests harbour important sites for biodiversity including bats, birds, reptiles and other small mammals of conservation value in Tanzania and internationally. Unfortunately, knowledge about the community structure and diversity of arthropods is still limited to a few studies. This study was conducted as part of a biodiversity action plan (BAP), the aim of which was to examine the effects of landscape intensification due to SGR construction on the abundance and diversity of different arthropod groups. The study sought to promote sustainable management of biodiversity resources, to maintain the ecological integrity of protected areas and the benefits from ecosystem services through the adoption of practices that integrate conservation needs and development priorities. Therefore, arthropod surveys are envisaged to ensure that the SGR line causes no net loss of arthropods through mapping their distribution in the area, designing monitoring protocols, and suggesting future conservation measures that can be taken.

5.2 Methods

5.2.1 Description of study sites

The field study was conducted within Pugu Hills and Ruvu South Forest Reserves, within the SGR project's footprint. For each forest reserve, three transects were set perpendicular to the SGR line (Figs. 5.1 and 5.2). For each transect, a 3km transect was set perpendicular to the SGR line, within each transect there were 6 sampling

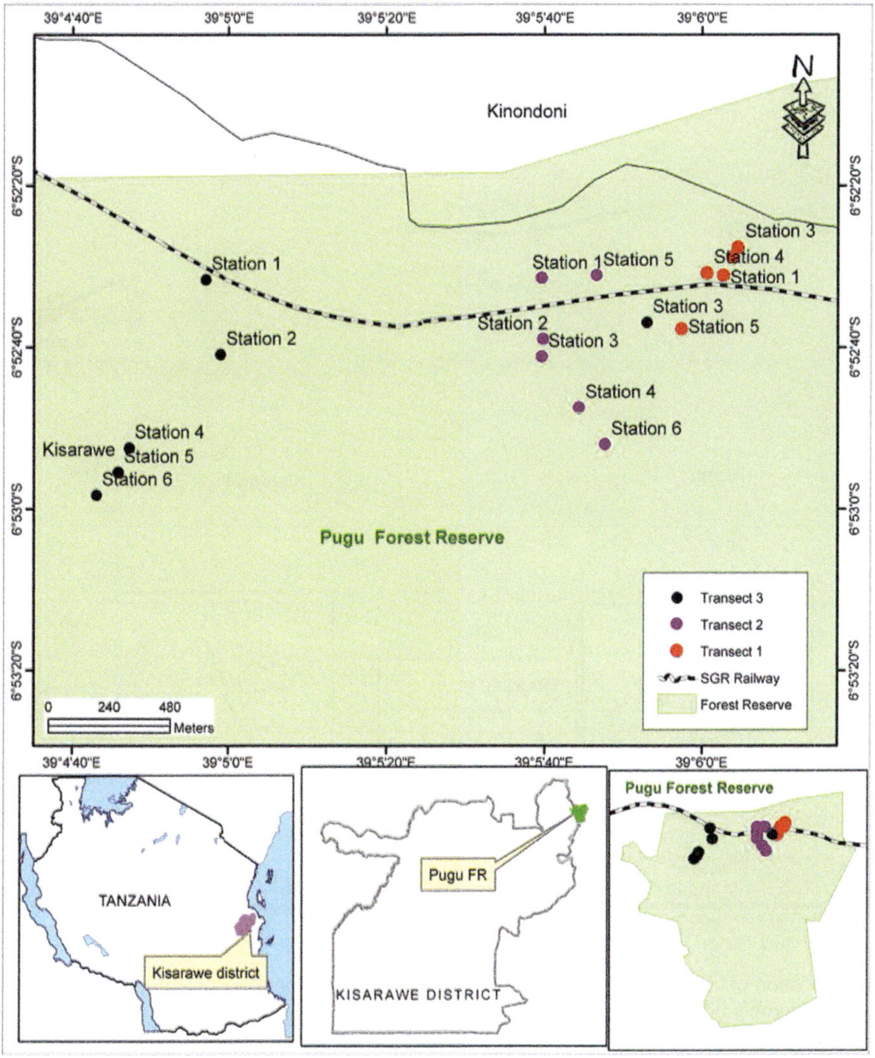

Fig. 5.1. Location of transects for arthropod sampling points (stations) in Pugu Hills Forest Reserve (April and November 2021).

stations which were located 500m apart. The study was conducted in two seasons, April and November 2021, covering both the rainy and dry seasons.

5.2.2 Sample collection

Arthropods were collected from the two forests by employing three different methods: pitfall trapping, sweep netting and physical searching. Pitfall trapping is the most frequently used method for sampling ground-dwelling arthropods (Phillips and Cobb, 2005). Pitfall traps can be used to evaluate the distribution of arthropods in diverse ecosystems at different scales, to describe activity patterns and habitat associations, as well as to establish relative species abundances, or the effects that disturbance can have on biodiversity (Mazía *et al.*, 2006). In this study, plastic containers were used for pitfall trapping; a one-litre plastic container, half-filled

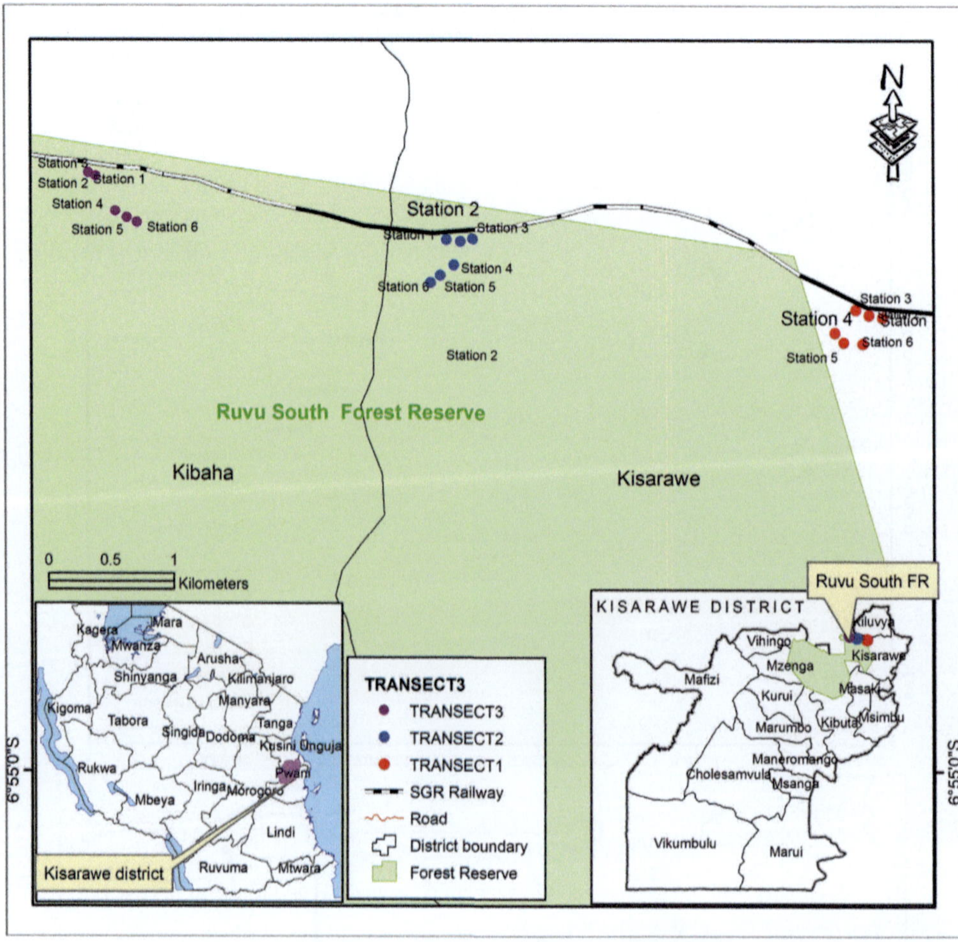

Fig. 5.2. Location of transects and arthropod sampling points (stations) in Ruvu South Forest Reserve (April and November 2021).

with water with a few drops of soap solution, was used to trap ground-foraging invertebrates. The traps were arranged in a 5 × 3 grid and placed 10 m apart, as described in Duelli *et al.* (1999). The stations were at least 500 m apart, and each station contained 15 plastic containers.

A standard 45 cm-diameter sweep net was used to collect flying arthropods and invertebrates on herbs, grass and other vegetation. Arthropods were swept from vegetation by vigorously brushing flowers, foliage or grasses with a strong-framed sweep net as described by Upton and Chapman (2010). Each five-minute sweeping constituted a sample according to established protocol (Schauff, 1997; Upton and Chapman, 2010).

The other method was direct searching of arthropods from discrete habitats such as under bark, rocks or logs, in ant or termite nests, at the base of grass tussocks, in flood debris and in leaf litter. While not as productive as other methods of collecting, direct searching provides information about the habits and life histories of species when collecting in this manner. A total of five searching samples was collected from each sampling station, being 15 min of continuous searching. The process was repeated for 2 days at each station. The collected specimens were morphologically identified using published identification keys and field guidebooks (Dijkstra and Clausnitzer, 2014; Enghoff, 2016).

5.2.3 Data analysis

Data were compiled using MS Excel into a species-sample-abundance matrix, summarized into tables and presented in graphs and charts. Abundance between sites was compared using the Mann-Whitney U test. Diversity was calculated using species richness and the Shannon-Wiener diversity Index. A similarity index between pairs of sites was calculated based on the Jaccard Index.

5.3 Findings and Discussion

5.3.1 Preliminary remarks

The two forests were rich in arthropods. However, the richness and diversity were more or less the same between the two reserves and between different transects at varying distances from the railway line. This is probably a reflection of similar vegetation and more or less the same level of disturbance experienced by the two forests. The geographical closeness of the two forest reserves may further explain the high similarity in arthropod taxa.

5.3.2 Athropod abundance

During the rainy season, 420 samples were collected. From these, 1876 and 1401 invertebrates were identified from Pugu and Ruvu South forest reserves, respectively. The number of samples collected during the dry season was 1307 from the two study sites (561 from Pugu Hills and 746 Ruvu South). The samples yielded 4837 total individual arthropods during the dry season (2077 (46%) and 2760 (54%) from Pugu Hills and Ruvu South, respectively). The mean abundance at the two sites was 3.70 (±0.24) and 3.70 (±0.57) individuals per sample, respectively. The data were not normally distributed (Shapiro-Wilk test, $W = 0.9302$ and 0.8714, respectively, $p < 0.05$). The Mann-Whitney test revealed that the mean abundance of arthropods was significantly higher in Ruvu South Forest Reserve compared to Pugu Hills Forest Reserve ($U = 1.93 \times 10^5$, $p < 0.05$). Data are summarized and presented in Table 5.1.

5.3.3 Arthropod richness and diversity

The morphological identification yielded the following classes of arthropods: Arachnida, Insecta, Malacostraca, Diplopoda and Chilopoda (Table 5.2). Class Insecta was the richest taxon with 14 orders, 91 families and 147 species. Class Arachnida was represented by three orders (Acari, Araneae and Scorpiones), 13 families and 31 species. Class Malacostraca was represented by one order (Isopoda), four families and six species, whereas the two classes Diplopoda and Chilopoda were each represented by one order and one family, respectively. The richest insect orders were Coleoptera, Hemiptera, Orthoptera and Hymenoptera, and for the Arachnids, Araneae was the richest group. Over 50 species of arthropods collected were represented by single individuals (singletons). Similar observations have been previously reported (Pizarro-Araya *et al.*, 2022) and were attributed to the presence of arthropod communities with rare species restricted to specific environments and with specialized ecological requirements, thus limiting their breeding and proliferation. Some important arthropod groups such as Lepidopterans were poorly represented; only three butterflies and ten butterflies and moths were recorded from Pugu and Ruvu forest reserves, respectively. This is a conservation concern and could be a reflection of the

Table 5.1. Statistical analyses of the abundance of arthropods collected from Pugu Hills and Ruvu South forest reserves (2021).

Statistic	Pugu Hills	Ruvu South
Number of samples	561	746
Sum	2077	2760
Mean	3.702	3.699
Standard error	0.241	0.573
Standard deviation	5.698	15.637

Table 5.2. Taxonomic composition of arthropods collected from Ruvu and Pugu forests.

Class	Order	Family	Species	Abundance Ruvu	Abundance Pugu	Total abundance	% abundance
Arachnida	Acari	1	1	32	0	32	0.66
	Araneae	11	29	197	107	304	6.32
	Scorpiones	1	1	0	2	2	0.04
Insecta	Blattodea	7	10	281	192	473	9.81
	Coleoptera	30	41	1429	172	1601	33.26
	Diplura	1	1	0	1	1	0.02
	Diptera	6	11	21	9	30	0.62
	Hemiptera	15	35	36	58	94	1.95
	Hymenoptera	9	21	402	1297	1699	35.61
	Mantodea	1	1	2	10	12	0.25
	Neuroptera	2	2	19	2	21	0.44
	Odonata	2	3	2	7	9	0.19
	Orthoprera	10	12	180	114	294	6.11
	Phasmatodea	1	2	2	13	15	0.31
	Trichoptera	1	1	4	7	11	0.23
	Isoptera	2	2	1	36	37	0.04
	Lepidoptera	4	5	10	3	13	0.27
Malacostraca	Isopoda	4	6	54	39	93	1.93
Diplopoda	Julida	1	1	79	8	87	1.81
Chilopoda	Scolopendromorpha	1	1	9	0	9	0.19
Totals		111	188	2760	2077	4837	100

abiotic or biotic state of the habitat as a result of constant disturbance and the impact of environmental change on the arthropod community or the ecosystem as a whole (Hodkinson and Jackson, 2005).

The two coastal forests are known as biodiversity hotspots for insects and particularly the Lepidopterans. In 2013, Howell and others recorded an estimated 200 species of insects in Pugu Forest Reserve (Howell *et al.*, 2013). Earlier, in 2001, a survey of 140 species of butterflies constituting 54% of the total collection was documented in the same area (Hall *et al.*, 2002). However, in the current study, members of Lepidoptera constituted a negligible percentage. Lepidopterans have an aesthetic value, which makes them popular among amateur naturalists and collectors. As a consequence, they are a relatively well studied group throughout the world and are used to monitor environmental changes and forest disturbance; thus, their diversity is considered as a righteous indicator of the strength of any given terrestrial biotope (Thomas, 2005; Tiple *et al.*, 2007; Nordqvist, 2009). Lepidopterans, and especially butterflies, are very vulnerable to habitat loss; this is because their larval stages are dependent on a few specific plant species as their host plants, and their reduction or loss, can lead to local extinction of the species (Laghude *et al.*, 2019). Furthermore, few earthworms (47 individuals) were recorded. These are important organisms that play a role in soil mixing and nutrient recycling and whose disappearance may ultimately affect the plant community. These results partly indicate that the tremendous decrease in the number of lepidopterans and earthworms is attributable to the ongoing SGR construction activities. However, an increased scope of sampling beyond the SGR alignment in the areas would confirm these findings.

5.3.4 Similarity index

The taxonomic composition of specimens collected shows similarity in terms of species collected from Pugu and Ruvu forest reserves. A total of 66 out of 188 species recorded were found in both habitats. Jaccard's similarity index was J = 0.36. The species richness and diversity were more or less similar between the two reserves and in different transects (Table 5.3) – Kruskal-Walli's test for equal medians – $Chi^2 = 10.15$; $p = 0.001465$; df = 3 and $Chi^2 = 9.828$; $p = 0.001786$; df = 3. This is probably a reflection of similar vegetation and more or less the same level of disturbance. The geographical closeness of the two forest reserves may further explain the high similarity in arthropod taxa. Further analysis of the data to establish the impact of disturbance with reference to the distance from the railway line indicated that there was no significant difference in abundance of arthropods between the sampling stations of Pugu forests; $Chi^2 = 5.489$; $p = 0.2119$; df = 6; whereas there was a significant difference in the abundance of arthropod assemblages between the sampling stations in Ruvu forest – $Chi^2 = 8.319$; $p = 0.04981$; df = 6. Notwithstanding the statistical analyses, the abundance of arthropods increased towards areas of high disturbances across all transects along the SGR line (Fig. 5.3).

Shannon-Wiener index indicates there was no significant difference in insect species diversity between Pugu and Ruvu forests ($t = 1.9296$, df = 188; $P = 0.053724$).

5.3.5 Arthropod distribution trends towards the SGR line

The abundance of arthropods decreased with increasing distance from the railway line. The assumption is that the level of disturbance is relatively high towards the railway line (Fig. 5.3). Anthropogenic activities can influence both positively and negatively the distribution of insect species and subsequently species richness (Chowdhury *et al.*, 2017). Human activities such as population growth, infrastructural development, livestock expansion, bush fires, timber and charcoal logging, mining, commercial firewood collection and climate change have a negative impact on diversity and species abundance of different organisms (Nordqvist, 2009). On the other hand, human activities such as agriculture maintain some open habitats and encourage environmental heterogeneity that can represent promising habitats for the growth and proliferation of invertebrates including butterflies (Tiple *et al.*, 2007).

The two forests (Pugu and Ruvu) have undergone constant disturbance due to anthropogenic activities over the years. The activities include agriculture, timber harvesting, firewood, charcoal, poles, clearance for cultivation, collection of medicines and bush meat (Rocky and Mligo, 2012). The ongoing SGR construction exerts more pressure on the already constantly disturbed forests. Many tropical forests worldwide are in constant disequilibrium due to ongoing human disturbance and degradation (Laghude *et al.*, 2019), which makes continued monitoring of biodiversity crucial to accurately document the impacts of habitat degradation on biological communities. Continuous surveillance and monitoring of species is of paramount importance because of their role in the ecosystem. This survey adds to the existing data on monitoring the trend of invertebrates in the two forests for conservation purposes.

5.3.6 Ecological functions of the collected arthropods

Tropical coastal forests are recognized as hotspots of global biodiversity and are highly productive, providing important ecosystem services. These ecosystems regulate nutrient cycles by retaining, transforming or removing nutrients and pollutants and enhancing primary productivity that helps support trophic levels along the food chain (Novoa *et al.*, 2020; Pizarro-Araya *et al.*, 2022). However, for several years now, the two coastal forests (Ruvu and Pugu), being close to Dar es Salaam City, have been subjected to many pressures of human nature (Boussougou *et al.*, 2021). Within these ecosystems, arthropods play an important ecological role (Table 5.4), including pollination, soil decomposition and nutrient recycling. Some species act as bio-indicators of the ecosystem's health; other species are

Table 5.3. Comparative species richness and similarity of arthropods of Ruvu and Pugu forest reserves.

Ruvu South Forest Reserve			Pugu Hills Forest Reserve		
Species name	Abundance	Percentage	Species name	Abundance	Percentage
Agelenopsis spp.	17	0.83	*Agelenopsis* spp.	19	0.88
Armadillum spp.	40	1.95	*Allodapulla* spp.	16	0.74
Armitermes hastus	31	1.51	*Amitermes hastatus*	81	3.77
Brachyiulus spp.	79	3.86	*Blepharodera* spp.	30	1.40
Catantopis spp.	46	2.25	*Componetus* spp.	13	0.60
Cicindela spp.	11	0.54	*Crematogaster* spp.	48	2.23
Cophogryllus spp.	49	2.39	*Euophrys* spp	14	0.65
Copris spp.	46	2.25	*Exochomus* spp.	26	1.21
Ctenus spp.	20	0.98	*Eyprepocnemis* spp.	15	0.70
Cyclocephala spp.	67	3.27	*Lepisiota* spp.	127	5.91
Dinothrombium spp.	30	1.46	*Linepithema humule*	38	1.77
Euophrys spp.	149	7.28	*Locustana* spp.	11	0.51
Eyprepocnemis spp.	32	1.56	*Macrotermes natalensis*	44	2.05
Graphipterus spp.	15	0.73	*Macrotermes* spp.	17	0.79
Holotermes spp.	231	11.28	*Maransis rufolieatus*	12	0.56
Hydrophilus spp.	13	0.63	*Messor* spp.	39	1.81
Leptacris spp.	18	0.88	*Oecophylla longinoda*	141	6.56
Megacephala spp.	161	7.86	*Omorgus* spp.	13	0.60
Oecophyla longinoda	17	0.83	*Onosandrus* spp.	22	1.02
Omorgus spp.	13	0.63	*Pachycondyla tarsatus*	264	12.28
Onosandrus spp.	17	0.83	*Paratrechina* spp.	25	1.16
Onthophagus spp.	337	16.46	*Polyrhachis gagates*	214	9.96
Pachycondyla tarsata	73	3.56	*Porcellio* spp.	31	1.44
Paratrechina spp.	26	1.27	*Rhabdomiris* spp.	10	0.47
Polyrhachis gagates	59	2.88	*Schizonycha* spp.	19	0.88
Solenopsis punctaticeps	35	1.71	*Scyphophorus acupuntatus*	12	0.56
Syngenes spp.	11	0.54	*Streblognathus aethiopicus*	143	6.65

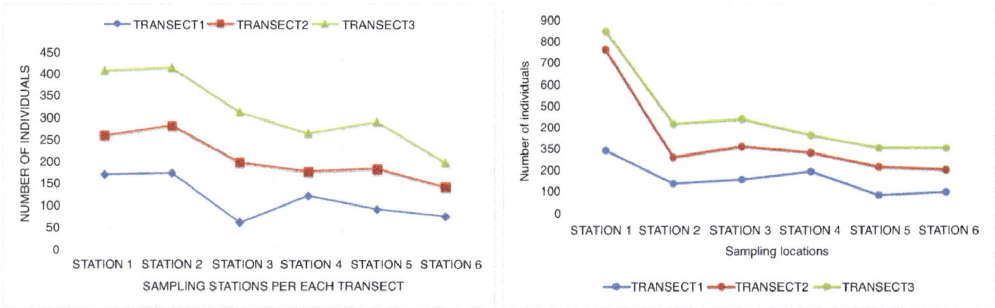

(A). Pugu Hills forest reserve (B). Ruvu South forest reserve

Fig. 5.3. Trend of invertebrate abundance from 50 m (station 1) – 3 km (station 6) away from the SGR line under construction in Pugu Hills and Ruvu South forest reserves, respectively.

predators to other invertebrates and they help to control agricultural pests by keeping them in check to balance the ecosystem. Species such as *Armadillidium* spp., *Brachyiulus* spp., *Amitermes hastus* and *Porcellio* spp. have been reported to play a nutrient recycling role (Dias *et al.*, 2012; Nervo *et al.*, 2017). Several other species recorded in this study – *Agelenopsis* spp., *Ctenus* spp., *Graphipterus* spp., *Euophrys* spp., *Oecophyla longinoda*, *Crematogaster* spp., *Streblognathus aethiopicus*, *Dinothrombium* spp., etc. are predators and play a significant role as control agents of crop pests (Zhang, 1999).

Furthermore, indicator species of disturbance (*Eyprepocnemis* spp. and *Paratrechina* spp.) were recorded from both forests, and the presence of these species attest to that. Anthropogenic activities such as logging, charcoal production, urban advancement, agriculture and infrastructure development (Boussougou *et al.*, 2021), and the ongoing SGR construction within the two forests has led to serious disturbances and alteration of their structure, functioning and diversity. Other species of ecological importance recorded are *Allodapula* spp. and *Copris* spp. for pollination and seed dispersal, respectively (Nichols *et al.*, 2008; Ley and Claßen-Bockhoff, 2009). Although arthropods play an important role in the ecosystem, knowledge of the potential of the rich and diverse arthropod fauna in the two forests is fragmented, whereas birds and other vertebrates have been widely studied (Howell, 1993; Hall *et al.*, 2002; Howell *et al.*, 2013).

5.4 Conclusion and Recommendations

The implications of this study for management and conservation of the two forest reserves is limited by lack of previous studies on arthropods in the studied area. However, disturbance in natural reserves resulting from construction activities directly removes important species requirements, such as food, cover, and resting and breeding sites. For arthropods, their large surface area-to-body ratio further imposes great danger to even subtle changes in habitat conditions such as light penetration. Removal of host larval plants reduces nectars and nectar plants that provide a source of energy to pollinators and other phytophagous invertebrates. This in turn may have significant impact on their populations. Therefore, arthropods assessment in perturbed systems such as this area under SGR construction can ensure habitat quality through improved habitat management strategies informed by research. Continuous monitoring of the diversity, abundance and distribution of arthropods post-construction is recommended to establish the trend for possible habitat restoration measures.

Table 5.4. Ecological functions of the key arthropod species collected in Ruvu and Pugu forests.

Species name	Ecological functions/services
Agelenopsis spp., *Ctenus* spp., *Graphipterus* spp., *Euophrys* spp., *Oecophyla longinoda*, *Crematogaster* spp., *Streblognathus aethiopicus* *Dinothrombium* spp.	Predators and generalist feeders serve a significant role in keeping populations of many insect pests in check, thus balancing an ecosystem. Control agents of pests and fungal diseases in tropical agroforestry systems.
Armadillidium spp., *Brachyiulus* spp., *Amitermes hastus Porcellio* spp.	These are detritivores and play an initial role in nutrient cycling by breaking down organic matter in the soils.
Allodapula spp.	These play a key role in the pollination of plants, especially 'small (horizontal)' flowers.
Componotus spp., *Omorgus* spp.	These are decomposers and clean the environment by feeding on organic waste, aerating the soil. *Componotous* spp. also serve as prey for bird species and other wild animals in the ecosystem.
Copris spp.	Secondary seed dispersal, nutrient cycling and parasite suppression
Cyclocephala spp.	These function in the cycling of organic matter, act as bio-indicators of environmental quality, and some species are considered agricultural pests.
Hydrophilus spp.	Water beetles are good indicators of biodiversity in aquatic ecosystems.
Leptacris spp., *Euophrys* spp.	Sensitive indicators of the state of disturbance in ecosystems
Lepisiota spp.	They play an important role in the seed bank, seedling establishment and plant composition of arid ecosystems.
Eyprepocnemis spp., *Linepithema humule*, *Messor* spp.	Disruptive communities of ants may cause shift in composition of the plant community, owing to a disproportionate reduction in the densities of large-seeded plants.
Locustana spp.	They play an important role in food chains; some grasshopper species are ecological indicators of ecosystem qualities.
Macrotermes spp., *Megacephala* spp., *Onthophagus* spp.	These are a source of heterogeneity on tree diversity and conversion and are widely advocated as ecological indicators.
Pachycondyla tarsata	These contribute a large biomass of tropical rainforest fauna with high biomass and individual density.
Paratrechina spp.	Agricultural and household pests, adapted to survive in highly disturbed habitats; used as indicator of disturbances.
Solenopsis punctaticeps	Sensitive to land management and serve as important regulators of key soil processes; can be used as indicators of soil-based ecosystem services

References

Andersen, A.N., Fisher, A., Hoffmann, B.D., Read, J.L. and Richards, R. (2004) Use of terrestrial invertebrates for biodiversity monitoring in Australian rangelands, with particular reference to ants. *Austral Ecology* 29(1), 87–92.

Boussougou, G.B., Brou, Y.T., Valimba, P, *et al*. (2021) Vulnerability of Pugu and Kazimzumbwi forest reserves under anthropogenic pressure in southeast Tanzania. In: Gervasi, O., Murgante, B., Misra, S., Stankova, E. and Torre, C.M. (eds), *Computational Science and Its Applications. ICCA*, Springer International Publishing AG, pp. 231–240.

Chowdhury, S., Hesselberg, T., Böhm, M., Islam, M.R. and Aich, U. (2017) Butterfly diversity in a tropical urban habitat (lepidoptera: Papilionoidea). *Oriental Insects* 51(4), 417–430.

Collen, B., Ram, M., Zamin, T. and McRae, L. (2008) The tropical biodiversity data gap: Addressing disparity in global monitoring. *Tropical Conservation Science* 1, 97–110.

Congdon, C. and Collins, S. (1997) *Kiellands's Butterflies of Tanzania: Supplement*. Hill House/ABRI-Lambillionea, Nairobi.

Dias, N., Hassall, M. and Waite, T. (2012) The influence of microclimate on foraging and sheltering behaviours of terrestrial isopods: Implications for soil carbon dynamics under climate change. *Pedobiologia* 55(3), 137–144.

Dijkstra, K.-D.B. and Clausnitzer, V. (2014) *The Dragonflies and Damselflies of Eastern Africa: Handbook for All Odonata from Sudan to Zimbabwe*. In: Musee Royale de l'Afrique Centrale. Tervuren, Belgium.

Duelli, P., Obrist, M.K. and Schmatz, D.R. (1999) Biodiversity evaluation in agricultural landscapes: Above-ground insects. In: Pooletti, M.G. (ed.) *Invertebrate Biodiversity as Bioindicators of Sustainable Landscapes: Practical Use of Invertebrates to Assess Sustainable Land Use*. Elsevier.

Enghoff, H. (2016) Checklist of the millipedes (diplopoda) of Tanzania. *Journal of East African Natural History* 105, 51–113.

Espinosa, R., Andino, P., Cauvy-Fraunié, S., Dangles, O., Jacobsen, D. *et al*. (2020) Diversity patterns of aquatic Macroinvertebrates in a tropical high-Andean catchment. *International Journal of Tropical Biology* 68(2).

Fleishman, E. and Murphy, D.D. (2009) A realistic assessment of the indicator potential of butterflies and other Charismatic Taxonomic groups. *Conservation Biology* 23, 1109–1116.

Gallou, A., Baillet, Y., Ficetola, G.F. and Despres, L. (2017) Elevational gradient and human effects on butterfly species richness in the French Alps. *Ecology and Evolution* 7(11), 3672–3681.

Hall, S.M., Staddon, S., Howell, K.M. and Fanning, E. (2002) *Kazimzumbwi Forest Reserve: A Biodiversity Survey* [Frontier Tanzania Technical Report 26]. Coastal Forest Research Programme. Society for Environmental Exploration/Universti of Dar es Salaam.

Hemp, C. (2017) Annotated checklist of Orthoptera from Kazimzumbwi forest reserve, Tanzania with the description of new species and discussion of the Biogeographic patterns of threatened species. *Zootaxa* 4226, 151–193.

Hodkinson, I.D. and Jackson, J.K. (2005) Terrestrial and aquatic invertebrates as Bioindicators for environmental monitoring, with particular reference to mountain Ecosystems. *Environmental Management* 35, 649–666.

Howell, K.M. (1993) Herpetofauna of the eastern African forests. In: Lovett, J.C. and Wasser, S.K.*Biogeography and Ecology of the Rain Forests of Eastern Africa*Cambridge University Press, Cambridge, UK.

Howell, K.M., Msuya, C., Perkin, A., Ngalasson, W., Clarke, P, *et al*. (2013) Tanzania's coastal forests are a biodiversity hotspot. *The Arc Journal* 7–10.

Laghude, R., Rhobragade, B. and Irungbam, J.S. (2019) Species diversity of butterflies in moderately disturbed forests and along forest edges: a case study of Karjat district, Ahmednagar, Maharashtra, India. *Uttar Pradesh Journal of Zoology* 40(4), 214–221.

Lesch, V. and Bouwman, H. (2018) Terrestrial Arthropods as Indicators of Environmental Pollution. In: SETAC Conference 2018, Rome.

Ley, A.C. and Claßen-Bockhoff, R. (2009) Pollination syndromes in African Marantaceae. *Annals of Botany* 104(1), 41–56.

Mazía, C.N., Chaneton, E.J. and Kitzberger, T. (2006) Small-scale habitat use and assemblage structure of ground-dwelling beetles in a Patagonian shrub Steppe. *Journal of the Arid Environment* 67, 177–194.

McKinney, M.L. (2008) Effects of urbanization on species richness: A review of plants and animals. *Urban Ecosystems* 11, 161–176.

Nervo, B., Caprio, E., Celi, L., Lonati, M., Lombardi, G., *et al.* (2017) Ecological functions provided by dung beetles are interlinked across space and time: Evidence from 15n isotope tracing. *Ecology* 98(2).

Nichols, E., Spector, S., Louzada, J., Larsen, T., Amezquita, S. *et al.* (2008) Ecological functions and Ecosystem services provided by Scarabaeinae dung beetles. *Biological Conservation* 141, 1461–1474.

Nordqvist, E. (2009) *Butterfies as indicators of forest quality in miombo woodlands, Tanzania*. MSc thesis, University of Uppsala.

Novoa, V., Rojas, O., Ahumada-Rudolph, R., Sáez, K., Fierro, P. *et al.* (2020) Coastal wetlands: Ecosystems affected by urbanization? *Water* 12(3).

Phillips, I.D. and Cobb, T.P. (2005) Effects of habitat structure and lid transparency on pitfall catches. *Environmental Entomology* 34, 875–882.

Pizarro-Araya, J., Alfaro, F.M., Gómez, F.A. and Villablanca, R. (2022) Arthropod fauna of the urban Coastal Wetland of Aguada La Chimba (Antofagasta region, Chile): a Wetland in an arid matrix. *Anthropocene Coasts* 5.

Ramirez, M.I., Azcárate, J.G. and Luna, L. (2003) Effects of human activities on monarch butterfly habitat in protected mountain forests of Mexico. *Forestry Chronicle* 79(2), 242–246.

Rocky, J. and Mligo, C. (2012) Regeneration pattern and size-class distribution of indigenous woody species in exotic plantation in Pugu forest Reserve Tanzania. *International Journal of Biodiversity and Conservation* 4(1), 1–14.

Schauff, M.E. (1997) *Collecting and Preserving Insects and Mites: Techniques and Tools*. National Museum of Natural History, Washington, DC.

Thomas, J. (2005) Monitoring change in the abundance and distribution of insects using butterflies and other indicator groups. *Philosophical Transactions of the Royal Society of London. Series B: Biological Sciences* 360, 339–357.

Tiple, A.D., Khurad, A.M. and Dennis, R.L. (2007) Butterfly diversity in relation to a human-impact gradient on an Indian University campus. *Nota Lepidopterologica* 30(1), 179–188.

Townsend, C.R., Scarsbrook, M.R. and Doledex, S. (1997) Quantifying disturbance in streams: Alternative measures of disturbance in relation to Macroinvertebrate species traits and species richness. *Journal of the North American Benthological Society* 16, 531–544.

Uehara-Prado, M., Fernandes, J.O., Bello, A.M., Machado, G., Santos, A.J. *et al.* (2009) Selecting terrestrial Arthropods as indicators of small-scale disturbance: A first approach in the Brazilian Atlantic forest. *Biological Conservation* 142, 1220–1228.

Upton, S.M. and Chapman, L.B. (2010) *Methods for Collecting, Preserving and Studying Insects and Other Terrestrial Arthropods*. Australian Entomological Society, Fitzroy, Australia.

Zhang, Z.Q. (1999) Biology and ecology of trombidiid mites (acari: Trombidioidea). In: Bruin, J., Van der Geest, L.P.S. and Sabelis, M.W. (eds) *Ecology and Evolution of the Acari*. Springer, Dordrecht, The Netherlands, pp. 277–289.

6 Macroinvertebrate and Fish Assemblages as Indicators of Wetland Modification within the SGR Footprint

Edmond Alavaisha*
Institute of Resource Assessment, University of Dar es Salaam, Tanzania

Abstract
The wetlands outside of conservation areas are threatened by human activities. Tanzania is currently constructing a standard-gauge railway (SGR) from Dar es Salaam to Makutupora, which is expected to impact the area's ecological condition and biodiversity. Monitoring and preserving these areas' integrity and ecosystem services require cost-effective tools. This study investigated whether macroinvertebrates and fishes could indicate predictable changes due to construction activities. Using a modified standardized kick-net method and the Fish-based Assessment Method for Ecological Status (FEME), we collected macroinvertebrates at 13 wetlands and fishes at eight selected wetlands. Despite varying degrees of modification, all wetlands supported biodiversity, hosting macroinvertebrates from ten orders and 22 families, and 1139 individual fishes from 23 species, with *Tilapia niloticus* and *Astatotilapia* spp. being the most abundant. Both macroinvertebrates and the fishes appeared to reflect the wetlands' modification levels, suggesting their potential as reliable indicators for assessing human impacts on wetlands, including those affected by SGR construction in Tanzania.

Keywords: wetland, macroinvertebrates, fish assemblages, wetland modification

6.1 Background Information

Globally, wetlands suffer from numerous anthropogenic forces that usually act concurrently on biota (Ramsar Convention on Wetlands, 2018). Human populations and the resultant pressure from the increasing consumption of wetland resources and infrastructure development (Ramsar Convention on Wetlands, 2018; Bridgewater and Kim, 2021) increase these forces. The consequence is excess of nutrients (Bruinsma, 2003; Feio *et al.*, 2023), landscape modification (Tscharntke *et al.*, 2012; Feio *et al.*, 2023), altered streams flow and connectivity (Bobbink *et al.*, 2006; Demars *et al.*, 2012), physical habitat structure, loss of riparian vegetation and variation in species dynamic (Alavaisha *et al.*, 2019; Masese *et al.*, 2023). The anthropogenic stressors have instigated greater losses of biodiversity in wetlands (Bridgewater and Kim, 2021). They have degraded biological assemblage structure and composition, reduced functional diversity, and altered ecosystem functioning (Alavaisha, 2020; Frazier and Schriever, 2022; Henriques *et al.*, 2022) leading to losses in potential ecosystem services. Such services consist of food and water provisioning, regulation of local climate, and regulation of water quantity and quality.

*edmondalavaish@yahoo.com; alavaisha.edmond@udsm.ac.tz

DOI: 10.1079/9781800626942.0006

With current expansion of transport infrastructures, wetlands are increasingly investigated from an ecosystems and ecological integrity perspective (Allan, 2004). Estimations of environmental status are done by looking at the influence of disturbance on aquatic organisms and the living environment. Measurement of physico-chemical parameters and monitoring using modern experiments are usually time-consuming, expensive and dependent on special instruments and expertise (Alavaisha *et al.*, 2019). Furthermore, majority measurements show quality of environment at the time of measurement and these can change over time. Currently, indicators based on the presence or absence of aquatic organisms have been developed to assess environmental quality and ecological status (Rubach *et al.*, 2010; Frazier and Schriever, 2022). For example Mustapha and Aris (2012) found the river Getsi to be modified and its water highly polluted, but Alavaisha *et al.* (2019) and Elias *et al.* (2014) reported the water of rivers in Kilombero and Pangani to be modified due to nutrients leaching using invertebrates scores.

State-of-the-art methodologies and bio-assessment programmes have been developed for wetlands across several continents and nations (Kaufmann *et al.*, 2022; Feio *et al.*, 2023). The assessment of wetland environment is a critical tool for evaluating status and trends in health, demonstrating biological risk, elaborating cost-effective restoration/rehabilitation measures, and evaluating the results of environmental policy and management actions. Nevertheless, few projects, particularly in developed nations, have effectively implemented statutory bio-assessment tools (Dallas, 1997; Kaaya *et al.*, 2015; Kaufmann *et al.*, 2022). Moreover, the biological condition of wetlands is not regularly monitored to capture the improvement from proposed measures. These limitations have hindered rigorous assessments at local scale to prevent the wide spread of pollution, meanwhile safeguarding the integrity of the environment and ecosystem.

Macroinvertebrates and fishes are often used for bioassessment as they respond to modification of wetland environments and are relatively easy and quick to sample (Kaaya *et al.*, 2015; Fierro *et al.*, 2021; Kaufmann *et al.*, 2022). Macroinvertebrates are important for the recycling of organic matter and a necessary food resource for higher trophic levels; i.e. their fluctuations may affect the ecosystem function and environment. Moreover, their mobility, richness, diversity and life cycles indicate the degree of modification or improvement of wetlands. Norris and Thoms (1999) suggested that the effect on macroinvertebrates and fishes is usually the final point of environmental degradation and thus an important indication of overall ecosystem health. This gives confidence that they reveal local conditions that point out landscape change. Macroinvertebrates and fishes have been widely used in North and South America, Australia, Europe and South Africa (Hering *et al.*, 2004; Ollis *et al.*, 2006; Frazier and Schriever, 2022; Feio *et al.*, 2023; Sinche *et al.*, 2023) where most studies were connected to peri-urban catchments and nutrient pollution and modification of rivers due to intensive agriculture. So far, only a few studies have used macroinvertebrates and fishes (e.g. Minaya *et al.*, 2013; Elias *et al.*, 2014; Kilonzo *et al.*, 2014; Shimba and Jonah, 2016; Alavaisha *et al.*, 2019; Tumusiime *et al.*, 2019; Masese *et al.*, 2023) in East African countries with limited studies investigating the response of them to land modification due to construction of infrastructure.

With the current trend of infrastructure development in Tanzania, concern is growing on the importance of conserving wetlands. This is very crucial as 40% of ecosystem services are coming from the wetlands (Zedler and Kercher, 2005). For this to be effectively managed, assessment and monitoring must be in place and easily adopted at local scale. The construction of the standard-gauge railway (SGR) from Dar es Salaam to Makutupora passes through various fringed and flooded wetlands, some being seasonal while others are permanent. Among the important large wetlands are the Ruvu, Mkondoa and Mkata rivers.

This chapter intends to demonstrate the environmental conditions of selected wetlands from Dar es Salaam to Dodoma. Ensuring compliance with requirements for local and international standards, including the Tanzania Environmental Management Act (EMA) of 2004 and the International Finance Corporation Performance (IFC) Standards, is an important step towards assessing the state of wetlands. Further, the chapter is expected to indicate the response of wetlands, using indicator species – macroinvertebrates and fishes – to inform the levels of modification of wetlands along the constructed SGR footprint.

6.2 Materials and Methods

6.2.1 Description of study area

The study involved selected wetlands between Dar es Salaam and Makutupora (see Fig. 6.1). Various types of wetlands (Table 6.1) associated with rivers were sampled within a buffer of 60m either side of the SGR line.

The wetlands involved are in a tropical climate with one to two rainfall seasons. The rainfall is seasonal with long rains between March and June and short rains between October and December (Rocky and Mligo, 2012) in the segment of Dar es Salaam to Morogoro. The rainfall distribution and vegetation cover varies with altitude and mean annual temperature is 28°C, and the variation throughout the year is relatively low. Even though sampling was conducted during the rainy season, most areas in Dar es Salaam and Morogoro had fewer rains; Dodoma and Singida regions were dry. The Dodoma segment of the SGR, on the other hand, receives little rainfall throughout the year with the driest months being June through August, whereas January, February and September are the months that receive most rainfall.

6.2.2 Data collection

Macroinvertebrates and fish species are indicators of water quality, river health and wetland modifications. Macroinvertebrates were collected along delineated water bodies within the project footprint area following the described methodology for a rapid bioassessment using the multi-habitat approach of Gerber and Gabriel

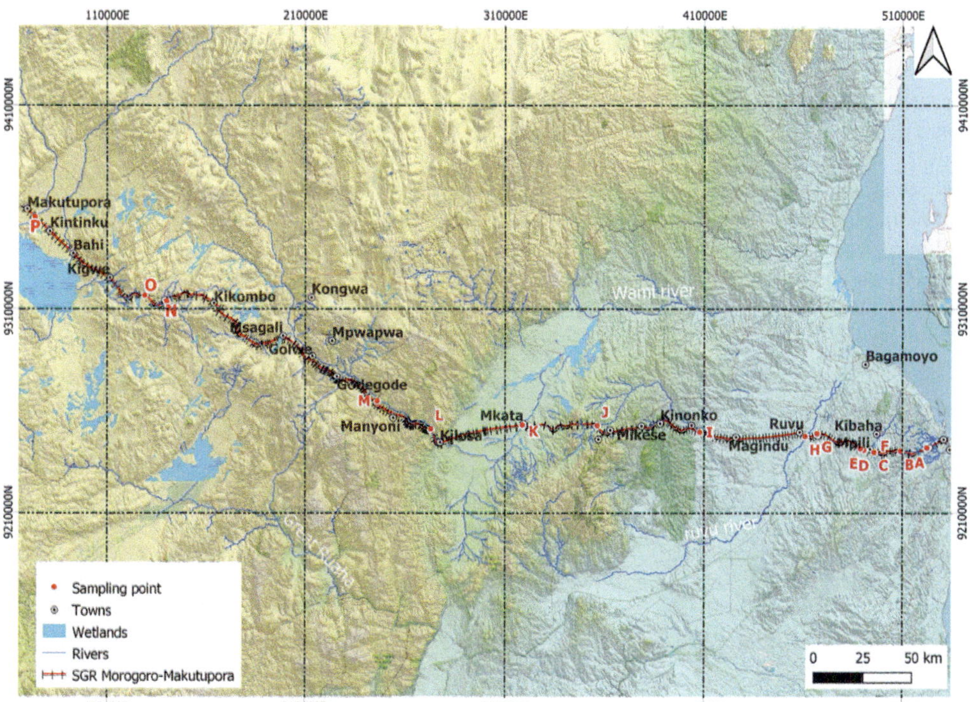

Fig. 6.1. Study area location and sampling points along the SGR line from Dar es Salaam to Makutupora.

Table 6.1. Study sites (sampling points) along the SGR line from Dar es Salaam to Makutupora via Dodoma.

Site	UTM Zone	Easting	Northing	Description	Name
A	37S	522846	9241679	Flood wetland	Vingunguti
B	37S	509162	9240016	Fringe wetland	Msimbazi river
C	37S	495642	9239619	Fringe wetland	Mpiji river
D	37S	490465	9240519	Fringe wet land	
E	37S	488751	9241307	Flood wet land	
F	37S	495601	9239469	Fringe wet land + stream	Kwamturuki swamp
G	37S	466928	9248698	Fringe wet land	
H	37S	460878	9247296	River	Ruvu river
I	37S	407936	9249221	Fringe	Ngerengere river 1
J	37S	356639	9252422	River + swamp	Ngerengere river 2
K	37S	318884	9252701	Ffringe	Mkata river
L	37S	272915	9250569	River	Mkondoa river
M	37S	246065	9264220	Fringe	Lumuma river
N	36S	804560	9313547	Fringe pond	Korongo idara ya maji
O	36S	793431	9316460	Flood wetland	Zuzu
P	36S	738351	9355673	Fringe	Bubu river

(2002). A modified kick-net method (using hand net mini-SASS with 1 mm mesh size) was used for collection of macroinvertebrates. The net was swept through vegetation, and stone and gravel + sand + mud (GSM) biotope in opposing directions for 2 minutes to ensure equivalent sampling effort and to cover spatial variation. Obtained samples were sorted and preserved in 70% alcohol. Identification and counting for abundance and diversity were done in the laboratory. To estimate the ecological health of wetland, the following indices were used:

i. average score per taxa (ASP) (Eqn 6.1) calculated from Tanzania River Scoring System (TARISS) score (Kaaya *et al.*, 2015); and
ii. Simpson's diversity index (Eqn 6.2) (Simpson, 1949).

$$\text{Average Score per Taxa (ASPT)} = \frac{\text{TARISS score}}{\text{No.of Taxa}} \quad \text{(Eqn 6.1)}$$

The ASPT expressed sensitivity score (TARISS score) of recorded taxon. TARISS score is the perceived sensitivity of taxon in the range of 1–15 with increasing order to water quality. The ASPT will be used to define the ecological category/health of the wetland ecosystem (Table 6.2)

$$\text{Simpson's diversity index (D)} = 1 - \sum_{i=1}^{s} \frac{n_i(n_i - 1)}{N(N - 1)} \quad \text{(Eqn 6.2)}$$

where n is the number of individuals of each species and N is the total number of individuals of all species.

Modified time-limited method of fishing net was based on FAME (Fish-based Assessment Method for the Ecological Status) (Schmutz *et al.*, 2007). This active sampling technique involved use of seine net of 6 mm mesh to sample fish in selected wetlands along the SGR. The techniques allow the sampler to actively target communities and provide immediate results; however, seine netting is difficult in areas with many snags in the water. The fishing net was dragged in the margin of the wetland to maximize the chances of capturing fish. We did 2 hours of fishing in each wetland to ensure similar fishing effort for every site. Most of the sampled fish were identified, counted and released *in situ*. Some fish that

Table 6.2. Ecological categories for interpretation of ASPT.

Ecological band	Water category	Range of ASPT	Description
A	Very good (Natural)	Greater than 7	No, or negligible, modification (relatively little human impact)
B	Good	6.0–6.9	Biodiversity and integrity largely intact (some human-related disturbance but ecosystems essentially in a good state)
C	Fair	5.0–5.9	Sensitive species may be lost, with tolerant or opportunistic species dominating (moderately modified).
D	Poor	4.0–4.9	Mostly only tolerant species present; alien species invasion; disrupted population dynamics; species are often diseased (largely modified)
E	Very poor	Less than 3.9	Critical modification; almost complete loss of natural habitat and indigenous species with severe alien invasion (seriously modified).

needed to be identified further were fixed in 70% alcohol and transferred to the laboratory. The identification was done using the field guidebook by Eccles (1992). Finally, the species diversity was computed using the equation (Eqn 6.2) above. The species were also grouped for each surveyed wetland.

6.3 Results and Discussion

6.3.1 Macroinvertebrates and fish occurrences and distributions

Wetland habitat varied by depth, water movement and type in sampling sites along the SGR line. Table 6.3 presents habitat features noted at each sampling site for macroinvertebrates and fish. Most wetlands were fringe, some associated with swamps and ponds. Some sites had emerging wetland due to water blockage and damming activities during construction of the standard-gauge service road and railway line. Eight sites were located in slow-moving-water areas dominated by small substrates and extensive aquatic macrophyte beds. These were indicative of most of the flooded areas, ponds and swamps. Seven sites were located in moderate-depth water. These sites were characteristic of the pelagic/profundal zone areas that extend to more than 3 m depth. At all these sites, major differences in habitat were associated with substrate composition and the presence of aquatic macrophytes.

6.3.2 Macroinvertebrate abundances and taxa richness

A total of 730 macroinvertebrate individuals belonging to 10 orders and 22 families from 13 sampled wetlands were collected, identified and counted (see Table A3.8, Ch. 3). All identified macroinvertebrates were common in Tanzanian rivers and wetlands (Gerber and Gabriel, 2002; Kaaya, 2014). The majority of taxa were from the groups Odonata (20%), Coleoptera (19%), Ephemeroptera (13%) and Trichoptera (8%;

Table 6.3. Information on sampling sites along the SGR line.

Sampling site	Description	Name	Sampling	Depth	Stream habitat
A	Flood wetland	Vingunguti	M	Shallow	Slow
B	Fringe wetland	Msimbazi river	M, F	Moderate	Slow
C	Fringe wetland	Mpiji river	M, F	Moderate	Slow
D	Fringe wetland		NIL		
E	Flood wetland		M	Moderate	Slow
F	Fringe wetland + swamp	Kwamturuki river	F		
G	Fringe wetland		M	Shallow	Slow
H	River	Ruvu river	M, F	Moderate	Fast
I	Fringe	Ngerengere river 1	M	Moderate	Fast
J	River + swamp	Ngerengere river 2	M, F	Moderate	Fast
K	Fringe	Mkata river	NIL		
L	River	Mkondoa river	M, F	Moderate	Fast
M	Fringe	Lumuma river	M	Shallow	Fast
N	Fringe + pond	Korongo idara yamaji	M, F	Shallow	Slow
O	Flood wetland	Zuzu	M	Shallow	Slow
P	Fringe	Bubu river	M, F	Shallow	Slow

Sampling: "M" for macroinvertebrates; "F" = for fish; "NIL", for no sampling

Fig. 6.2). The literature indicates that macroinvertebrates would have a marked gradient in heterogeneity as a function of the gradient in water conditions, and in this aspect it is similar to existing descriptions of other rivers (Elias *et al.*, 2014; Shimba and Jonah, 2016; Alavaisha *et al.*, 2019). Previous reports indicate that the taxa Coleoptera, Ephemeroptera and Trichoptera are abundant under moderate turbulent flow, associated with a high oxygen concentration and much particulate matter that makes a food source. In general, higher abundance and taxa richness appeared to be associated with littoral areas of the moderate water flow, sites H, J and L (Table 6.4; Table A3.8, Ch. 3). Distinct taxa at a station are organisms that were not observed at any other sampling station. For the case of wetland along the SGR line, all species were common; however they varied in abundance, probably affected slightly by the sampling effort and on-site disturbances. The low taxa richness observed at site N, a fringe pond in Dodoma, might be explained by the unstable shifting sands associated with the alluvial fan located near the confluence of the river. In comparison, taxa richness was moderate to higher ($D > 0.5$) at other stations where the substrate was more diverse (Table 6.4). The higher diverse site is more stable, even in face of perturbation, making these sites less vulnerable to environmental stress, such as that brought about through land intensification and increased human activities in the wetland scape. Sampling in shallow and moderate-depth water did not reveal any distinct taxa since most sites were cleared, leaving few riparian vegetations to provide sufficient forage and shadow for species.

6.3.3 Fish abundance and richness

The species of fish caught depends to a large extent on the method used. The percentage composition and Simpson's diversity index were used to characterize the species collected along the SGR line. A total of 1139 individual fish from 23 different species were caught in wetland (Table A3.9, Ch. 3). All fish species were common in the region. *Tilapia niloticus* (34.24%) and *Astatotilapia* spp. (32.75%) had high abundance and were found almost in all wetlands.

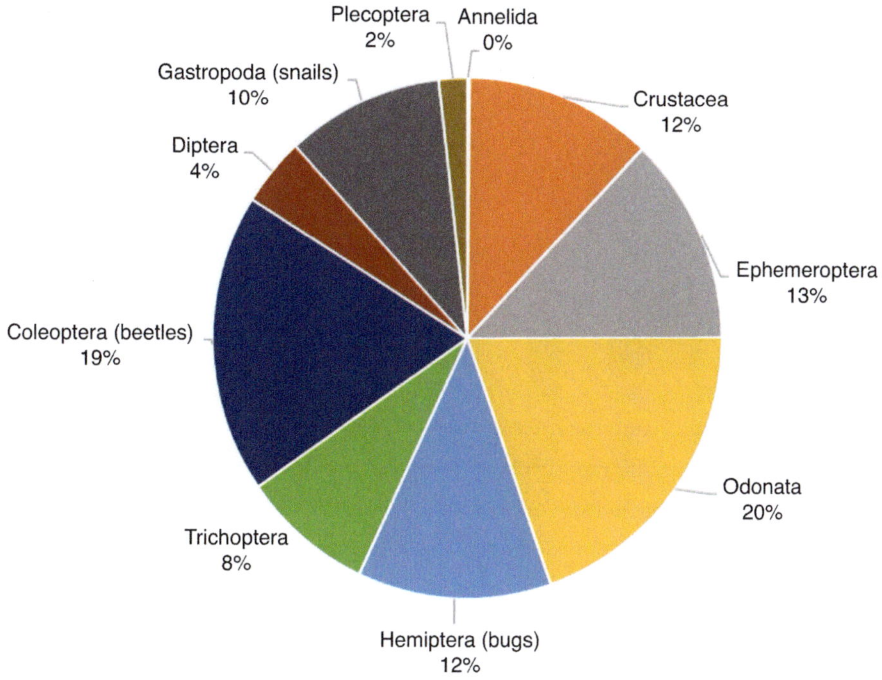

Fig. 6.2. Percentage distribution of macroinvertebrate taxa in the SGR line.

Table 6.4. Biological and ecological indices of wetland sampling sites along SGR line.

Indices	A	B	C	E	G	H	I	J	L	M	N	O	P
TARISS score	58	34	20	20	28	40	24	40	40	20	21	24	37
No. of taxa	10	7	4	4	6	6	3	7	7	4	4	4	5
ASPT	5.8	4.9	5	5	4.7	6.67	8	5.7	5.7	5	5.25	6	7.4
Ecological band	C	D	C	C	D	B	A	C	C	C	C	B	A
Diversity (D)	0.8	0.8	0.64	0.7	0.7	0.84	0.6	0.7	0.8	0.8	0.38	0.5	0.6

The diversity was moderate to slightly higher in all sites except new emerging wetland, site F (Fig. 6.3), where there was land transformation due to sand translocation for service road construction.

The fishery resources consist mainly of riverine genera. In flood plains, fishing was carried out during the rainy season and in pools which become isolated as the river levels drop. In such pools, fish are caught with ring nets. Although fishing activities may take place all year round, the most important fishing period was during the rainy season, as at this time fish ascend the rivers to spawn and can be caught easily.

Higher fish diversity was recorded in all sites, except site F (Kwamturuki). This was new emerging river and swamp (Fig. 6.4). Despite development of riparian vegetation and reeds, mixing of water was very limited. In addition, because of construction works, the environment was too unstable to allow colonization of new fish species. The site is, potentially, a fishing ground in the future, after natural restocking of the swamp.

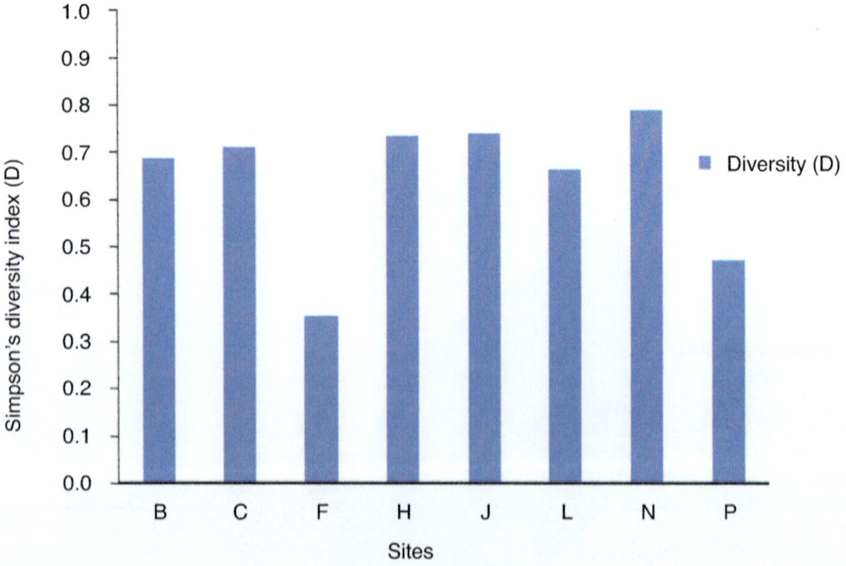

Fig. 6.3. Diversity of fish species among sampling sites.

Fig. 6.4. Emerging wetland following excavation of dry land at Kwamturuki.

6.3.4 Trophic interactions between macroinvertebrates and fish

There is a close link between macroinvertebrates and fish. Research of the trophic role of macroinvertebrates in wetland revealed that in those ecosystems there are herbivores, detritivores, shredders, carnivores, and those that feed on vegetal residuals that originated from the surrounding catchment basins (Tomanova and Usseglio-Polatera, 2007; Hauer and Resh, 2017; Pallottini *et al.*, 2017). For instance, Ephemeroptera, Trichoptera and Plecoptera larval stages are shredders and decapods feed on dead vegetal matter and predate on freshwater mollusc. In turn, mollusc and larval insect stages are prey for native fishes. The majority of fish feed on Arthropoda, Annelida, Mollusca, Porifera, Vertebrata and vegetables. Thus, in this scenario, the macroinvertebrates and fish are reported to have a unique prey–predator relationship that maintains biodiversity and ecosystem health.

6.3.5 Macroinvertebrates, fish and wetlands modification

Overall, 69% of the wetland assessed ranged from fair to very poor condition while 31% ranged from good to very good in terms of macroinvertebrates and fish indices (Table 6.4; Fig. 6.3). The fringe wetland, G, and part of Msimbazi river, B, were severely impaired. The fringe wetland was blocked from the main Ruvu river and streams, and therefore limited the water exchange and nutrient flow. Similar reasons were shared by Alavaisha *et al.* (2019), Seeteram *et al.* (2019) and Masese *et al.* (2023), reporting that excess nutrients and alteration of flow were causing the modification of the wetland. The assessments based on fish indices showed slightly lower modification among the selected wetlands. This is not surprising; it has been reported (e.g. Aazami *et al.*, 2015) that fish are not as sensitive as macroinvertebrates, and, further (Marzin *et al.*, 2012; Aazami *et al.*, 2015), that fish species can withstand the wetland landscape modification unless the threshold capacity is exceeded. Fish move across larger spatial scales and across habitats within wetland than do invertebrates to access food resources and complete their life cycles. Despite the slight variation, both macroinvertebrate and fish data are site-quality classifications of modification of wetlands. Our results support that human land disturbances and wetland modifications during the construction of the SGR would have a proportional impact on the ecological status of wetland condition. Similarly, Alavaisha *et al.* (2019), Seeteram *et al.* (2019) and Feio *et al.* (2023) noted low diversity and slightly lower quality of wetland in modified wetlands. Further, Shimba (2017) and Alavaisha *et al.* (2019) indicated lower water quality for rivers in all sites along the roads that signify modifications. We observed substantial differences in the biological quality of rivers that may also be associated with non-point source pollution, which we were unable to capture during the study.

6.4 Conclusion

The loss and change in structure of macroinvertebrates and fish communities in the study area reflect landscape modification. The modification altered species distributions, simplified aquatic community structure and composition, and increased the likelihood of non-native species. Such changes disrupt ecosystem functioning because of altered proportions of the available functional traits of species. Consequently, the ability of wetlands, along the SGR footprint, to provide ecosystem services to human populations is decreased, including losses in climate regulation, water quality, carbon regulation, water and food provisioning, recreation, and disease prevention. Therefore, the macroinvertebrates and fish can offer easy and cheap approaches for assessment of wetland modification to provide early warning before the wide spread of any impact.

Acknowledgements

We acknowledge Tanzania Railway Corporation for providing funds to conduct this study.

References

Aazami, J., Esmaili-Sari, A., Abdoli, A., Sohrabi, H. and Van den Brink, P.J. (2015) Monitoring and assessment of water health quality in the Tajan river, Iran using physicochemical, fish and macroinvertebrates indices. *Journal of Environmental Health Science and Engineering* 13(1), 1–12. DOI: 10.1186/s40201-015-0186-y.

Alavaisha, E. (2020) Agricultural expansion impacts on wetland ecosystem services from Kilombero Valley, Tanzania. Stockholm, Sweden. Dissertations in Physical Geography, no. 7. University of Stockholm.

Alavaisha, E., Lyon, S.W. and Lindborg, R. (2019) Assessment of water quality across irrigation schemes: a case study of wetland agriculture impacts in Kilombero valley, Tanzania. *Water (Switzerland)* 11(4). DOI: 10.3390/w11040671.

Allan, J.D. (2004) Influence of land use and landscape setting on the ecological status of rivers. *Limnetica* 23(3–4), 187–198. DOI: 10.1146/annurev.ecolsys.35.120202.110122.

Bobbink, R., Beltman, B., Verhoeven, T.A. and Whigham, D.F. (eds) (2006) *Wetlands: Functioning, Biodiversity Conservation, and Restoration*. Springer, Berlin.

Bridgewater, P. and Kim, R.E. (2021) The Ramsar Convention on Wetlands at 50. Nature Ecology & Evolution 5(3), 268–270. DOI: 10.1038/s41559-021-01392-5.

Bruinsma, J. (ed.) (2003) *World Agriculture: Towards 2015/2030: An FAO Perspective*. Routledge, London. DOI: 10.4324/9781315083858.

Dallas, H.F. (1997) A preliminary evaluation of aspects of SASS (South African scoring system) for the rapid bioassessment of water quality in rivers, with particular reference to the incorporation of SASS in a national biomonitoring programme. *Southern African Journal of Aquatic Sciences* 23(1), 79–94. DOI: 10.1080/10183469.1997.9631389.

Demars, B.O.L., Kemp., J.L., Friberg, N., Usseglio-Polotero, P. and Harper, D.M. (2012) Linking biotopes to invertebrates in rivers: Biological traits, taxonomic composition and diversity. *Ecological Indicators* 23, 301–311. DOI: 10.1016/j.ecolind.2012.04.011.

Eccles, D.H. (1992) *Field Guide to the Freshwater Fishes of Tanzania*. FAO Species Identification Sheets for Fishery Purposes, FAO, Rome.

Elias, J.D., Ijumba, J.N., Mgaya, Y.D. and Mamboya, F.A. (2014) Study on freshwater macroinvertebrates of some Tanzanian rivers as a basis for developing biomonitoring index for assessing pollution in tropical African regions. *Journal of Ecosystems*. DOI: 10.1155/2014/985389.

Feio, M.J., Hughes, R.M., Serra, S.R.Q., Nichols, S.J., Kefford, B.J, *et al*. (2023) Fish and macroinvertebrate assemblages reveal extensive degradation of the world's rivers. *Global Change Biology* 29(2), 355–374. DOI: 10.1111/gcb.16439.

Fierro, P., Valdovinos, C., Lara, C. and Saldias, G. (2021) Influence of intensive agriculture on benthic macroinvertebrate assemblages and water quality in the Aconcagua river basin (Central Chile). *Water (Switzerland)* 13(4). DOI: 10.3390/w13040492.

Frazier, C.F. and Schriever, T.A. (2022) Patterns of invertebrate community composition and functional structure across a dune succession gradient. *Wetlands* 42(7). DOI: 10.1007/s13157-022-01596-w.

Gerber, A. and Gabriel, M.J.M. (2002) *Aquatic Invertebrates of South African Rivers Field Guide*. Department of Water Affairs and Forestry, South Africa. Available at: http://scholar.google.com/scholar?hl=en&btnG=Search&q=intitle:Aquatic+Invertebrates+of+South+African+Rivers+Field+Guide#0 (accessed 29 April 2024).

Hauer, F.R. and Resh, V.H. (2017) Macroinvertebrates. In: Hauer, F.R. and Lamberti, G. (eds) *Methods in Stream Ecology*, 3rd edn. Elsevier, pp. 297–319. DOI: 10.1016/B978-0-12-416558-8.00015-9.

Henriques, M., McVicar, T.R., Holland, K.L. and Daly, E. (2022) Riparian vegetation and geomorphological interactions in anabranching rivers: A global review. *Ecohydrology* 15(2), e2370. DOI: 10.1002/ECO.2370.

Hering, D., Moog, O., Sandin, L. and Verdonschot, P.F.M. (2004) Overview and application of the AQEM assessment system. *Hydrobiologia* 516(1–3), 1–20. DOI: 10.1023/B:HYDR.0000025255.70009.a5.

Kaaya, L.T. (2014) Biological assessment of tropical riverine systems using aquatic macroinvertebrates in Tanzania, East Africa. PhD thesis, University of Cape Town. Available at: https://open.uct.ac.za/bitstream/item/8977/thesis_sci_2014_kaaya_lt.pdf?sequence=1 (accessed 29 April 2024).

Kaaya, L.T., Day, J.A. and Dallas, H.F. (2015) Tanzania River Scoring System (TARISS): a macroinvertebrate-based biotic index for rapid bioassessment of rivers. *African Journal of Aquatic Science* 40(2), 109–117. DOI: 10.2989/16085914.2015.1051941.

Kaufmann, P.R., Hughes, R.M., Paulsen, S.G., Peck, D.V., Seeliger, C.W, *et al.* (2022) Physical habitat in conterminous US streams and rivers, part 1: Geoclimatic controls and anthropogenic alteration. *Ecological Indicators* 141, 109046. DOI: 10.1016/j.ecolind.2022.109046.

Kilonzo, F., Masese, F.O., Griensven, A.V., Bauwens, W., Obando, J, *et al.* (2014) Spatial-temporal variability in water quality and macro-invertebrate assemblages in the upper Mara river Basin, Kenya. *Physics and Chemistry of the Earth* 67–69, 93–104. DOI: 10.1016/j.pce.2013.10.006.

Marzin, A., Archaimbault, V., Belliard, J., Chauvin, C., Delmas, F, *et al.* (2012) Ecological assessment of running waters: Do Macrophytes, Macroinvertebrates, diatoms and fish show similar responses to human pressures? *Ecological Indicators* 23, 56–65. DOI: 10.1016/j.ecolind.2012.03.010.

Masese, F.O., Wanderi, E.W., Nyakeya, K., Achieng, A.O., Fauchy, K, *et al.* (2023) Bioassessment of multiple stressors in Afrotropical rivers: evaluating the performance of a Macroinvertebrate-based index of biotic integrity, diversity, and regional biotic indices. *Frontiers in Environmental Science* 11(March), 1–27. DOI: 10.3389/fenvs.2023.1015623.

Minaya, V., McClain, M.E., Moog, O., Omengo, F. and Singer, G.A. (2013) Scale-dependent effects of rural activities on benthic macroinvertebrates and physico-chemical characteristics in headwater streams of the Mara river, Kenya. *Ecological Indicators* 32, 116–122. DOI: 10.1016/j.ecolind.2013.03.011.

Mustapha, A. and Aris, A.Z. (2012) Multivariate statistical analysis and environmental modelling of heavy metals pollution by the industries. *Polish Journal of Environmental Studies* 21(5), 1353–1361. Available at: https://www.academia.edu/2159133/Multivariate_Statistical_Analysis_and_Environmental_Modeling_of_Heavy_Metals_Pollution_by_Industries (accessed 5 December 2018).

Norris, R.H. and Thoms, M.C. (1999) What is river health? *Freshwater Biology* 41(2), 197–209. DOI: 10.1046/j.1365-2427.1999.00425.x.

Ollis, D.J., Dallas, H.F., Esler, K.J. and Boucher, C. (2006) Bioassessment of the ecological integrity of river ecosystems using aquatic macroinvertebrates: An overview with a focus on South Africa. *African Journal of Aquatic Science* 31(2), 205–227. DOI: 10.2989/16085910609503892.

Pallottini, M., Cappelletti, D., Fabrizi, A., Gaino, E., Selvaggi, R. *et al.* (2017) Macroinvertebrate functional trait responses to chemical pollution in agricultural–industrial landscapes. *River Research and Applications* 33(4), 505–513. DOI: 10.1002/rra.3101.

Ramsar Convention on Wetlands (2018) *Global Wetland Outlook: State of the World's Wetlands and Their Services to People*. Ramsar Convention on Wetlands, Gland, Switzerland. Available at: https://static1.squarespace.com/static/5b256c78e17ba335ea89fe1f/t/5b9ffd2e0e2e7277f629eb8f/1537211739585/RAMSAR+GWO_ENGLISH_WEB.pdf (accessed 29 April 2024).

Rocky, J. and Mligo, C. (2012) Regeneration pattern and size-class distribution of indigenous woody species in exotic plantation in pugu forest reserve, tanzania. *International Journal of Biodiversit Conservation* 4, 1–14.

Rubach, M.N., Baird, D.J. and Van Den Brink, P.J. (2010) A new method for ranking mode-specific sensitivity of freshwater arthropods to insecticides and its relationship to biological traits. *Environmental Toxicology and Chemistry* 29(2), 476–487. DOI: 10.1002/etc.55.

Schmutz, S., Cowx, I.G., Haidvogl, G. and Pont, D. (2007) Fish-based methods for assessing European running waters: a synthesis. *Fisheries Management and Ecology* 14(6), 369–380. DOI: 10.1111/J.1365-2400.2007.00585.X.

Seeteram, N.A., Hyera, P.T., Kaaya, L.T., Lalika, M.C.S. and Anderson, E.P. (2019) Conserving rivers and their biodiversity in Tanzania. *Water (Switzerland)* 11(12), 2612. DOI: 10.3390/w11122612.

Shimba, M.J. (2017) Spatial and seasonal variations in surface water quality of the Mkondoa River, Tanzania. DOI: 10.5897/JSSEM2017.0640.

Shimba, M.J. and Jonah, F. (2016) Macroinvertebrates as bioindicators of water quality in the Mkondoa river, Tanzania, in an agricultural area. *African Journal of Aquatic Science* 41(4), 453–461. DOI: 10.2989/16085914.2016.1230536.

Simpson, E.H. (1949) Measurement of diversity. *Nature* 163, 688. DOI: 10.1038/163688a0.

Sinche, F., Cabrera, M., Vaca, L., Segura, E. and Carrera, P. (2023) Determination of the ecological water quality in the Orienco stream using benthic macroinvertebrates in the Northern Ecuadorian Amazon. *Integrated Environmental Assessment and Management* 19(3), 615–625. DOI: 10.1002/ieam.4666.

Tomanova, S. and Usseglio-Polatera, P. (2007) Patterns of benthic community traits in neotropical streams: Relationship to mesoscale spatial variability. *Fundamental and Applied Limnology* 170(3), 243–255. DOI: 10.1127/1863-9135/2007/0170-0243.

Tscharntke, T., Clough, Y., Wanger, T.C., Jackson, L., Motzke, I, *et al*. (2012) Global food security, biodiversity conservation and the future of agricultural intensification. *Biological Conservation* 151(1), 53–59. DOI: 10.1016/j.biocon.2012.01.068.

Tumusiime, J., Tolo, C.U., Dusabe, M.-C. and Albrecht, C. (2019) Reliability of the Tanzania river scoring system (TARISS) macroinvertebrate index of water quality: a case study of the river Mpanga system, Uganda. *Journal of Freshwater Ecology* 34(1), 541–557. DOI: 10.1080/02705060.2019.1631895.

Zedler, J.B. and Kercher, S. (2005) Wetland resources: Status, trends, ecosystem services, and restorability. *Annual Review of Environment and Resources* 30(1), 39–74. DOI: 10.1146/annurev.energy.30.050504.144248.

7 Herpetofauna within the Standard-gauge Railway Footprint from Dar es Salaam to Makutupora, Singida, Tanzania

Wilirk Ngalason* and John Valentine Lyakurwa
College of Natural and Applied Sciences, University of Dar es Salaam, Tanzania

Abstract
The standard-gauge railway (SGR) between Dar es Salaam and Singida regions in Tanzania traverses in varied ecological zones from coastal to semi-arid areas in the central part of the country. This study assesses herpetofauna species richness, composition and conservation status along the SGR. It assumes species richness varies between the coastal and semi-arid areas. Audio-visual surveys were supplemented with information from literature resulting in 51 amphibian and 107 reptilian species. One amphibian and three reptilian species are endemic to Tanzania, whereas one amphibian and one reptilian species are considered 'critically endangered'. The similarity index in species composition between the lowlands and the semi-arid regions is 52.9% for amphibians and 48.6% for reptiles. Possible potential impacts of the project to the herpetofauna include mortality due to train collisions, barrier effects and habitat disturbances. Construction of adequate passes in the form of tunnels, underpasses and drainage culverts is recommended to facilitate movements.

Keywords: amphibia, reptile, richness, impact, conservation, mitigation

7.1 Background Information

Railways are among the major global transportation systems and are promoted by several governments due their economic and environmental advantages (Borda-de-Água *et al.*, 2017). Due to these benefits, the Government of Tanzania embarked on the construction of an electrified standard-gauge railway (SGR) between Dar es Salaam and Mwanza in different phases, with the first two running between Dar es Salaam and Makutupora in Singida region. Despite the attached benefits, railway projects have associated impacts on biodiversity including barrier effects, mortality, species invasions and environmental disturbances caused by chemical pollution and noise (Borda-de-Água *et al.*, 2017). This chapter aims at documenting the existing herpetofauna along the SGR between Dar es Salaam and Dodoma regions which will form a basis for monitoring and management of herpetofauna.

The term herpetofauna refers to the assemblage of amphibians and reptiles in a given area (Karns, 1986; Howell, 2002). The two groups are normally included in one scientific discipline of herpetology despite being quite different. Many of them crawl along the ground and are commonly placed together in museums, and similar methods of collection and preservation are used (Karns, 1986). Amphibians, including

*Corresponding author: wilirk@udsm.ac.tz

 Biodiversity and Conservation along an East African Railway (eds C.A. Masao *et al.*)
DOI: 10.1079/9781800626942.0007

frogs, toads, newts, salamanders and caecilians, have moist, thin, glandular and permeable skins, whereas reptiles, including crocodiles, lizards, snakes, tortoises, terrapins, turtles and tuatarans, have dry, scaly, waterproof skins. Amphibians use their skins to respire and absorb water (termed cutaneous drinking) beside protection of internal organs, whereas the skins of reptiles are dry and covered with scales or scutes and thus cannot perform the same functions as those of amphibians. Due to their differences in biology, the two groups have different ecological requirements. Amphibians are prone to desiccation due to their permeable skins and because of that they are restricted to moist environments such that they are neither found in deserts nor in marine environments. Additionally, they limit their activities to at night or during rainy periods. Reptiles, on the other hand, due to their protective skins, are able to live in very dry places including deserts and marine environments. Another big difference between amphibians and reptiles is the way they reproduce. Most of both groups lay eggs, although a few species give birth to live offspring. The eggs of amphibians are of anamniotic type, consisting of a single layer of vitelline membrane that surrounds the yolk, and thus are laid in moist environments, whereas the eggs of reptilian fauna are of amniotic type with extraembryonic membranes, including the yolk sac, amnion, the allantois, the chorion and the shell that protects the egg, and thus they can be laid in any habitat, including in very dry places.

The word 'amphibia' comes from two Greek words *amphi* meaning 'double' and *bios* meaning 'life', indicating that they can spend part of their life in water and also in a terrestrial environment, although there are exceptions (du Preez and Caruthers, 2009). They are thought to have evolved from the lobe-fined fish sarcopterygians during the Palaezoic era during the Devonian period about 370 million years ago (Vitt and Caldwell, 2014). *Elginerpeton*, *Ichthyostega*, *Acanthostega* and *Ventastega* were the first fossils assigned to amphibians (Vitt and Caldwell, 2014). The fossil amphibians have features that resembled the fossils of lobe-fined fish, including radial fin rays that supported a tail fin, labyrinthodont teeth, a lateral line system, similar vertebra outline, and fish branchial arches supporting internal gills. Over evolutionary history, the earliest amphibians diversified and adapted to different ways of life to form the modern amphibians that are different from their ancestors. All the extant amphibians are grouped in the subclass Lissamphibia, with three orders including Anura or Salientia (frogs and toads), Caudata or Urodela (newts and salamanders), and Apoda or Gymnophiona (caecilians). Frost (2023) names 8673 amphibian species that are distributed worldwide in all the continents except Antarctica. Anura is the largest order, with representatives in all continents except Antarctica, and including Tanzania (Vitt and Caldwell, 2014). The second largest group is Order Caudata or Urodela that consists of newts and salamanders with their distribution in the northern hemisphere, with the exception of one group, Family Plethodontidae, a lungless salamander that was able to cross to the southern hemisphere in South America. Urodelans are not found in Tanzania or Africa south of the Sahara. In Africa they are restricted to the northern tip of the continent (Channing and Howell, 2006; Channing and Rödel, 2019). The third group of amphibians includes the caecilians, Order Apoda, that are worm-like with elongated bodies, legless, with reduced eyes and a solid and compact skull, being adaptations for subterranean life. The caecilians are distributed in the tropical forests of South America, Africa and South-east Asia (AmphibiaWeb, 2023).

Reptiles are thought to have evolved from a lineage of amphibians that became more terrestrial about 315 million years ago during the Carboniferous period (Branch, 1998; Vitt and Caldwell, 2014). The evolution of reptiles involved the development of scaly skin to avoid desiccation; evolution of amniotic eggs to reduce the dependence on water for reproduction and the development of lungs to support respiration on land (Vitt and Caldwell, 2014). The evolution of these features enabled reptiles to diversify and use both aquatic and terrestrial resources that the amphibians were not able to use because of their biology. Currently 11,940 reptile species are known (Uetz *et al*., 2023), which include the crocodiles in the Order Crocodilia (Crocodylia); terrapins, tortoises and turtles in the Order Testudines; worm lizards, lizards and snakes in the Order Squamata; and tuatarans in the Order Rhynchocephalia. The former three orders are represented in Tanzania whereas the tuatarans are only found in New Zealand (Vitt and

Caldwell, 2014). The modern amphibians and reptiles are very much diversified and adapted to different lifestyles including terrestrial, fossorial, arboreal and aquatic species.

Amphibians and reptiles are among the threatened vertebrate groups mainly through habitat alternation that is caused by deforestation, degradation, logging, uncontrolled fires and clearance of vegetation for cultivation, urbanization and for some developmental projects (Heatwole and Wilkinson, 2009; Menegon *et al.*, 2020). Other factors leading to biodiversity loss include the effect of climate change, pollution, overharvest, and diseases (Heatwole and Wilkinson, 2009; Hof *et al.*, 2011). Amphibians and reptiles are among the less-studied groups of vertebrates in Tanzania. Much of the known literature on the amphibian and reptile richness of the Dar es Salaam-Makutupora SGR is contained in the field guides of amphibians and reptiles of the region (e.g. Schiøtz, 1975, 1999; Spawls *et al.*, 2002, 2006, 2018; Channing and Howell, 2006; Pickersgill, 2007; Channing and Rödel, 2019). Literature on the influence of linear infrastructures, such as railway projects, on amphibians and reptiles is very limited and in Tanzania such information does not exist. The SGR that starts in the coastal areas with coastal climates in Dar es Salaam to the semi-arid areas in Singida region in central Tanzania provides an opportunity for such a study. Two attempts to assess herpetofauna species richness from coastal lowlands to highlands were conducted in southern Africa and in southern Tanzania (Poynton, 1992, Poynton, 2003). There has not been such an attempt in other regions, including central Tanzania. This chapter is an attempt to explore the herpetofauna species richness along the SGR which starts in the coastal areas with coastal climates near sea level in Dar es Salaam to the semi-arid areas in highland areas of central Tanzania in Singida region. The study further reviews the potential impacts of the SGR on the amphibian and reptilian fauna. The information generated will form a basis for monitoring and mitigating the impacts of such projects on the amphibian and reptilian fauna.

7.2 Methodology

7.2.1 Study area

This study was conducted along the SGR footprint from coastal Tanzania in Dar es Salaam through Coast, Morogoro, Dodoma and Singida regions. The footprint covers a distance of 541 km and it traverses different eco-regions. Important eco-regions traversed by the SGR include the northern Zanzibar-Inhambane coastal forest mosaic, eastern miombo woodlands, and *Acacia-Commiphora* bushlands and thickets. These different eco-regions can generally be classified into modified and natural habitats. Forest reserves traversed by the SGR route include Pugu-Kazimzumbwi Nature Forest Reserve (PKNFR) and Ruvu South Forest Reserve (RSFR). Within the SGR footprint are wetlands that are scattered throughout the line. The elevation in the study area increases gently from near sea level in Dar es Salaam to 1190 masl in Dodoma in central Tanzania before decreasing to Makutupora at 850 masl. Climate within the footprint varies considerably from coastal to tropical with variable rainfall patterns ranging from 956 mm per annum in the coastal area in Dar es Salaam to 539 mm per annum in the semi-arid areas in Dodoma (Cauldwell *et al.*, 2019). Bimodal rainfall pattern is experienced in the majority of the stretch with the exception of areas in Dodoma region that experience unimodal pattern (ESIA, 2019). Temperature ranges between 15°C and 32°C with the coastal areas being the warmest. Typically, long rains fall between March and May; short rains between October and December; and dry seasons are experienced between January and February and between June and September.

7.2.2 Study sites

A study of the herpetofauna species richness along the SGR footprint was conducted between 22 April 2021 and 5 May 2021 during the rainy season. Several standard herpetofauna sampling techniques are available; they have been used successfully to obtain baseline information on the diversity of herpetofauna. They range from live-trapping (e.g. pitfall and funnel traps) to cover boards, plots, transects, time-constrained searches, to general or opportunistic searches (Grant *et al.*, 1992; Heyer *et al.*, 1994). There is no single method that is adequate to sample all the herpetofauna communities due to the diverse lifestyles of herpetofauna including arboreal, subterranean, terrestrial and aquatic styles. Arboreal species include the tree frogs, chameleons, some geckoes and some snakes, whereas fossorial species include caecilians, some frogs, worm lizards, some skinks and some snakes. Other species live an aquatic lifestyle including some frogs, lizards, terrapins, turtles, snakes and crocodiles. Several other herpetofauna species live on the terrestrial environment including some of frogs and toads, lizards, snakes and testudines.

The surveys of the herpetofauna within the SGR footprint were conducted both during the day and at night. This is due to the fact that some species are active during the day whereas others are active at night. Amphibians are mostly active at night when it is cool, whereas some reptiles are active during the day and others at night. The study involved searching for herpetofauna along the footprint by audio-visual encounter surveys (Grant *et al.*, 1992; Heyer *et al.*, 1994; Doan, 2003). The method is one of the appropriate techniques for both inventory and monitoring studies (Crump and Scott Jr, 1994; Heyer *et al.*, 1994). Most herpetofauna species can be easily detected by the method with less effort. The surveys involved driving along the SGR and stopping for herps searches in different habitat types occurring along the footprint including wetlands. The area within 60 m (i.e. 30 m either side of the SGR) was searched. All herpetofauna species encountered or heard calling were identified and recorded. The standard field guides to the amphibians and reptiles of East Africa (Channing and Howell, 2006; Pickersgill, 2007; Harper *et al.*, 2010; Spawls *et al.*, 2018) were used. In addition to field surveys, literature on the amphibian and reptilian fauna occurring within the SGR geographic region was reviewed. This includes Loveridge (1957), Tilbury (2010), Schiøtz (1975, 1999), Spawls *et al.* (2002, 2006, 2018), Channing and Howell (2006), Pickersgill (2007), Harper *et al.* (2010), Barratt (2017), and Channing and Rödel (2019). The taxonomy of the amphibian species followed that of Amphibian Species of the World (Frost, 2023) and that of reptiles followed that of East Africa (Spawls *et al.*, 2018; Uetz *et al.*, 2023).

A study by Poynton (2003) shows a reduction or high turnover of the lowland and highland amphibian species between 600m and 1200 m. Additionally, Harper *et al.* (2010) indicate the coastal forests to range from sea level to about 500 masl, and in some places they extend to as high as 1040 m. These coastal forests are considered one of the globally important regions for biodiversity conservation (Mittermeier *et al.*, 1998, 1999; Myers *et al.*, 2000). At least two coastal forests are traversed by the SGR including PKNFR and RSFR which form one of the important habitats for flora and fauna. Due to the subtraction observed, this study attempts to group the herpetofauna species into lowland species that occur from near sea level in Dar es Salaam to about 500m in Kilosa (37 M 0280000 UTM 9244000) and highland species that occur from Kilosa to Makutupora (36 M 0733285 UTM 9360186).

7.3 Findings and Discussion

7.3.1 Amphibian species richness

Fifty-one amphibian species were found to occur along the SGR footprint from Dar es Salaam to Singida region. Seventeen species were found during the surveys and the other 34 species are known to occur in the SGR geographic region, based on literature. The low numbers obtained during the surveys reflect low sampling effort. The species obtained belong to 12 anuran families including Arthroleptidae (4 species), Brevicipitae (1 species), Bufonidae (8 species), Hemisotidae (1 species), Hyperoliidae (14 species), Microhylidae (1 species), Phrynobatrachidae (5 species), Pipidae (1 species), Ptychadenidae (8 species), Pyxicephalidae (5 species), Ranidae (1 species) and Rhacophoridae (2 species) (Table 7.1). Of the 51 species, 27 (52.7%) occur throughout the SGR from Dar es Salaam to Makutupora, Singida; 20 species occur in the lowlands between Dar es Salaam and Kilosa; and only

Table 7.1. Conservation status of amphibian fauna within the SGR footprint from Dar es Salaam to Dodoma region.

#	Taxonomic group	Common name	Conservation status	Coast	Upland
	Amphibians, Order Anura, Family Arthroleptidae				
1	*Arthroleptis stenodactylus* Pfeffer, 1893*	Long-fingered Squeaker	Least Concern	x	x
2	*Arthroleptis xenodactyloides* Hewitt, 1933	Dwarf Squeaker	Least Concern	x	-
3	*Leptopelis argenteus* (Pfeffer, 1893)	Silvery Tree Frog	Least Concern	x	-
4	*Leptopelis flavomaculatus* (Günther, 1864)*	Yellow-spotted Tree Frog	Least Concern	x	x
	Order Anura, Family Brevicipitidae				
5	*Breviceps mossambicus* Peters, 1854	Mozambique Rain Frog	Least Concern	x	-
	Order Anura, Family Bufonidae				
6	*Mertensophryne lindneri* (Mertens, 1955)	Lindner's Forest Frog	Least Concern	x	x
7	*Mertensophryne loveridgei* (Poynton, 1991)	Loveridge's Forest Frog	Least Concern	x	-
8	*Mertensophryne micranotis* (Loveridge, 1925)	Woodland Forest Frog	Least Concern	x	-
9	*Mertensophryne taitana* (Peters, 1878)	Taita forest frog	Least Concern	x	x
10	*Schismaderma carens* (Smith, 1848)	Red Toad	Least Concern	x	x
11	*Sclerophrys gutturalis* (Power, 1927)*	Guttural Toad	Least Concern	x	x
12	*Sclerophrys pusilla* (Mertens, 1937)	Merten's Striped Toad	Least Concern	x	x
13	*Sclerophrys xeros* (Tandy, Tandy, Keith & MacKay, 1976)	Desert Toad	Least Concern	x	-
	Order Anura, Family Hemisotidae				
14	*Hemisus marmoratus* (Peters, 1854)	Marbled Piglet Frog	Least Concern	x	x
	Order Anura, Family Hyperoliidae				
15	*Afrixalus delicatus* Pickersgill, 1984	Delicate Spiny Reed Frog	Least Cconcern	x	x
16	*Afrixalus fornasini* (Bianconi, 1849)*	Fornasini's Spiny Reed Frog	Least Concern	x	-
17	*Afrixalus stuhlmanni* (Pfeffer, 1893)	Stuhlimann's Spiny Reed Frog	Least Concern	x	-
18	*Hylambates maculatus* Duméril, 1853*	Red-legged Wot-wot	Least Concern	x	-
19	*Hyperolius argus* Peters, 1854*	Argus Reed Frog	Least Concern	x	-

Continued

Table 7.1. Continued

#	Taxonomic group	Common name	Conservation status	Coast	Upland
20	*Hyperolius mariae* Barbour and Loveridge, 1928	Mary's Reed Frog	Least Cncern	x	-
21	*Hyperolius mitchelli* Loveridge, 1953	Mitchell's Reed Frog	Least Concern	x	-
22	*Hyperolius parkeri* Loveridge, 1933	Parker's Reed Frog	Least Concern	x	-
23	*Hyperolius pusillus* (Cope, 1862)*	Water Lily Reed Frog	Least Concern	x	-
24	*Hyperolius ruvuensis* Barratt *et al.*, 2017	Ruvu Reed Frog	Critically Endangered	x	-
25	*Hyperolius substriatus* Ahl, 1931	Mainland Reed Frog	Least Concern	x	-
26	*Hyperolius tuberilinguis* Smith, 1849*	Tinker Reed Frog	Least Concern	x	-
27	*Hyperolius viridiflavus* (Duméril & Bibron, 1841)*	Common Reed Frog	Least Concern	x	x
28	*Kassina senegalensis* (Duméril & Bibron, 1841)	Senegal Running Frog	Least Concern	x	x
	Order Anura, Family Microhylidae				
29	*Phrynomantis bifasciatus* (Smith, 1847)*	Banded Rubber Frog	Least Concern	x	x
	Order Anura, Family Phrynobatrachidae				
30	*Phrynobatrachus acridoides* (Cope, 1867)*	East African Puddle Frog	Least Concern	x	x
31	*Phrynobatrachus mababiensis* FitzSimons, 1932	Mababe Puddle Frog	Least Concern	x	x
32	*Phrynobatrachus natalensis* (Smith, 1849)*	Natal Puddle Frog	Least Concern	x	x
33	*Phrynobatrachus pallidus* Pickersgill, 2007	Pale Puddle Frog	Least Concern	x	-
34	*Phrynobatrachus scheffleri* (Nieden, 1911)	Scheffler's Puddle Frog	Least Concern	x	x
	Order Anura, Family Pipidae				
35	*Xenopus muelleri* (Peters, 1844)*	Müller's Clawed Frog	Least Concern	x	x
	Order Anura, Family Ptychadenidae				
36	*Hildebrandtia ornata* (Peters, 1878)	Common Ornate Frog	Least Concern	x	x
37	*Ptychadena anchietae* (Bocage, 1868)*	Anchieta's Grass Frog	Least Concern	x	x
38	*Ptychadena mascareniensis* (Duméril & Bibron, 1841)*	Mascarene Grass Frog	Least Concern	x	x
39	*Ptychadena mossambica* (Peters, 1854)	Mozambique Grass Frog	Least Concern	x	x

Continued

Table 7.1. Continued

#	Taxonomic group	Common name	Conservation status	Coast	Upland
40	*Ptychadena nilotica* (Seetzen, 1855)	Nile Grass Frog	Least Concern	x	x
41	*Ptychadena oxyrhynchus* (Smith, 1849)	Sharp-nosed Grass Frog	Least Concern	x	x
42	*Ptychadena schillukorum* (Werner, 1908)	Schilluk Grass Frog	Least Concern	x	x
43	*Ptychadena taenioscelis* Laurent, 1954	Southern Dwarf Grass Frog	Least Concern	x	x
	Order Anura, Family Pyxicephalidae				
44	*Pyxicephalus adspersus* Tschudi, 1838	African Bullfrog	Least Concern	-	x
45	*Pyxicephalus angusticeps* Parry, 1982	Beira Bullfrog	Least Concern	x	-
46	*Pyxicephalus edulis* Peters, 1854*	Edible Bullfrog	Least Concern	x	x
47	*Tomopterna luganga* Channing, Moyer and Dawood, 2004	Red Sand Frog	Least Concern	-	x
48	*Tomopterna tandyi* Channing and Bogart, 1996	Tandy's Sand Frog	Least Concern	-	x
	Order Anura, Family Ranidae				
49	*Amnirana galamensis* Duméril & Bibron, 1841	Galam White-lipped Frog	Least Concern	x	x
	Order Anura, Family Rhacophoridae				
50	*Chiromantis xerampelina* Peters 1854	Southern Foam-nest Frog	Least Concern	x	-
51	*Chiromantis petersii* Boulenger, 1882	Peters' Foam-nest Frog	Least Concern	-	x
	Total number of species				

Note: * = species detected by audio-visual surveys, x = present, - = no record within the vicinity of the area four species were found above 500 m, including *Pyxicephalus edulis*, *Tomopterna luganga*, *T. tandyi* and *Chiromantis petersi*. Based on the Jaccard Similarity Index (S_J) (Ludwig and Reynolds, 1988), the similarity in species between the lowland and highland areas was 52.9% indicating a moderate similarity in amphibian species between the two areas. This observation can be accounted for by low elevation gradient between the two sites (0–500 m against 500–~1000 m). Studies on elevation distribution of amphibians and reptiles in higher mountains have shown very low similarity in species in distantly located sites, e.g. on Mount Kilimanjaro and Uluguru (Ngalason and Mkonyi, 2011; Ngalason, 2016).

7.3.2 Conservation status of amphibians

One amphibian species, the Ruvu spiny reed frog, *Hyperolius ruvuensis*, that occurs within the SGR footprint, is endemic to Tanzania in RSFR. This species is further assigned a conservation status of Critically Endangered (CR) by the International Union for Conservation of Nature (IUCN). The IUCN establishes criteria for assessing the threat status of organisms that are based on species population size, rate of decline, area of geographical distribution and occurrence, degree of population and distribution fragmentation, and the probability of extinction in the wild (IUCN, 2023). All the species assigned a CR category are facing an extremely high risk of extinction, reduction of the population size and decline of the geographical range, and the risk of extinction is more than 50% within three generations (IUCN 2023). The forest reserve in which the frog occurs is traversed by the SGR network. One potential threat is related to increased human pressure on the forest resources due to improved accessibility caused by the railway route. Such human pressure may involve deforestation or logging for poles and charcoal production, as well as increased wildfire incidences, that may impact the species. Other amphibian species that occur within the SGR footprint are categorized as Least Concern by IUCN indicating that they are currently not considered to be of conservation concern because they are widespread, have sufficient populations and are not currently threatened with extinction.

7.3.3 Reptilian species richness

A total of 108 reptilian species in 24 families occur along the SGR footprint. Eleven species were detected during the surveys and the remaining 97 species are known to occur in the geographic area based on literature review. These include one family of crocodiles, Family Crocodylidae (1 species); two families of testudines, Families Testudinidae (5 species) and Pelomedusidae (5 species); ten families of lizards, including Family Eublepharidae (1 species), Gekkonidae (17 species), Scincidae (12 species), Lacertidae (7 species), Cordylidae (1 species), Gerrhosauridae (3 species), Agamidae (3 species), Chamaeleonidae (4 species), Varanidae (2 species), Amphisbaenidae (2 species); and eleven families of snakes, including Family Typhlopidae (3 species), Leptotyphlopidae (1 species), Booidae (1 species), Pythonidae (2 species), Lamprophiidae (5 species), Prosymnidae (1 species), Psammophiidae (8 species), Atractaspididae (5 species), Colubridae (14 species), Elapidae (4 species), and Viperidae (3 species) (Table 7.2).

7.3.4 Reptilian species composition and conservation status

Of the 107 reptilian species occurring within the SGR footprint, 52 are widespread from Dar es Salaam to Makutupora, whereas 35 species occur in the coastal lowlands and 20 in upland areas (Table 7.2). The species similarity index based on the Jaccard Similarity Index (S_J) (Ludwig and Reynolds, 1988) between the lowland coastal and the highland semi-arid areas was 48.6%. Three species are endemic to Tanzania including *Ancylodactylus barbouri* (occurring in RSFR and a few other forests in Tanzania), *Lygodactylus inexpectatus* (occurring in Dar es Salaam), and *Scolecoseps acontias* (occurring in Dar es Salaam and Kilwa). These species are known from few localities (Spawls *et al.*, 2018; Uetz *et al.*, 2023) and their distribution may extend to the SGR footprint. One parthenogenic

Table 7.2. Conservation status of reptilian fauna within the SGR footprint from Dar es Salaam to Dodoma region.

#	Taxonomic group	Common name	Coastal	Highland	Conservation status
	Order Crocodilia, Family Crocodylidae	Crocodiles			
1	*Crocodylus niloticus* Laurenti, 1768	Nile Crocodile	x	x	Least Concern
	Order Testudines, Family Testudinidae	Tortoises			
2	*Stigmochelys pardalis* (Bell, 1828)	Leopard Tortoise	x	x	Least Concern
3	*Kinixys belliana* Gray, 1831	Bell's Hingeback Tortoise	x	-	Not Evaluated
4	*Kinixys spekii* Gray, 1863*	Speke's Hingeback Tortoise	-	x	Not Evaluated
5	*Kinixys zombensis* Hewitt, 1931	Eastern Hinged Tortoise	x	-	Not Evaluated
6	*Malacochersus tornieri* (Siebenrock, 1903)	Pancake Tortoise	-	x	Critically Endangered
	Order Testudines, Family Pelomedusidae	Terrapins			
7	*Pelomedusa subrufa* (Bonnaterre, 1789)	Helmeted Terrapin	x	x	Not Evaluated
8	*Pelomedusa neumanni* Petzold, Vargas-Ramírez, Kehlmaier *et al.*, 2014	Neumann's Marsh Terrapin	x	x	Not Evaluated
9	*Pelusios castanoides* Hewitt, 1931	Yellow-bellied Hinged Terrapin	x	-	Least Concern
10	*Pelusios sinuatus* (Smith, 1838)	Serrated Hinged Terrapin	x	-	Not Evaluated
11	*Pelusios subniger* (Bonnaterre, 1789)	Pan-hinged Terrapin	-	x	Least Concern
	Order Squamata, Family Eublepharidae	Clawed Geckoes			
12	*Holodactylus africanus* Boettger, 1893	Boettger's Strong-clawed Gecko	-	x	Least Concern
	Squamata, Sauria, Gekkonidae	Geckoes			
13	*Ancylodactylus barbouri* (Perret, 1986)	Uluguru Forest Gecko	x	-	Least Concern
14	*Hemidactylus angulatus* Hallowell, 1854	East African House Gecko	x	x	Least Concern
15	*Hemidactylus mabouia* (Moreau De Jonnès, 1818)*	Tropical House Gecko	x	x	Least Concern

Continued

Table 7.2. Continued

#	Taxonomic group	Common name	Coastal	Highland	Conservation status
16	*Hemidactylus platycephalus* Peters, 1854*	Tree Gecko	x	x	Least Concern
17	*Hemidactylus squamulatus* Tornier, 1896	Nyika Gecko	x	x	Least Concern
18	*Lygodactylus angolensis* Bocage, 1896	Angolan Dwarf Gecko	-	x	Least Concern
19	*Lygodactylus capensis* (Smith, 1849)	Cape Dwarf Gecko	x	x	Least Concern
20	*Lygodactylus grotei* Sternfeld, 1911*	Grote's Dwarf Gecko	x	-	Least Concern
21	*Lygodactylus inexpectatus* Pasteur, 1965	Dar es Salaam Dwarf Gecko	x	-	Data Deficient
22	*Lygodactylus picturatus* (Peters, 1870)*	Yellow-headed Dwarf Gecko	x	x	Least Concern
23	*Chondrodactylus turneri* (Gray, 1864)	Turner's Thick-toed Gecko	x	x	Least Concern
24	*Elasmodactylus tuberculosus* Boulenger, 1895	Tuberculate Thick-toed Gecko	x	x	Least Concern
25	*Homopholis fasciata* (Boulenger, 1890)	Banded Velvet Gecko	-	x	Least Concern
26	*Phelsuma dubia* (Boettger, 1881)	Dull Green Day Gecko	x	-	Least Concern
	Order Squamata, Sub-order Sauria, Family Scincidae	Skinks			
27	*Acontias percivali* Loveridge, 1935	Percival's Legless Skink	-	x	Least Concern
28	*Trachylepis dichroma* Günther, Whiting & Bauer, 2005	Two-coloured/ Zebra Skink	-	x	Least Concern
29	*Trachylepis maculilabris* (Gray, 1845)*	Speckle-lipped Skink	x	-	Least Concern
30	*Trachylepis margaritifera* (Peters, 1854)	Rainbow Skink	-	x	Least Concern
31	*Trachylepis megalura* (Peters, 1878)	Grass Top/Long-tailed Skink	x	-	Least Concern
32	*Trachylepis planifrons* (Peters, 1878)	Tree Skink	x	x	Least Concern
33	*Trachylepis striata* (Peters, 1844)*	Striped Skink	x	x	Least Concern

Continued

Table 7.2. Continued

#	Taxonomic group	Common name	Coastal	Highland	Conservation status
34	*Trachylepis varia* (Peters, 1867)*	Variable Skink	x	x	Least Concern
35	*Mochlus sundevalli* (Smith, 1849)	Sundevall's Writhing Skink	x	x	Least Concern
36	*Cryptoblepharus africanus* (Sternfeld, 1918)	Coral Rag Skink	x	-	Least Concern
37	*Panaspis massaiensis* (Angel, 1924)	Maasai Snake-eyed Skink	x	x	Not Evaluated
38	*Scolecoseps acontias* (Werner, 1913)	Legless Sand Skink	x	-	Data Deficient, Tanzania endemic
	Sub-order Sauria, Family Lacertidae	Typical Lizards			
39	*Gastropholis vittata* Fischer, 1886	Striped Keel-bellied Lizard	x	-	Least Concern
40	*Holaspis laevis* Werner, 1896	Eastern Blue-tailed Gliding Lizard	x	-	Least Concern
41	*Nucras boulengeri* Neumann, 1900	Boulenger's Scrub Lizard	x	x	Least Concern
42	*Latastia johnstonii* Boulenger, 1907	Johnston's/Malawi Long-tailed Lizard	-	x	Least Concern
43	*Latastia longicaudata* (Reuss, 1834)	Southern Long-tailed Lizard	-	x	Least Concern
44	*Heliobolus neumanni* (Tornier, 1905)	Neumann's Sand Lizard	x	-	Data Deficient
45	*Heliobolus spekii* (Günther, 1872)*	Speke's Sand Lizard	x	x	Least Concern
	Sub-order Sauria, Family Cordylidae	Girdled Lizards			
46	*Cordylus tropidosternum* (Cope, 1869)	Tropical Girdled Lizard	x	x	Least Concern
	Sub-order Sauria, Family Gerrhosauridae, Plated Lizards	Plated lizards			
47	*Broadleysaurus major* (Duméril, 1851)	Great Plated Lizard	x	x	Least Concern
48	*Gerrhosaurus flavigularis* Wiegmann, 1828	Yellow-throated Plated Lizard	x	x	Least Concern
49	*Gerrhosaurus nigrolineatus* Hallowell, 1857	Black-lined Plated Lizard	x	x	Least Concern
	Sub-order Sauria, Family Agamidae	Agamas			

Continued

Table 7.2. Continued

#	Taxonomic group	Common name	Coastal	Highland	Conservation status
50	*Agama armata* Peters, 1855	Peter's Ground Agama	-	x	Least Concern
51	*Agama dodomae* Loveridge, 1923	Dodoma Rock Agama	-	x	Least Concern
52	*Agama mossambica* Peters, 1854	Mozambique Agama	x	-	Least Concern
	Sub-order Sauria, Family Chamaeleonidae	Chameleons			
53	*Rieppeleon brachyurus* (Günther, 1893)	Beardless Pygmy Chameleon	x	-	Least Concern
54	*Rieppeleon brevicaudatus* (Matschie, 1892)	Bearded pygmy chameleon	x	-	Least Concern
55	*Chamaeleo dilepis* Leach, 1819	Flap-necked Chameleon	x	x	Least Concern
56	*Trioceros melleri* (Gray, 1865)	Giant One-horned Chameleon	x	-	Least Concern
	Sub-order Sauria, Family Varanidae	Monitor Lizards			
57	*Varanus albigularis* (Daudin, 1802)	White-throated Savannah Monitor	x	x	Least Concern
58	*Varanus niloticus* (Linnaeus, 1766)*	Nile/Water Monitor	x	x	Least Concern
	Sub-order Amphisbaenia, Family Amphisbaenidae	Worm Lizards			
59	Geocalamus acutus Sternfeld, 1912	Voi Wedge-snouted Worm Lizard	-	x	Least Concern
60	*Geocalamus modestus* Günther, 1880	Mpwapwa Wedge-snouted Worm Lizard	-	x	Data Deficient
	Sub-order Ophidia (Serpentes), Family Typhlopidae	Blind Snakes			
61	*Afrotyphlops lineolatus* (Jan, 1864)	Lineolate Blind Snake	x	-	Least Concern
62	*Afrotyphlops mucruso* (Peters, 1854)	Zambezi Giant Blind Snake	x	x	Least Concern
63	*Indotyphlops braminus* (Daudin, 1803)	Flowerpot Blind Snake	x	-	Least Concern, Exotic/Introduced species
	Sub-order Ophidia (Serpentes), Family Leptotyphlopidae	Worm/Thread Snakes			

Continued

Table 7.2. Continued

#	Taxonomic group	Common name	Coastal	Highland	Conservation status
64	*Leptotyphlops merkeri* (Werner, 1909)	Merker's Worm Snake	x	x	Least Concern
	Sub-order Ophidia (Serpentes), Family Booidae	Boas			
65	*Eryx colubrinus* (Linnaeus, 1758)	Kenya Sand Boa	-	x	Least Concern
	Sub-order Ophidia (Serpentes), Family Pythonidae	Pythons			
66	*Python natalensis* Smith, 1840	Southern African Rock Python	x	x	Least Concern
67	*Python sebae* (Gmelin, 1789)	Central African Rock Python	x	-	Near Threatened
	Sub-order Ophidia (Serpentes), Family Lamprophiidae	African Snakes			
68	*Boaedon fuliginosus* (Boie, 1827)	Brown House Snake	x	x	Least Concern
69	*Lycophidion capense* (Smith, 1831)	Cape Wolf Snake	x	x	Least Concern
70	*Lycophidion depressirostre* Laurent, 1968	Flat-snouted Wolf Snake	x	x	Least Concern
71	*Gracililima nyassae* (Günther, 1888)	Dwarf/Black File Snake	x	-	Least Concern
72	*Limaformosa capensis* (Smith, 1847)	Cape File Snake	x	-	Least Concern
	Sub-order Ophidia (Serpentes), Family Prosymnidae				
73	*Prosymna stuhlmanni* (Pfeffer,1893)	East African Shovel-snout	x	x	Least Concern
	Sub-order Ophidia (Serpentes), Family Psammophiidae				
74	*Hemirhagerrhis nototaenia* (Günther, 1864)	South-eastern Bark Snake	x	x	Least Concern
75	*Psammophis angolensis* (Bocage, 1872)	Dwarf Sand Snake	x	x	Least Concern
76	Psammophis biseriatus Peters, 1881	Link-marked Sand Snake	-	x	Least Concern

Continued

Table 7.2. Continued

#	Taxonomic group	Common name	Coastal	Highland	Conservation status
77	*Psammophis mossambicus* Peters, 1882	Olive/Hissing Sand Snake	x	x	Least Concern
78	*Psammophis orientalis* Broadley, 1977	Eastern Stripe-bellied Sand Snake	x	x	Least Concern
79	*Psammophis sudanensis* Werner, 1919	Northern Stripe-bellied Sand Snake	x	x	Least Concern
80	*Psammophis tanganicus* Loveridge, 1940	Tanganyika Sand Snake	-	x	Least Concern
81	*Rhamphiophis rostratus* Peters, 1854	Rufous Beaked Snake	x	x	Least Concern
	Sub-order Ophidia (Serpentes), Family Atractaspididae	Burrowing asps			
82	*Aparallactus capensis* Smith, 1849	Cape Centipede Eater	x	-	Least Concern
83	*Aparallactus guentheri* Boulenger, 1895	Black Centipede Eater	x	-	Least Concern
84	*Aparallactus lunulatus* (Peters, 1854)	Plumbeous Centipede Eater	-	x	Least Concern
85	*Amblyodipsas polylepis* (Bocage, 1873)	Common Purple-glossed Snake	x	-	Least Concern
86	*Atractaspis bibronii* Smith, 1849	Bibron's Burrowing Asp	x	x	Least Concern
	Sub-order Ophidia (Serpentes), Family Colubridae	Colubrid Snakes			
87	*Meizodon semiornatus* (Peters, 1854)	Semi-ornate Snake	x	x	Least Concern
88	*Philothamnus hoplogaster* (Günther, 1863)	South-eastern Green Snake	x	-	Least Concern
89	*Philothamnus punctatus* Peters, 1867	Speckled Green Snake	x	-	Least Concern
90	*Philothamnus semivariegatus* (Smith, 1840)	Spotted Bush Snake/Spotted Wood Snake	x	x	Least Concern
91	*Dispholidus typus* (Smith, 1828)	Boomslang	x	x	Least Concern
92	*Thelotornis mossambicanus* (Bocage, 1895)	Eastern Vine Snake	x	x	Least Concern
93	*Thelotornis usambaricus* Broadley, 2001	Usambara Vine Snake	x	-	Vulnerable

Continued

Table 7.2. Continued

#	Taxonomic group	Common name	Coastal	Highland	Conservation status
94	*Dasypeltis medici* Bianconi, 1859	Rufous/East African Egg Eater	x	x	Least Concern
95	*Dasypeltis scabra* (Linnaeus, 1758)	Common or Rhombic Egg Eater	x	x	Least Concern
96	*Telescopus semiannulatus* Smith, 1849	Tiger Snake	x	x	Least Concern
97	*Crotaphopeltis hotamboeia* (Laurenti, 1768)	White-lipped Snake	x	x	Least Concern
98	*Dipsadoboa flavida* (Broadley and Stevens, 1971)	Cross-Barred Tree-Snake	x	-	Least Concern
99	*Scaphiophis albopunctatus* Peters, 1870	Hook-nosed Snake	-	x	Least Concern
100	*Natriciteres olivacea* (Peters, 1854)	Olive Marsh Snake	x	-	Least Concern
	Sub-order Ophidia (Serpentes), Family Elapidae	Cobras, Mambas and Relatives			
101	*Naja nigricollis* Reinhardt, 1843	Black-necked Spitting Cobra	x	x	Least Concern
102	*Naja subfulva* Laurent, 1955	Brown Forest Cobra	x	-	Least Concern
103	*Dendroaspis angusticeps* (Smith, 1849)	Green Mamba	x	-	Least Concern
104	*Dendroaspis polylepis* (Günther, 1864)	Black Mamba	x	x	Least Concern
	Sub-order Ophidia (Serpentes), Family Viperidae	Vipers			
105	*Causus defilippii* (Jan, 1863)	Snouted Night Adder	x	x	Least Concern
106	*Bitis arietans* (Merrem, 1820)	Puff Adder	x	x	Least Concern
107	*Bitis gabonica* (Duméril, Bibron & Duméril, 1854)	Gaboon Viper	x	-	Vulnerable

Note: * = species detected by audio-visual surveys, x = present, - = no record within the vicinity of the area

species, the flowerpot blind snake, *Indotyphlops braminus*, is exotic to the coastal eastern Africa including Dar es Salaam. Females of this species can produce an egg that develops and forms an offspring without the need for fertilization by males.

One reptile species, the pancake tortoise (*Malacochersus tornieri*), is considered Critically

Endangered by the IUCN. The distribution map of this species extends to the semi-arid areas in central Tanzania (Spawls *et al.*, 2018), the area that is traversed by the SGR. Two other species' geographical range that extends to the area include the Usambara vine snake (*Thelotornis usambaricus*), and the Gaboon viper (*Bitis gabonica*). They are considered Vulnerable; one species (*Python sebae*), is Near Threatened, three species are considered Data Deficient, six species are not yet evaluated and 94 species are considered Least Concern by the IUCN.

7.4 Effects of the SGR Project on Herpetofauna and Mitigations

The SGR traverses through important habitats for amphibians and reptiles including wetlands and forest reserves such as PKNFR and RSFR is home to one endemic and Critically Endangered amphibian species, the Ruvu spiny reed frog (*Hyperolius ruvuensis*). Most amphibian species breed in wetlands such that key impacts of the project may include loss of terrestrial and wetland habitats as well as sedimentation and pollution of aquatic habitats. Herpetofauna species are more vulnerable to population reduction following the construction of the SGR due to the potential increase in human activities in the reserves caused by the improved accessibility, which may lead to unsustainable utilisation of forest resources, e.g. increased charcoal production and extraction of poles, all of which will impact the natural habitats of the species. The Ruvu spiny reed frog inhabits and breeds in swampy open grassland (Barratt *et al.*, 2017). The construction and operation of the SGR project may lead to its habitat disturbance, ranging from pollution to alteration of its habitat.

Impacts of railways to herpetofauna can be divided into mortality, barrier effects, species invasions and environmental disturbances through noise and pollution (Borda-de-Água *et al.*, 2017). Few studies exist on the mortality of herpetofauna due to train collision (Santos *et al.*, 2017). The studies show higher frequency of encountering amphibian casualties than in any other vertebrate group (Heske, 2015; Santos *et al.*, 2017). Reptilian fauna is another vertebrate group being affected by train collision although to a lesser extent than amphibians. The studies show the events to be higher, particularly after rainy periods when herpetofauna are more active (Heske, 2015).

Another possible impact of the railway line is formation of barriers to herpetofauna movements. Physical barriers imposed by the railway route are among the big problems facing mainly smaller animals. Amphibian fauna are among the vertebrates most affected by these linear infrastructures (Lucas *et al.*, 2017). Other slow-moving species, including tortoises, are also affected by failing to cross the railway or ending trapped between the rails leading to increased railway kills (Barrientos and Borda-de-Água, 2017). Mitigation measures for these problems may involve fencing the railway route to avoid mortality of small animals through train collision, although this may create a barrier effect. Barrier effects can be minimised by creation of wildlife passes in the form of tunnels, underpasses and drainage culverts, to reduce the effects (Barrientos and Borda-de-Água, 2017).

Despite these negative effects, railway infrastructures are known to provide habitats for some herpetofauna, including rocky substrates, vegetation that grows along the railway line, and the wetlands that are formed by river or flood plain inundations. The wetlands formed by blockage of drainage systems by the railway infrastructure may attract species such as some amphibians andreptiles, e.g. monitor lizards. These secondary habitats may lead to a local increase in herpetofauna species richness and abundance. The formation and expansion of wetlands may also contribute to the spread of species (both native and invasive) that may eliminate or displace some species, reducing biodiversity and negatively affecting ecological communities. Monitoring of habitats and herpetofauna within the SGR footprint is recommended to assess short- and long-term effects of the SGR.

7.5 Conclusion

This study revealed the occurrence of 51 amphibian and 107 reptilian species within the SGR footprint. Some species are distributed throughout the route from Dar es Salaam to Makutupora in Singida region whereas others are restricted to the coastal lowlands or in the highland semi-arid areas. Of all the amphibian species, one is endemic to Tanzania and is also considered Critically Endangered by the IUCN, whereas three reptile species within the footprint are endemic to Tanzania. The conservation status of the reptiles includes one species that is considered Critically Endangered, two species that are Vulnerable, one species that is Near Threatened, three species that are Data Deficient, six species that are not evaluated and 94 species that are considered Least Concern. Potential impacts of the railway infrastructure on the herpetofauna may range from barrier effect, habitat disturbances and mortality from train collision. Additionally, blockage of drainage systems may lead to formation of temporary or permanent water bodies that may attract some herpetofauna species as well as displacing others.

7.6 Recommendations

A well-designed railway infrastructure is recommended to reduce the impacts of the project on the herpetofauna. The infrastructure should be constructed with adequate herpetofauna passes in the form of tunnels, underpasses and drainage culverts, to prevent interruption of the movements of animals between the two sides of the railway. Additionally, construction of fences is recommended to reduce herpetofauna–train collisions. Adequate underpasses are also recommended to prevent blockage of drainage systems that can affect species richness and composition as well as abundance.

References

AmphibiaWeb (2023). University of California, Berkeley, California. Available at: https://amphibiaweb.org (accessed 27 July 2023).

Barratt, C.D. (2017) Biodiversity patterns and conservation of the coastal forests of Eastern Africa. PhD thesis, University of Basel, Switzerland.

Barratt, C.D., Lawson, L.P., Bittencourt-Silva, G.B., Doggart, N., Morgan-Brown, T, *et al.* (2017) A new, narrowly distributed, and critically endangered species of spiny-throated reed frog (Anura: Hyperoliidae) from a highly threatened coastal forest reserve in Tanzania. *Herpetological Journal* 27(1), 13–24.

Barrientos, R. and Borda-de-Água, L. (2017) Railways as barriers for wildlife: Current knowledge. In: Borda-de-Água, L., Barrientos, R., Beja, P. and Pereira, H.M. (eds) *Railway Ecology*. Springer Open, Switzerland, pp. 68–89.

Borda-de-Água, L., Barrientos, R., Beja, P. and Pereira, H.M. (2017) Railway ecology. In: Borda-de-Água, L., Barrientos, R., Beja, P. and Pereira, H.M. (eds) *Railway Ecology*. Springer Open, Switzerland, pp. 29–35.

Branch, B. (1998) *Field Guide to Snakes and Other Reptiles of Southern Africa*, 3rd edn. Stuik, Cape Town, South Africa, pp. i–xxv, 287.

Cauldwell, A., Lawrence, S. and Nahonyo, C.L. (2019) Standard Gauge Railway, Biodiversity Action Plan, Dar es Salaam to Makutupora, Tanzania. Environmental Resources Management (ERM) Southern Africa (Pty) Ltd, Johannesburg, South Africa, pp. i–xiv, 91.

Channing, A. and Howell, K.M. (2006) *Amphibians of East Africa*. Cornell University Press, Ithaca, New York.

Channing, A. and Rödel, M.O. (2019) *Field Guide to the Frogs and Other Amphibians of Africa*. Struik Nature, Cape Town, South Africa.

Crump, M.L. and Scott Jr, N.J. (1994) Visual encounter surveys. In: Heyer, W.R., Donnelly, M.A., McDiarmid, R.W., Hayek, L.A.C. and Foster, M.S. (eds) *Measuring and Monitoring Biological Diversity Standard Methods for Amphibians*. Smithsonian Institution Press, Washington & London, pp. 84–92.

Doan, T.M. (2003) Which methods are most effective for surveying rain forest Herpetofauna? *Journal of Herpetology* 37(1), 72–81.

du Preez, L. and Caruthers, V. (2009) *A Complete Guide to the Frogs of Southern Africa*. Stuik Nature, Cape Town, South Africa.

ESIA (Environmental and Social Impact Assessment) (2019) Final Report. A consultant report for Tanzania Railway Corporation, Dar es Salaam, Tanzania.

Frost, D.R. (2023) Amphibian Species of the World: An Online Reference. Version 6.2. Electronic database . American Museum of Natural History. Available at: https://amphibiansoftheworld. amnh.org/index.php (accessed 26 July 2023).

Grant, B.W., Tucker, A.D., Lovich, J.E., Mills, A.M., Dixon, P.M, *et al.* (1992) The use of coverboards in estimating patterns of reptile and amphibian biodiversity. In: Siegel, R. and Scott, N. (eds) *Wildlife*. Elsevier Science, London, pp. 379–403.

Harper, E.B., Measey, G.J., Patrick, D.A., Menegon, M. and Vonesh, J.R. (2010) *Field Guide to the Amphibians of the Eastern Arc Mountains and Coastal Forests of Tanzania and Kenya*. Camerapix Publishers International, Nairobi.

Heatwole, H. and Wilkinson, J. (eds) (2009) *Amphibian Biology: Amphibian Decline, Diseases, Parasites, Maladies and Pollution*, Series: Amphibia Biology, Vol. 8. Surrey Beatty & Sons PTY Limited, Australia.

Heske, E.J. (2015) Blood on the tracks: track mortality and scavenging rate in urban nature preserves. Urban Naturalist 4, 1–13.

Heyer, W., Donnelly, M., MacDiarmid, W., Hayek, L.C. and Foster, M.S. (1994) Measuring and monitoring biological diversity: Standard methods for amphibians. Smithsonian Institution Press, Washington, DC.

Hof, C., Araújo, M.B., Jetz, W. and Rahbek, C. (2011) Additive threats from pathogens, climate and land-use change for global amphibian diversity. *Nature* 480(7378), 516–519.

Howell, K.M. (2002) Amphibians and reptiles: the herptiles. In: Davies, G. (ed.) *African Forest Biodiversity: A Field Survey Manual for Vertebrates*. Earthwatch, pp. 17–44.

IUCN (2023) The IUCN Red List of Threatened Species. Version 2022-2. Available at: https://www.iucnredlist.org (accessed 26 July 2023).

Karns, D.R. (1986) Field herpetology: methods for the study of amphibians and reptiles in Minnesota. Occasional Paper 18. James Ford Bell Museum of Natural History, University of Minnesota.

Loveridge, A. (1957) Checklist of the reptiles and amphibians of East Africa (Uganda, Kenya, Tanganyika, Zanzibar). *Bulletin of the Museum of Comparative Zoology* 117, 153–362.

Lucas, P., de Carvalho, R.G. and Grilo, C. (2017) Railway disturbances on wildlife: types, effects, and mitigation measures. In: Borda-de-Água, L., Barrientos, R., Beja, P. and Pereira, H.M. (eds) *Railway Ecology*. Springer Open, Switzerland, pp. 81–99.

Ludwig, J.A. and Reynolds, J.F. (1988) *Statistical Ecology: A Primer on Methods and Computing*. John Wiley, New York.

Menegon, M., Lyakurwa, J. and Loader, S. (2020) *The Amphibians of the Tanzanian Forests*. A free downloadable visual guide. Version 1.0.

Mittermeier, R.A., Myers, N., Gil, P.R. and Mittermeier, C.G. (1999) *Hotspots: Earth's Biologically Richest and Most Endangered Terrestrial Ecoregions*. CEMEX, Mexico City.

Mittermeier, R.A., Myers, N., Thompsen, J.B., da Fonseca, G.A.B. and Olivieri, S. (1998) Global biodiversity hotspots and major tropical wilderness areas. *Conservation Biology* 12(3), 516–520.

Ngalason, W. (2016) Ecology and altitudinal distribution of herpetofauna on Mount Kilimanjaro, Tanzania. Unpublished PhD thesis, University of Dar es Salaam, Tanzania.

Myers, N., Mittermeier, R.A., Mittermeier, C.G., Fonseca, G.A.B. and Kent, J. (2000) Biodiversity hotspots for conservation priorities. *Nature* 403, 853–858.

Ngalason, W. and Mkonyi, F.J. (2011) Herpetofauna of montane areas of Tanzania. 2. Altitudinal distribution of amphibians on the Uluguru South mountains. *Fieldiana: Life and Earth Sciences* 4, 81–89.

Pickersgill, M. (2007) *Frog Search: Results of Expedition to Southern and Eastern Africa*. Edition Chimaira, Frankfurt am Main, Germany.

Poynton, J.C. (1992) Amphibian diversity and species turnover in southern Africa: Investigation by means of a Bloemfontein-Durban transect. *Journal of the Herpetological Association of Africa* 40(1), 2–8.

Poynton, J.C. (2003) Altitudinal species turnover in southern Tanzania shown by anurans: Some zoogeographical considerations. *Systematics and Biodiversity* 1(1), 117–126.

Santos, S.M., Carvalho, F. and Mira, A. (2017) Current knowledge on wildlife mortality in railways. In: Borda-de-Água, L., Barrientos, R., Beja, P. and Pereira, H.M. (eds) *Railway Ecology*. Springer Open, Switzerland, pp. 11–22.

Schiøtz, A. (1975) *The Tree Frogs of Eastern Africa*. Steenstrupia, Copenhagen.

Schiøtz, A. (1999) *Tree Frogs of Africa*. Edition Chimaira, Frankfurt am Main, Germany.

Spawls, S., Howell, K.M. and Drewes, R.C. (2006) *Pocket Guide to the Reptiles and Amphibians of East Africa*. A. and C. Black, London.

Spawls, S., Howell, K.M., Drewes, R.C. and Ashe, J. (2002) *A Field Guide to the Reptiles of East Africa*. A. and C. Black, London.

Spawls, S., Howell, K.M., Hinkel, H. and Menegon, M. (2018) *Field Guide to East African Reptiles*, 2nd edn. Bloomsbury, London.

Tilbury, C.R. (2010) *Chameleons of Africa: An Atlas Including the Chameleons of Europe, the Middle East and Asia*. Edition Chimaira/Serpent's Tale NHBD, Frankfurt, Germany.

Uetz, P., Freed, P., Aguilar, R., Reyes, F. and Hošek, J. (eds) (2023) The Reptile Database. Available at: http://www.reptile-database.org (accessed 26 July 2023).

Vitt, L.J. and Caldwell, J.P. (2014) *Herpetology: An Introductory Biology of Amphibians and Reptiles*, 4th edn. Academic Press, San Diego, California.

8 The Avifauna of Pugu Hills and Ruvu South Coastal Forests amidst Railway Development

Jasson John*
College of Natural and Applied Sciences, University of Dar es Salaam, Tanzania

Abstract
Pugu and Ruvu South are biologically important coastal forests, with several endemic species and host a significant number of bird species restricted to the Eastern Africa Coastal Forest biodiversity hotspot. The avifauna of the area is characterized by forest-dependent species, some of which are globally threatened. In this chapter, baseline information on birds along the electrified standard-gauge railway alignment within the two forests is provided. The chapter further provides an account of the avifaunal richness and status of the forest reserves. This compilation presents a total of 267 (29% forest-dependent) and 201 (26% forest-dependent) bird species known for Pugu and Ruvu South forest reserves, respectively. Nonetheless, both forests are threatened by unsustainable utilization of forest resources and the effects associated with linear infrastructure development. The rate of habitat modification in the reserves is worrying and requires immediate attention to preserve forest avian populations.

8.1 Background Information

Pugu Hills and Ruvu South forests represent an important ecosystem of the coastal forests with high and unique biological resources, some of which are not found anywhere else in the world. The two forests are part of the Kisarawe District Coastal Forest Important Bird Area (Baker and Baker, 2002). Important Bird Areas are networks of habitats that support significant numbers of globally threatened, endemic bird species and a high population of congregatory birds, especially waterbirds. The identification of Important Bird and Biodiversity Areas (IBAs, formerly called Important Bird Areas until 2013) has a long history (Donald *et al.*, 2019). In general, the regular presence of a Critical or Endangered bird species, irrespective of population size, at a site may be enough for a site to qualify as an IBA. For Vulnerable species, the presence of more than threshold numbers at a site is necessary to trigger selection. The selection of IBAs is achieved through the application of quantitative ornithological criteria, grounded in up-to-date knowledge of the sizes and trends of bird populations. The criteria ensure that the sites selected as IBAs have true significance for the international conservation of bird populations and provide a common currency that all IBAs adhere to, thus creating consistency and enabling comparability at global levels using criteria listed in Box 8.1. Sites should ideally be assessed against all relevant criteria for which data are available, but meeting thresholds under any of the criteria

*wildornithology@udsm.ac.tz

 Biodiversity and Conservation along an East African Railway (eds C.A. Masao *et al.*)
DOI: 10.1079/9781800626942.0008

Box 8.1. Four major criteria used to identify Important Bird and Biodiversity Areas. (From Donald *et al.*, 2019)

A1: Globally Threatened species criterion: the site is known or regularly thought to hold significant numbers of a Globally Threatened species.
A2: Restricted Range species criterion: the site is known or thought to hold a significant population of at least two range-restricted species.
A3: Bioregion-restricted assemblages criterion: the site is known or thought to hold a significant component of the group of species whose distributions are largely or wholly confined to one biome-realm.
A4: Congregations criterion: the site is known or thought to hold congregations of ≥ 1% of the global population of one or more species on a regular or predictable basis.

is enough for a site to be recognized as an IBA. The main aim of the IBA concept is to secure the long-term conservation of sites that are significant for birds and biodiversity.

Pugu and Ruvu South forests are especially important for the two globally threatened bird species: Spotted Ground-thrush *Geokichla guttata* and Sokoke Pipit *Anthus sokokensis* (BirdLife International, 2023). Throughout their range in the coastal forests in Kenya and Tanzania, numbers are declining due to habitat loss and degradation. The East African subspecies of Spotted Ground-thrush, *fischeri*, is migratory that breeds in the southern part of Tanzania and winters in the Arabuko Sokoke forests in the coast of Kenya (Ndang'ang'a *et al.*, 2005), thus utilizing the forests in between (including Pugu and Ruvu South) as stopping over during migration, whereas the Sokoke Pipit is a resident. In addition to being globally threatened, both species occur in a very low density in the two forests (Mlingwa, 1996a; Baker and Baker, 2002).

Other Afro-tropical migratory forest species that utilize the Kisarawe District Coastal Forest IBA include the African Pitta, *Pitta angolensis*, and Red-capped Robin-chat, *Cossypha natalensis* (Baker and Howell, 1992). The Red-capped Robin-chat has a small resident breeding population in Kisarawe district coastal forests. Both African Pitta and Spotted Ground-thrush could be expected during April–May and October–November on their movements from and to breeding areas in the southern part of the country. These forests are also important for higher-altitude visitors that have included Buff-spotted Flufftail *Sarothrura elegans*, Lemon Dove *Aplopelia larvata*, Orange Ground-thrush *Geokichla gurneyi* (Baker and Baker, 1992; Mlingwa *et al.*, 2000) and Olive-headed Bulbul *Arizelocichla striifacies* (Mlingwa, 1996b). Many long-distance migrants, Palearctic (northern Africa, Europe and parts of Asia), use these forests during northern winters (Harvey and Howell, 1987).

More importantly, these forests host populations of the closed-canopy evergreen forest species. The list in this category includes the Eastern Green Tinkerbird *Pogoniulus simplex*, Red-tailed Antithrush *Neocossyphus rufus*, Fischer's Greenbul *Phyllastrephus fischeri*, Kretschemer's Longbill *Macrosphenus kretschmeri*, Little Yellow Flycatcher *Erythrocercus holochlorus* and Uluguru Violet-backed Sunbird *Anthreptes neglectus*. The near-threatened East Coast Akalat *Sheppardia gunning*, is a resident bird of forest interior and Ruvu South and is suspected to host a large population for this species in Tanzania (Baker and Baker, 2002).

Like many of the East Africa coastal forests, Pugu and Ruvu South are facing serious threats, mainly from anthropogenic activities leading to habitat degradation and fragmentation (Burgess *et al.*, 1992; Rocky and Mligo, 2012; Mligo, 2019). Their proximity to Dar es Salaam, the most populated region in Tanzania, poses threats due to the high demands of the forest products such as charcoal, poles and timber. In addition, the two forests continue to be affected by linear infrastructure developments, namely railways, roads and power lines. The Central Railway System (CRS), built in 1912, traverses these forests. Furthermore, recently, the government, via the Tanzanian Railways Corporation (TRC), has constructed a single-track, electrified standard-gauge railway (SGR) from Dar es Salaam to Mwanza. Within the two forests, the constructed SGR runs almost parallel to the CRS. The direct footprint of the SGR represents about 3.6 and 0.14% of the Pugu Hills and Ruvu South forests, respectively (ERM, 2019). The development of SGR does not only result in loss of bird habitats, but also it will likely cause long-term impacts due to habitat fragmentation and interference of movements of local birds.

Thus, based on this development, this chapter presents an assessment of the ornithological importance of the Pugu and Ruvu South forests, with detailed field observations along the SGR alignment within the area. The observations on SGR corridor provide baseline information for future reference. The chapter intends to raise ecological understanding of the avifauna of the critical coastal habitat in relation to the SGR development.

8.2 Methodology

8.2.1 Scope of the assessment

The field study was conducted in Pugu and Ruvu South forests within the SGR project's footprint (TRC, 2020) located in the northern part of the two forests. However, the survey radiated from SGR alignment into the forest reserves; this provided opportunity for recording species that are sensitive to edge effects. The field surveys were conducted in both wet (April 2021) and dry (November–December 2021) seasons. The same team and equal sampling efforts were employed in both surveys. To account for ornithological status of the two forests, a detailed review of scientific and grey literature was done.

8.2.2 Sampling in reference to SGR footprints

Two different methods were employed in the field to sample birds. These were mist netting (Ralph *et al.*, 1993) and transect using fixed-width point counts (Bennun and Howell, 2002) (including playback for Sokoke Pipit and Spotted Ground-thrush). Mist netting and playback were employed as suitable methods for understorey and skulking bird species. All sampling points were geo-referenced using a hand-held Garmin GPS.

Sampling stations for Pugu and Ruvu South

Two general sampling locations were established in Pugu, the first site at Mambisi (SGR 24 km), popularly known as 'Shimo la Popo', and a second station was set at SGR 22 km near a bird-watching site trail. The first site was located near the riverine vegetation along Msimbazi River. In Ruvu South, birds were also sampled at two different sites. The first station was set at SGR 38 km and the second station was set further south-west about 19 km from the first camp. This site was named as a forest site identified from satellite images, because the area near SGR is more of *Acacia-Commiphora* scrub, and with intermixed woodlands, and highly disturbed by anthropogenic activities, charcoal harvesting, grazing and fire. This lowered the possibility of finding typical forest species or birds of closed canopy, including Spotted Ground-thrush and Sokoke Pipit in the areas near SGR. Note also that only 25% of Ruvu South is considered forest (Clarke and Dickinson, 1995; Baker and Baker, 2002).

Mist netting

At each station within the SGR corridor, nets were erected between 50 m and 100 m, and 300–400 m from the SGR line. All mist nettings were done in the southern part of the SGR; the northern part of the SGR line for Pugu has a river and CRS, and for Ruvu South the area was highly unsuitable for putting nets, and at times livestock wandered around. In each season (wet and dry), a total of 20 nets of 12 m length were erected at each camp, ten close to the SGR line (within 50–100 m – hereafter 'near SGR') and another ten were set within 300–400 m (hereafter 'away from SGR') from the SGR line. The same netting hours for 20 nets were followed at a forest at Ruvu South. Nets were checked every 30–40 min and run for 30 diurnal hours (on three consecutive days). Birds trapped were ringed with metal rings obtained from the National Museum of Kenya, Nairobi, and released.

Point counts and search using playbacks

The fixed-width point count method (Bibby *et al.*, 2000; Bennun and Howell, 2002) was employed. Points were established along the SGR (within 50 m) and perpendicular to the

SGR going outwards on both the southern and northern side (Figs. 8.1 and 8.2). Points were set at least 200 m from each other. In most cases, transects had two to three points, i.e. at 200 m, 400 m and 600 m from the SGR. Point census was also done at a forest site in Ruvu South. Extra points were set randomly, especially in Pugu forest, to increase the likelihood for recording the rare and skulking species. At all census points with suitable habitats – a canopy

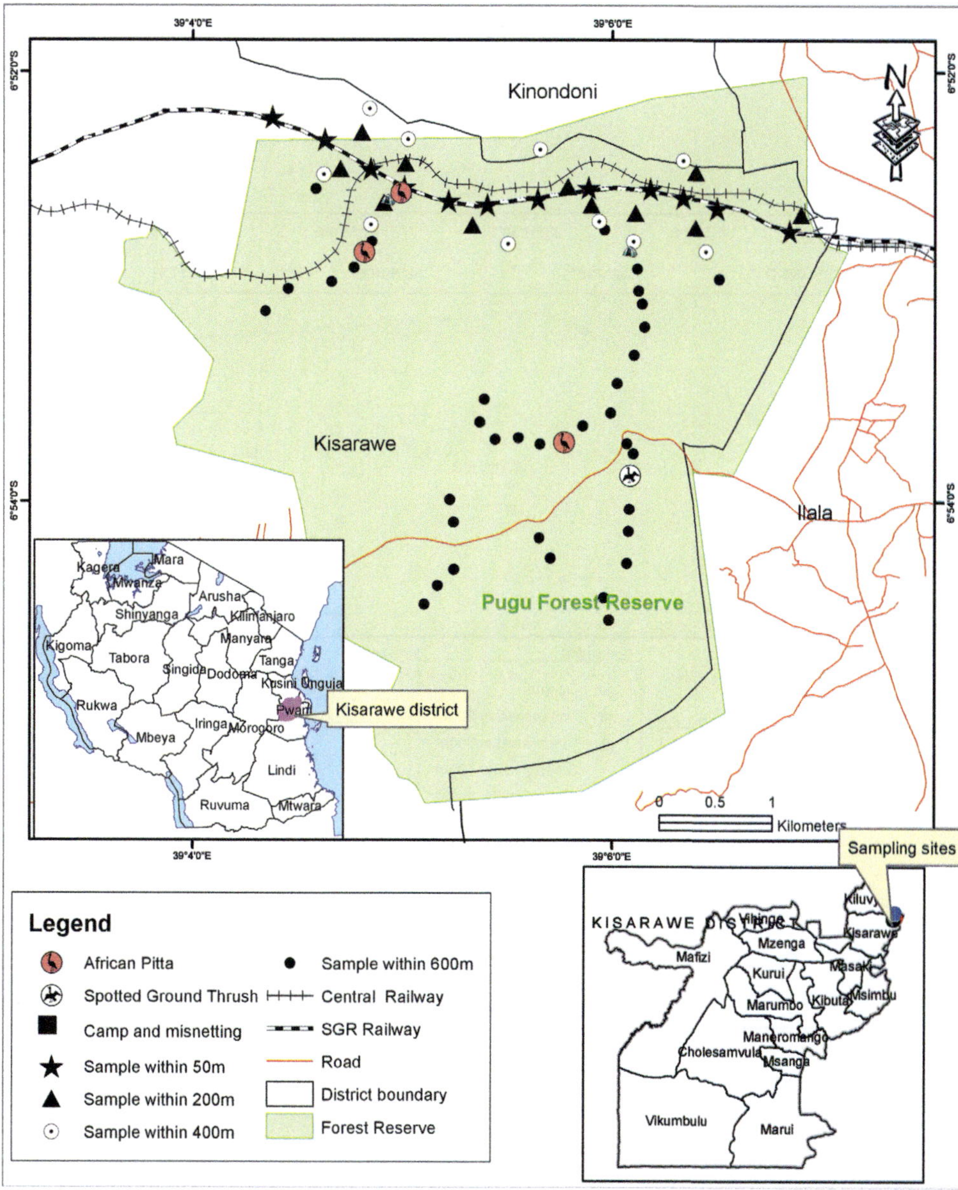

Fig. 8.1. Bird sampling locations in Pugu forest in relation to SGR alignment in April, and November–December 2021.

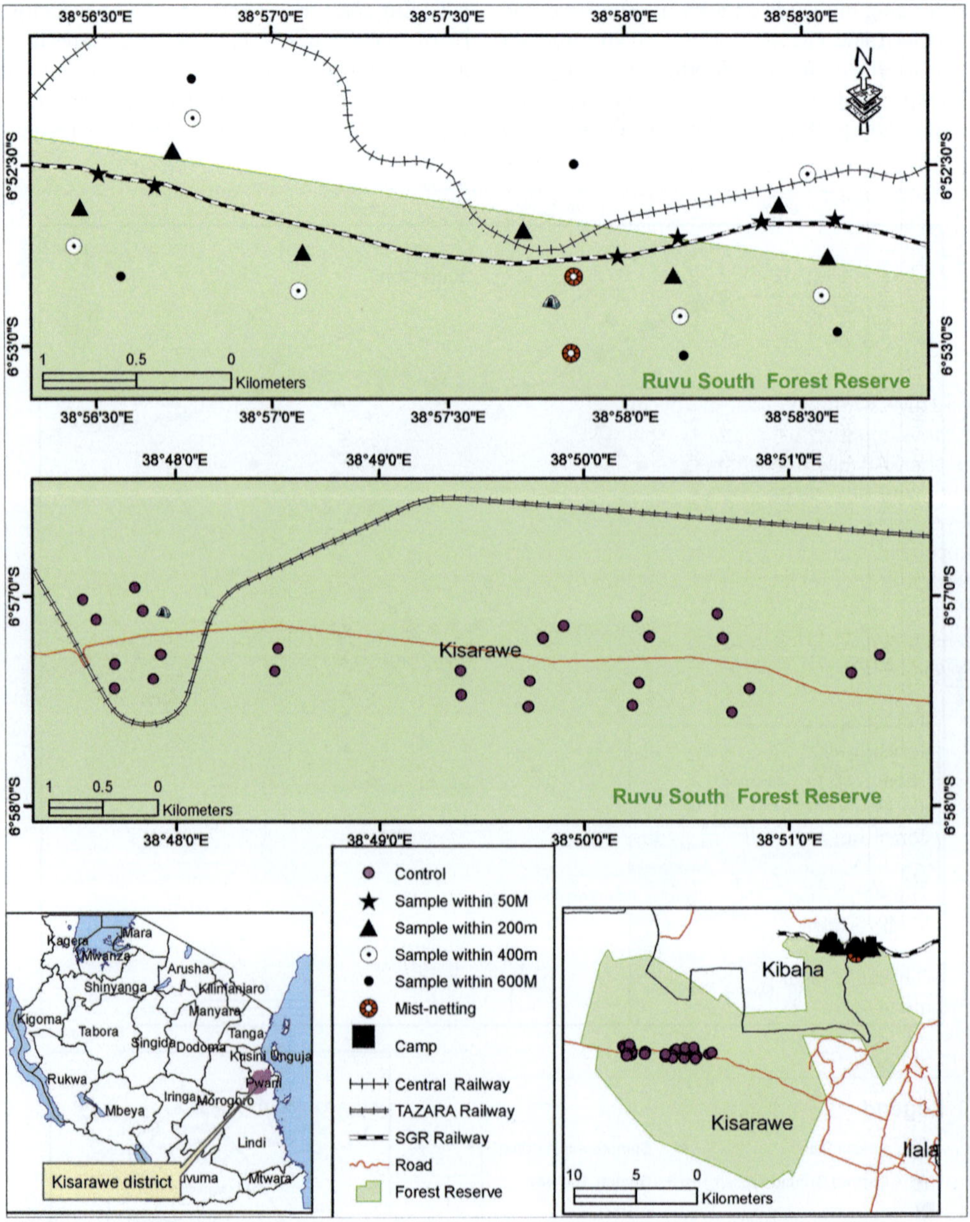

Fig. 8.2. Bird sampling locations in Ruvu South Forest in relation to SGR alignment and at a forest (control) site in April, and November–December 2021.

height above 5 m and an open floor with leaf litters – playbacks for Spotted Ground-thrush and Sokoke Pipit were done. The same protocol was applied both for wet and dry seasons. For playbacks, bird calls from the Birds of East Africa App were played on smartphones using Dodocool Min wireless speaker (DA84-1).

8.2.3 Literature and unpublished data

An assessment of both scientific and grey literature on birds of Pugu Hills and Ruvu South was done to pull out all birds known to occur there. Most old publications for the two forests were accessed from the digital Biodiversity Heritage Library (BHL). Tanzania Bird Atlas (TBA, n.d), accessible online (http://tanzaniabirdatlas.net) offered another great source of information for both sites. All my bird records (2021–July 2023) and reliable sightings from field assistants and visitors are included. Mlingwa *et al.* (2000) provided a basis for forest dependency at Pugu.

8.2.4 Data presentation

Bird diversity analyses were done to compare different sampling clusters and seasons in reference to the SGR alignment, and Shannon-Winener and diversity t-test was used. Significance levels were considered at ≤ 0.05. Tables and an appendix summarize important information. Bird forest dependence categorization follows Bennun *et al.* (1996), Mlingwa *et al.* (2000) and Jensen *et al.* (2005) and was confined to forest specialist (FF) and generalist (F):

- Forest specialists (FF): are typical of the forest interior and likely to disappear when the forest is modified.
- Forest generalists (F) occur in undisturbed forest but can exist at the forest edge or in modified and fragmented forests, but continue to depend upon the forest for some of their resources, such as nesting sites.

Threat categories are assigned to globally threatened bird species following the IUCN Red List of 2023 (BirdLife International, 2023). The threat category is based on the definitions provided in Collar *et al.* (1994) and Mace and Stuart (1994): Critically Endangered (CR) – facing an extremely high risk of extinction in the wild with a rapid population decline of 80–90% over the previous ten years (or three generations); Endangered (EN) – facing a very high risk of extinction in the wild with a 20% chance of extinction in 20 years; Vulnerable (VU) – facing a high risk of extinction in the wild with a 10% chance of extinction in 100 years; Near Threatened (NT) – species very close to VU. Globally Threatened species are those found in the three categories CR, EN and VU. The IUCN Red List is a critical indicator of the health of the world's biodiversity.

Taxonomy and nomenclature follow the *Handbook of the Birds of the World* (HBW) and Birdlife taxonomic checklist version 7 (BirdLife International, 2023). The Appendix provides information on the checklist of birds known to occur in Pugu Hills and Ruvu South, which should be referred to for scientific names henceforth. Maps were generated to show the sampling location in reference to the SGR alignment.

8.2.5 The study limitations for the SGR corridor

Sampling within the SGR corridor was conducted in April and November–December. The timing for Afro-tropical migratory forest species was perfect and Paleartic migrants were covered in April (those that move north late) and in November–December surveys. However, many Paleartic migrant species just pass by on the way to the southern Africa region and on their way back north; very few will remain there for a significant time; hence, sampling at only two specific short periods of time could miss many passers-by. Some rare resident species such as Sokoke Pipit require more time as they occur in low density in the area (Harvey and Howell, 1987; Mlingwa, 1996a) and are elusive. Construction activities were in an advanced stage during the bird survey and most disturbances such as habitat clearance had already occurred; thus, the current assessment along the SGR alignment refers to this stage. Additionally, the presence of a fence along the SGR was not considered in the design of field methods because the survey team became aware of it during the repeated survey when it was being erected.

8.3 Findings and Discussion

8.3.1 Bird diversity in Pugu forest

Seasonal patterns in bird diversity in reference to SGR alignment

From 59-point counts, 1466 individuals were recorded comprising 115 species in the wet season, and 974 individuals from 115 species in the dry season (Table 8.1; Appendix 8.1). Although the same number of species richness was recorded for both wet and dry season, there were changes in species diversity and composition. Significant seasonal variation in species diversity was only noted along the SGR line (0 m) and at 400 m from the SGR line (Table 8.2), which can be partly ascribed to the local habitat conditions. The area near SGR is close to Msimbazi river, which may attract many bird species in the dry season, whereas the 400 m is an interface between grassland-bush and thicket. Interestingly, in both seasons, the diversity increased progressively from SGR (Table 8.2) with significance levels in the wet seasons and the species diversity at ≥ 600 m for the dry season differing significantly from the rest of the clusters.

Some species were recorded only in single season, mostly because they are rare not recorded in the area before (Half-collared Kingfisher); restricted to habitats or migratory (Great Snipe, European Roller, Red-backed Shrike); species with erratic movements (African Pygmy-kingfisher, Garden Warbler, Spotted Flycatcher, Mangrove Kingfisher, Spotted Ground-thrush, Jacobin's Cuckoo, and Green-backed Twinspot).

For the same reasons, habitat conditions, local movements and migration, discussed previously, birds netted showed seasonal variation. The area near SGR had higher diversity in the the wet season than in the dry season while the area away from the SGR showed a reversed pattern (Tables 8.1 and 8.3). Additionally, within seasons, the two areas showed a significant level in bird diversity only in the wet season. The wet season attracted many of the seed-eaters and insectivores. Birds were observed feeding on insects that come with rain near the SGR line. The area away from SGR is a forested area, and dry season may have resulted in an increase in forest visitors seeking cover. Forest cover acts as cool refugia when temperatures are high enough to cause thermal stress (Seavy, 2006). A small part near SGR at birdwatching trail site (SGR 22 km) had also been burnt during the dry survey session.

Table 8.1. Summary of the species richness and abundance (number of individuals counted) recorded during the field surveys in Pugu Hills, wet and dry seasons, 2021.

Sampling clusters	Species richness		No. of individuals	
	Wet	Dry	Wet	Dry
0–50 m (SGR line)	57	57	383	211
At 200 m (from SGR line)	56	57	260	197
At 400 m (from SGR line)	56	48	231	136
At ≥ 600 m (from SGR line)	84	81	592	430
Mist netting (near SGR, 50–100 m)	35	20	79	33
Mist netting (away from SGR, 300–400 m)	21	22	103	69

Table 8.2. Seasonal differences in species diversity with varying distance from the SGR alignment in Pugu forest.

Sampling clusters	Wet (*H*)	Dry (*H*)	Significance status
0 m	3.05	3.6314	**
200 m	3.5897	3.6775	*
*400 m	3.7553	3.5488	**
≥ 600 m	3.9217	3.9815	*

* = not significant, ** = significant, *H* = Shannon-Weiner Species Diversity Index

Some of the species were re-trapped in the dry season, meaning that the nets were set within their home ranges where they either forage, roost or move around within the area. However, a few of these (Brown-hooded Kingfisher, African Paradise Flycatcher, Red-capped Robin-chat, Red-throated Twinspot) are partial migrants whose movements are poorly understood. For example, 72 individuals of Red-throated Twinspot were netted in the wet season while only one bird was captured in the

Table 8.3. Seasonal differences in species diversity indices for birds netted in areas near and away from SGR in Pugu Hills, wet and dry seasons, 2021.

Sampling clusters	Wet (*H*)	Dry (*H*)	Significance level
Near SGR (50–100 m)	3.2563	2.8608	**
Away from SGR (300–400 m)	1.7982	2.6379	**
Significance level	**	*	

* = not significant, ** = significant, H = *Shannon-Weiner Species Diversity Index*

dry season. Many Red-throated Twinspots had a brood patch, and a nest with two eggs was found in the area in the rainy season indicating that they breed in the area and disperse afterwards.

Overall, the study found almost the same number of species for areas near to the SGR and away from the SGR, but with high species turnover (see Appendix 8.1). The area near SGR was dominated by open-country species as compared to birds netted away from the SGR. The northern part of Pugu Hills, where SGR traverses, has been, historically, subjected to anthropogenic activities including linear infrastructure development, especially CRS, charcoal harvesting and logging, resulting in large areas of open grassland that are susceptible to periodic fire outbreaks (Baker and Baker, 2002; Mligo, 2019).

Bird diversity of Pugu forests – literature and unpublished data

Pugu forest is one of the most studied areas for birds of the Tanzanian coastal forests and until the mid-1980s it was already listed among the 80 most important forests for bird conservation in Africa (Collar and Stuart, 1988). The birdlife of this forest is well known from specimen collection during the first half of the 20th century (Fuggles-Couchman, 1939, 1986; Ripley and Heinrich, 1966, 1969; Britton, 1978; Britton, 1981), which are now held in a number of museums including the British Natural History Museum, the Tanzania National Museum and the University of Dar es Salaam. Using information documented about the collections and reports from several zoologists (Mackworth-Praed and Grant, 1960; Gerhart, 1977), Howell (1981) lists only 81 bird species for the area. Extensive bird-ringing activity in Pugu Hills during the 1980s and 1990s was partly published, but not in the form of a checklist; and include altitudinal migration (Baker, 1984; Mlingwa, 1996b), Afro-tropical migration (Baker and Howell, 1992), breeding and moult phenology (Mlingwa, 2000). Mlingwa *et al.* (2000) provide a list of 63 forest-dependent species known in Pugu Hills. Other ringing data, especially from Neil Baker and William Newmark (1980–1997) remain unpublished.

The Pugu Hills, now being isolated, once formed a large and extensive coastal forest engulfing the Dar es Salaam area, and thus bird records of the forest were included in the birds of Dar es Salaam area in the late 1980s (Harvey and Howell, 1987) but mostly were those listed in Howell (1981). Recenly, Gwegime *et al.* (2013a) listed 88 species for Pugu forest, while the Tanzania Bird Atlas (http://tanzaniabirdatlas.net) gives about 60 species for this area. The atlas provides an interactive online database for the squares falling within an area of interest. In a period of 1 year (May 2021–May 2022), for monthly visits at a private forest adjacent to Pugu Hills, Jane Goodall's Pugu Nature Centre documented 131 species for the centre (John, 2022). In compiling the checklist (Appendix 8.1), additional records were drawn from my field observations and bird-ringing programme in the forests since 2019 (author's unpublished data).

All the known bird records from different sources and recent field surveys have been pulled together to produce a comprehensive bird list for Pugu Hills. A total of 267 species can now be accepted to have occurred in Pugu forest. About 158 species were observed during the recent systematic survey (point count and mist nets) while 109 come from the other sources (see Appendix 8.1). The birds recorded in Pugu forest represent 67 families. Particularly well represented groups include the hawks and eagles (28 species), Estrildid finches (15 species), Old-world flycatchers and chats (14 species), cuckoos (14 species), cisticolas and allies (12 species), and pigeons and doves (11 species). However, a handful of eagle species were encountered in recent visits at Pugu (see Appendix 8.1); they might be in decline due to ongoing urbanization pressure surrounding the forest and lack of nesting emergent trees that were removed for charcoal and timber.

It is also important to note that some of the species, such as the White-eared Barbet (Baker, 1993) and White-chested Alethe are very rare, whereas others such as the Hooded Vulture, Rüppell's Vulture, Lesser Seedcracker, Sokoke Pipit and Spotted Ground-thrush are probably undergoing local extinction in the area. Harvey and Howell (1987) reported that Hooded Vultures were roosting in Pugu forest and Rüppell's Vultures were thought to breed in rocky hills near Pugu forest; but both species have not been seen in recent years. Most vultures are now globally threatened to extinction, and populations of many species have collapsed worldwide (Ogada *et al.*, 2016, 2022), especially near to urban areas.

8.3.2 Bird diversity in Ruvu South

Seasonal patterns in reference to SGR alignment

From 23 census points (Fig. 8.2) established along the SGR corridor in Ruvu South, 438 individuals were recorded from 84 species in wet season (April), and 355 individuals from 69 species in dry season. Information on the species richness and individuals obtained in both seasons for specific locations are presented in Table 8.4.

Table 8.5. Seasonal differences in species diversity with varying distance from the SGR alignment in Ruvu South forest during wet and dry seasons, 2021.

Sampling clusters	Wet (*H*)	Dry (*H*)	Significance status
0 m	3.843	2.5232	**
200 m	2.932	3.4211	**
400 m	3.5448	3.2884	**
≥ 600 m	3.4781	3.4476	*

* = not significant, ** = significant; H=Shannon-Weiner Species Diversity Index

Seasonal analyses and comparisons for species diversity are shown in Table 8.5. Although no patterns can be deduced from the analysis, all clusters except one (≥ 600 m) showed a significant seasonal difference in bird diversity. This difference can only be attributed to local movements, behavioural differences during the breeding season when birds are very active, and breeding plumage, pairing, and singing, which may influence detection. The frequent records of the birds in the genuses *Euplectus* and *Vidua* in the wet season attest to this (see Appendix 8.1).

A total of 279 individuals from 62 species were trapped within the SGR footprint. The number of birds captured at each location in both seasons are presented in Table 8.4. The most fascinating capture was that of an endangered Palearctic migrant species, the Basra

Table 8.4. Summary of bird species richness and abundance (number of individuals counted) from field surveys at Ruvu South during wet and dry seasons, 2021.

	Species richness		No. of individuals	
Sampling clusters	Wet	Dry	Wet	Dry
0–50 m (SGR line)	37	22	75	47
At 200 m (from SGR line)	38	41	141	115
At 400 m (from SGR line)	45	35	113	94
At ≥ 600 m (from SGR line)	40	40	109	99
Mist netting (near SGR, 50–100 m)	37	25	104	47
Mist netting (away from SGR, 300–400 m)	29	26	61	67

Table 8.6. Seasonal differences in species diversity indices for birds netted in areas near and away from the SGR during wet and dry seasons, 2021.

Cluster	Wet (*H*)	Dry (*H*)	Significance level
Near SGR (50–100 m)	3.0081	2.95547	*
Away from SGR (400–500 m)	3.1689	2.846	**
Significance level	*	*	*

* = not significant, ** = significant, *H* = Shannon-Weiner Species Diversity Index

Reed-warbler, on 27 November 2021. Britton (1978) notes a specimen from Soga, located not far from the 2021 capture, collected by Andersen on 30 November 1960.

In the wet season, the area near SGR recorded many birds in mist nets as compared to the one set away from the SGR (Table 8.4) because groups of Red-billed Quelea flocked into the mist nets, but this did not affect species diversity between seasons (Table 8.6). The general area becomes dry in November, which is the end of the dry season, and this may influence species diversity in some areas without water sources. The areas near SGR had seasonal swamps and temporary water bodies (because of ongoing construction) to attract birds, especially during the dry season. There were also humans making charcoal in the same areas in the dry season.

Birds at a forest site

A total of 912 individuals from 112 species were registered. Of these, only 32 species (99 individuals) came from netting. The following species were recorded from mist netting only: Pygmy-kingfisher, East Coast Batis, Miombo Wren-warbler, Northern Brownbul, Pale-breasted Illadopsis, Green-backed Twinspot and Black-and-white Mannikin. Although the wet season had 64 species from point counts compared with 80 species in the dry season, from equal sampling effort of 26 census points, species diversity did not show a significant seasonal difference ($H_{wet} = 3.834$ vs $H_{dry} = 3.918$, $p = 0.1768$). Similarly, species diversity for birds captured in mist nets did not differ significantly ($H_{wet} = 2.7917$ vs $H_{dry} = 2.6541$, $p = 0.4677$).

This site registered many forest bird species (about 50% of all the known forest-dependent in Ruvu South) which were not found in areas close to the SGR (see Appendix 8.1). The 26 forest-dependent bird species in this site included a first-time record for Pale-breasted Illadopisis in Ruvu South. Additionally, over 20 forest visitors preferring woodlands and bushy habitats were also recorded only from this area. It was therefore important to survey this habitat in order to enrich the species checklist of the reserve.

Bird diversity of Ruvu South – literature and unpublished data

Ruvu South has received little attention for bird studies as compared to the Pugu forest. With the exception of early collections by ornithologist Fuggles-Couchman in 1955 (Clarke and Dickinson, 1995), no bird list has ever been published for Ruvu South. However, the reserve was surveyed by ornithologists during the IBA project (Baker and Baker, 2002). Between 2010 and 2013, Gwegime *et al.* (2013b) recorded 88 species from Ruvu South. The IBA survey assessment was mainly done in the woodland areas to the north-east of the reserve with a brief visit to the main forest block (Baker and Baker, 2002) and not many forest species were known before this study. The Pale-breasted Illadopsis, previously thought to be endemic to Pugu (Howell, 1981) and Kazimzumbwi forests (Mlingwa *et al.*, 1993), recorded in Ruvu South for the first time during this survey, is an indication on how poorly the avifauna of the area is.

At present, at least 201 species can now be accepted to have occurred in Ruvu South Forest Reserve. Most species were recorded during the systematic field survey, and only 26 species were drawn from existing literature while another six species were added from opportunistic observation. Species recorded only once and those added from literature sources (see Appendix 8.1) may be among those exhibiting the following characteristics: habitat specialist or localized and rare (e.g. Sokoke Pipit, Lesser Seedcracker); skulking (Common Button-quail (see Baker and

Baker, 1994)); exhibit movements and migration (e.g. Lemon Dove, Madagascar Cuckoo and Lesser Cuckoo). It can also be suspected that some resident species, e.g. Sokoke Pipit, have declined in the recent past due to continued human threats. The current bird checklist of this forest represents 57 families. Particularly well represented groups include the cuckoos (12 species), cisticolas and allies (12 species), hawks and eagles (11 species), weavers (11 species) and Old-world flycatchers and chats (11 species).

8.3.3 Forest-dependent and threatened species in Pugu and Ruvu South

Forest-dependent and globally threatened species

The presence of forest-dependent bird species is a strong reason why both Pugu and Ruvu South forests are considered important for bird conservation. These species are sensitive to the alterations of the forest habitats and the efforts put in in the region by different scholars at different periods (Bennun *et al.*, 1996; Mlingwa *et al.*, 2000; Jensen *et al.*, 2005) allowing ornithologists to assign many of the Coastal Forest species to different forest dependence categories.

Pugu and Ruvu South host many of the FF and F species (see Section 2.4), which are characteristic of the East African Coastal Forests Biodiversity Hotspots (see Appendix 8.1) reflecting on their importance for avian conservation in the region. Pugu holds 31 and 48 species of FF and F, respectively; likewise, Ruvu South hosts 20 and 33 species of FF and F species, respectively. The race *T.r. puguensis*, a forest specialist, previously thought to be endemic in Pugu and in the adjacent forest of Kazimzumbwi (Howell, 1981; Mlingwa *et al.*, 1993), is among the forest specialists for this area. This subspecies is isolated from the nearest populations in the East Usambaras and Zanzibar. Because of the isolation, scholars suspect that this may prove to have sufficiently diverged to warrant a specific status (Baker and Baker, 2002).

Although some of the forest species recorded for Pugu and Ruvu South forests may not represent typical coastal forest birds – notably, several are essentially montane, with small temporary or permanent populations (Mlingwa, 1996; Mlingwa *et al.*, 2000; Baker and Baker, 2002) – the two forests are still important for them: White-starred Robin, Black-fronted Bush-shrike, Lemon Dove, Orange Ground-thrush, Eastern Bronze-naped Pigeon, Olive Pigeon, Buff-spotted Flufftail, Barred Long-tailed Cuckoo and Olive-headed Bulbul. The latter has recently been split from the Stripe-cheeked Bulbul *Arizelocichla milanjensis* found in Malawi and Zimbabwe. The two were previously lumped as *Andropadus milanjensis* (del Hoyo *et al.*, 2016).

The African Pitta, an intra-African migrant, is another important forest bird in the area, as it passes through approximately in May and November to and from breeding ground in the southern Tanzania coastal forests, Rondo Plateau in particular. Two birds were trapped in April at Pugu forest at Camp 1 (at SGR 24 km), both near SGR and 400 m from SGR. A small group of the African Pitta (with four birds) was also encountered in Pugu forest on 26 April 2021 about 2 km south of SGR. Before the 2021 field observations, about 20 records (nine in May and 11 in April) of African Pitta had been recorded in Pugu Hills between 1984 and 1993 (N. Baker, Iringa, 2023, personal communication). The African Pitta is probably still doing well along its migratory routes in coastal habitats. It seems a good number of this species pass through Pugu on their way north (April–May) and south (breeding ground) (October–December). During our ringing session at Pugu, on 30 May 2023, an adult bird was netted with a post-breeding brood patch, while a young bird was netted on 29 April 2022.

Only two forest-dependent species of the Pugu and Ruvu South forests are Globally Threatened: Spotted Ground-thrush (VU) and Sokoke Pipit (EN). Another three forest-dependent species are classified as Near-threatened (NT): Southern-banded Snake-eagle, Crowned Eagle and East Coast Akalat. Below is a detailed account on some of the conservation-concern forest-dependent species for which Pugu and Ruvu South forests are considered critical habitat.

Sokoke Pipit (EN). It has a very small range, within which the total area of its forest habitat is declining owing to the clearance and degradation of the species' forest habitat, mainly through charcoal burning, agricultural encroachment, logging and pole cutting.

Although known from Pugu and Ruvu South forests (Baker and Baker, 2002), its recent status remains unknown. A first specimen for this species in the Dar es Salaam area was collected from Pugu in 1938 (Harvey and Howell, 1987). There are no recent published records of the Sokoke Pipit in Pugu since its rediscovery in the mid-1990s (Mlingwa, 1996a). Our efforts to search for this species in both forests did not yield any sign of its presence in over 100 census points (including playback) and netting (in Pugu, since 2019–July 2023). Likewise, none was netted by Neil Baker (1980–1990), Bill Newmark (1996–1997) or Faustine Kirway (March–May 2019), and there are no records from Gwegime *et al.* (2013a, b) in both Pugu and Ruvu South forests. Therefore, sightings of 28 birds in Ruvu South reported by Mbogo *et al.* (2018) remain suspicious.

Spotted Ground-thrush (VU). It has a small (patchy) population (Ndang'ang'a *et al.*, 2005), which is undergoing a continuing decline because of destruction and degradation of its habitat. The East Africa race, *fisheri*, although its known breeding site is now found in a newly gazetted Rondo Nature Forest Reserve in southern Tanzania, is probably CR, for the same threats facing Sokoke Pipit in much of its range, especially passage, including Pugu and Ruvu South forests. A single bird was recorded on 16 April 2021 at Pugu in an intact forest segment about 1.5 km south of SGR. Before this study, only 16 records (two in April, 14 in May) had been made between 1984 and 1991 (N. Baker, Iringa, 2023, personal communication). Like the African Pitta, records on passage to the breeding sites are rare because birds move to their breeding grounds faster than to their wintering grounds. Note also that Pugu is halfway between Rondo Plateau and coastal Kenya wintering grounds.

East Coast Akalat (NT). It occurs in scattered forests in south-east Kenya, eastern Tanzania, northern Malawi and Mozambique. It inhabits closed, moist forest in the Tanzanian mountains (Fjeldså *et al.*, 2000), lowland and mid-altitude humid forest in Malawi (Dowsett-Lemaire, 1989) and Mozambique (Spottiswoode *et al.*, 2008), and tracts of mixed woodland and open, dry forest in coastal/lowland areas (Nemeth and Bennun, 2000). It prefers forest areas with dead logs and moss, which it uses for nesting material. Arabuko-Sokoke Forest (Kenya) has probably the largest and most stable population in Kenya (Matiku *et al.*, 1999; Banks *et al.*, 2012). Although Baker and Baker (2002) claim that Ruvu South has the largest population in Tanzania, risks are high due to ongoing human activities (charcoal production, fire and cattle grazing) in the reserve. However, from the field observations, including ongoing bird-ringing studies at Pugu, this species is still common at sites representing good or typical coastal habitats in both forests. It comprised about 7% of all captures (n = 483, 41 species) between February and July 2023 at Pugu forest.

Other globally threatened species in Pugu and Ruvu South

At least ten species that are not forest-dependent but threatened with extinction are found in Pugu and Ruvu South forests (see Appendix 8.1). They include Southern Ground-hornbill, Bateleur, Great Snipe, Hooded Vulture, White-backed Vulture, Rüppell's Vulture, Martial Eagle, Tawny Eagle and Steppe Eagle. The forests are therefore still very important for conservation of forest and threatened species although some of them, e.g. vultures, have continued to decline. With the exception of the Palm-nut Vulture, sightings of other vultures are very rare. The Hooded vultures used to roost in Pugu Hills (Harvey and Howell, 1987) but have now disappeared. The presence of the migratory endangered Basra Reed-warbler further reveals the importance of these forests for the conservation of globally threatened species.

8.4 Conclusion and Recommendations

8.4.1 Conclusion

In compiling the information for this chapter, it is hoped that others will come to appreciate the avifauna richness found in the area, and its local and international conservation value. However, the ongoing habitat modification due to human activities within the forest reserves calls for monitoring to determine probable impacts of the changes in the forest structure. For example, species recorded in the northern part, especially closer to the SGR line, were mostly those of open habitats and degraded environment including those attracted to human habitations such as the House Sparrow and House Crow. The information in this chapter provides a baseline for monitoring the effect of a recently constructed railway within Pugu and Ruvu South forest reserves. Futhermore, the importance of Pugu forest regionally may increase through time because it is now part of the Pugu-Kazimzumbwi Nature Forest Reserve, while other parts of the coastal forests in the region, including Ruvu South, are gradually losing their significance for birdlife due to heavy human pressure. Nonetheless, the two forests are important for many forest-dependents including eastern Africa coastal forest endemic bird species. Unfortunately, the destinations and precise routes of the passage migrants at Pugu and Ruvu forests, including the intra-African, are not well known. Lastly, Pugu and Ruvu South forests are excellent birding sites near Dar es Salaam and continue to offer opportunities for bird research and training to the universities and colleges around Dar es Salaam.

8.4.2 Recommendations

Based on the field observations and the literature, key recommendations to preserve forest birds of Pugu and Ruvu South forests are:

a) Forest enrichment planting to restore the degraded or modified habitats along the SGR alignment should use native plant species as forest-dependent bird species will not do better in exotic dominated habitats. Exotic habitats may not provide the most required cover and food resources for forest species.

b) Research and monitoring of the presence and use of Pugu and Ruvu South forests for Sokoke Pipit and Spotted Ground-thrush are recommended. This will require good timing and continous search because of their elusiveness. African Pitta, another important Afro-tropical migrant that follows almost the same timing as the Spotted Ground-thrush, is worth considering.

c) The potential impacts of the fence erected on both sides of the SGR line on the movements of understorey bird species within the reserves can be investigated using experimental capture-mark-recapture techniques and telemetry application.

d) The northern part of the two forests including SGR alignment experience periodic wildfire. Therefore, a well-thought-out fire-management strategy for this area is needed. While working on the fire-management plan, the reserve management is encouraged to increase forest patrols to combat other illegal anthropogenic activities in the forests and to stimulate natural regeneration and recovery.

Acknowledgements

I wish to thank the following: Edigar Apolinary, Maneno Mbilinyi, Abdulrahman Adam, Emmanuel William, Michael Munisi, Tumaini Rikanga, Hamis Musa Chakapu (Pengo), Nassor Said Jaswa, Jimminus Kakoko, Gabriel Nalaila, Iddrissa Faruku, Alex Manweki, Daudi Damas, Chande Awadh, Athumani Mahiga, George Msese, Abdallah Msuya, Magdalena Kitila, Catherine Masao, Eliapenda Pesa and Flora Magige for their field and logistical support. William Newmark provided his ringing data in Pugu forests (1995–1997) while Neil Baker shared information on key species from his unpublished data. Faustine Kirway shared his 2019 ringing data in Pugu. Tanzania Forest Services Agency granted entry and research permits for fieldwork. I am grateful to my family. This work was funded by TRC.

References

Baker, N.E. (1984) Recent coastal records of the white-starred forest Robin *Pogonocichla stellata* in Tanzania. *Scopus* 8, 51–52.

Baker, N.E. (1993) On the racial status of the white-eared barbet *Stactolaema leucotis* in coastal Tanzania. *Scopus* 16, 111–113.

Baker, N.E. and Baker, E.M. (1992) Four Afrotropical migrants on the East African coast: Evidence of a common origin. *Scopus* 15, 122.

Baker, N.E. and Baker, E.M. (1994) Recent records of the blue quail *Coturnix chinensis* from Tanzania. *Scopus* 18, 116.

Baker, N.E. and Baker, E.M. (2002) *Important Bird Areas in Tanzania: A First Inventory*. Wildlife Conservation Society of Tanzania, Dar es Salaam.

Baker, N.E. and Howell, K.M. (1992) Unusual movements of Afro-tropical birds in the Dar es Salaam area. In: *Proceedings of the Seventh Pan-African Ornithological Congress*, Nairobi.

Banks, J.E., Jackson, C., Baya, A., Minnella, H., Nitz, M. *et al.* (2012) Forest-type preference of an Afrotropical thrush (East Coast Akalat *Sheppardia gunningi sokokensis*) in Arabuko-Sokoke forest, Kenya. *Ostrich* 83, 105–108.

Bennun, L. and Howell, K. (2002) Birds. In: Davies, G. (ed.) *African Forest Biodiversity: A Field Survey Manual for Vertebrates*. Eathwatch Institute, UK, pp. 121–161.

Bennun, L., Dranzoa, C. and Pomeroy, D. (1996) The forest birds of Kenya and Uganda. *Journal of East Africa Natural History* 85, 23–48.

Bibby, C., Jones, M. and Marsden, S. (2000) *Expedition Field Techniques: Bird Surveys*. Royal Geographical Society, London.

BirdLife International (2023) IUCN Red List for Birds. Available at: http://www.birdlife.org (accessed 10 April 2023).

Britton, P.L. (1978) The Andersen collection from Tanzania. *Scopus* 2, 77–85.

Britton, P.L. (1981) Notes on the Andersen collection and other specimens from Tanzania housed in some West German museums. *Scopus* 5, 14–21.

Burgess, N.D., Mwasumbi, L.B., Hawthorne, W.J., Dickinson, A. and Doggett, R.A. (1992) Preliminary assessment of the distribution, status and biological importance of coastal forests in Tanzania. *Biological Conservation* 62, 205–218.

Clarke, G.P. and Dickinson, A. (1995) "*Status Reports for 11 Coastal Forests in Coast Region, Tanzania*" [Frontier Tanzania Technical Report no.17] [The Society for Environmental Exploration, London/UK & The University of Dar es Salaam]. The Society for Environmental Exploration, London/University of Dar es Salaam.

Collar, N.J. and Stuart, S.N. (1988) *Key Forests for Threatened Birds in Africa*. ICBP, Cambridge.

Collar, N.J., Crosby, M.J. and Stattersfield, A. (1994) *Birds to Watch 2: The World List of Threatened Birds*. BirdLife International, Cambridge, UK.

del Hoyo, J., Collar, N.J., Christie, D.A., Elliott, A., Fishpool, L.D.C. *et al.* (2016) *HBW and BirdLife International Illustrated Checklist of the Birds of the World. Volume 2: Passerines* Lynx Edicions and BirdLife International, Barcelona, Spain and Cambridge, UK.

Donald, P.F., Fishpool, L.D.C., Ajagbe, A., Bennun, L.A., Bunting, G. *et al.* (2019) Important Bird and Biodiversity Areas (IBAs): The development and characteristics of a global inventory of key sites for biodiversity. *Bird Conservation International* 29, 177–198.

Dowsett-Lemaire, F. (1989) Ecological and biogeographical aspects of forest bird communities in Malawi. *Scopus* 13, 1–80.

ERM (2019) *Standard Gauge Railway – Dar es Salaam to Makutopora, Tanzania: Biodiversity Action Plan*. Environmental Resource Management (ERM) Southern Africa (Pty) Ltd & Yapi Merkezi. Version 3.4. Available at: www.erm.com

Fjeldså, J., Roy, M.S. and Kiure, J. (2000) A new montane subspecies of *Sheppardia gunningi* (East-coast Akalat) from Tanzania. *Bulletin of British Ornithologists' Club* 120, 27–33.

Fuggles-Couchman, N.R. (1939) Notes on some birds of Eastern province of Tanganyika territory. *Ibis* 76, 106.

Fuggles-Couchman, N.R. (1986) Breeding records of some Tanzanian birds. *Scopus* 10, 20–26.

Gerhart, J. (1977) Pugu hills forest reserve, Dar es Salaam. *Bulletin of East Africa Natural History Society* 130–131.

Gwegime, J., Mwangoka, M., Said, H., Mulungu, E., Nowak, K. *et al.* (2013a) *Two Surveys of the Plants, Birds and Forest Condition of Pugu and Kazimzumbwi Forest Reserves in 2011 and 2012*. TFCG Technical Paper 36. Tanzania Forest Conservation Group, Dar es Salaam, Tanzania.

Gwegime, J., Mwangoka, M., Mulungu, E., Perkin, A. and Nowak, K. (2013b) *The Biodiversity and Forest Condition of Ruvu South Forest Reserve*. TFCG Technical Paper 37. Tanzania Forest Conservation Group, Dar es Salaam, Tanzania.

Harvey, W.G. and Howell, K.M. (1987) Birds of the Dar es Salaam area, Tanzania. *Le Gerfaut* 77, 205–258.

Howell, K.M. (1981) Pugu forest reserve: biological values and development. *African Journal of Ecology* 10, 73–81.

Jensen, F.P., Tøttrup, A.P. and Christensen, K.D. (2005) The avifauna of coastal forests in Southeast Tanzania. *Scopus* 25, 1–22.

John, J. (2022) *A Checklist of Birds of Jane Goodall's Pugu Nature Centre*. Unpublished report. University of Dar es Salaam and Jane Goodall Institute, Dar es Salaam.

Mace, G.M. and Stuart, S.N. (1994) Draft IUCN red list categories. *Species* 21–22, 13–24.

Mackworth-Praed, C.W. and Grant, C.H.B. (1960) *African Handbook of Birds.*, 2nd edn. Longmans Gree & Co., London.

Matiku, P.M., Bennun, L.A. and Nemeth, E. (1999) Distribution and population size of the threatened East Coast Akalat in Arabuko-Sokoke forest, Kenya. *Ostrich* 71, 282–285.

Mbogo, D., Baines, J. and Mwendwa, B. (2018) Potential distribution and conservation of endangered bird *Anthus Sokokensis* in Ruvu South Coastal forest reserve, Tanzania. *International Journal of Forest, Animal and Fisheries Research* 2, 123–127.

Mligo, C. (2019) Post fire regeneration of indigenous plant species in the Pugu forest reserve, Tanzania. *Global Ecology and Conservation* 18, e00611.

Mlingwa, C.O.F. (1996a) A note on the discovery the discovery of the Sokoke Pipit *Anthus sokokensisin* Pugu hills, Tanzania. *Bird Conservation International* 6, 293–294.

Mlingwa, C.O.F. (1996b) A first record of the stripe-cheeked Greenbul *Andropadus milanjensis* in the Pugu hills, Tanzania. *Bulletin of the British Ornithologists' Club* 116, 63–64.

Mlingwa, C.O.F. (2000) Breeding and moult phenology of an avian community in the Pugu hills, Tanzania. *Ostrich* 71, 87–90.

Mlingwa, C.O.F., Huxham, M.R. and Burgess, N.D. (1993) The Avifauna of Kazimzumbwe forest reserve,Tanzania: initial findings. *Scopus* 16, 81–88.

Mlingwa, C.O.F., Waiyaki, E.M., Bennun, L.A. and Burgess, N.D. (2000) Birds. In: Burgess, N.D. and Clarke, G.P. (eds) *Coastal Forests of Eastern Africa*. IUCN, Cambridge, pp. 149–171.

Ndang'ang'a, K., Sande, E., Evans, S.W., Buckley, P. and Nwebery, P., et al. (eds) (2005) *International Species Action Plan for the Spotted Ground Thrush* Zoothera guttata. BirdLife International, Nairobi, and Royal Society for the Protection of Birds, Sandy, UK.

Nemeth, E. and Bennun, L. (2000) Distribution, habitat selection and behaviour of the East Coast Akalat *Sheppardia Gunningi*. *Bird Conservation International* 10, 115–130.

Ogada, D., Shaw, P., Beyers, R.L., Buij, R., Murn, C, *et al.* (2016) Another continental vulture crisis: Africa's vultures collapsing towards extinction. *Conservation Letters* 9, 89–97.

Ogada, D., Virani, Z.M., Thiollay, J.M., Kendall, C.J., Thomsett, S, *et al.* (2022) Evidence of widespread declines in Kenya's raptor populations over a 40-year period. *Biological Conservation* 266. DOI: 10.1016/j.biocon.2021.109361.

Ralph, C.J., Geupel, G.R., Pyle, P., Martin, T.E. and De Sante, D.F. (1993) *Handbook of Field Methods for Monitoring Land Birds*. Pacific Southwest Research Station, Forest Services, US Department of Agriculture.

Ripley, S.D. and Heinrich, G.H. (1966) Comments on the Avifauna of Tanzania, 1. *Postilla* 96, 1–45.

Ripley, S.D. and Heinrich, G.H. (1969) Comments on the Avifauna of Tanzania, 2. *Postilla* 134, 1–21.

Rocky, J. and Mligo, C. (2012) Regeneration pattern and size-class distribution of indigenous woody species in exotic plantation in Pugu forest reserve, Tanzania. *International Journal of Biodiversity Conservation* 4, 1–14.

Seavy, N.E. (2006) Physiological correlates of habitat association in East African sunbirds (Nectariniidae). *Journal of Zoology* 270, 290–297.

Spottiswoode, C.N., Patel, I.H., Herrmann, E., Timberlake, J. and Bayliss, J. (2008) Threatened bird species on two little-known mountains (Chiperone and Mabu) in northern Mozambique. *Ostrich* 79, 1–7.

Tanzania Bird Atlas (n.d.). Available at: http://tanzaniabirdatlas.net/TZmap_squarelists.html (accessed 10 May 2021).

TRC (2020) *terms of references, Ornithologist. Dar es Salaam to Makutopora standard gauge railway project, Tanzania. TRC, Dar Es Salaam.*

Appendix 8.1: Checklist of birds known for Pugu Hills (part of Pugu-Kazimzumbwi Nature Forest Reserve) and Ruvu South coastal forest reserves. Systematic field surveys were conducted in April (wet), and November and December (dry) 2021

No. HBW	Family	English name	Species	0 m		200 m		400 m		≥600 m		Forest_ RS		Nets_ near SGR		Nets_ away from SGR		Net_ forest_ RS		Obs. & Lit. only
				W	D	W	D	W	D	W	D	W	D	W	D	W	D	W	D	
139	Numididae	Helmeted Guineafowl	*Numida meleagris*		p	pr	p		p	r	r									
145	Numididae	Eastern Crested Guineafowl F	*Guttera pucherani*							p	p		r							
237	Phasianidae	African Blue Quail	*Synoicus adansonii*																	r
273	Phasianidae	Red-necked Francolin	*Pternistis afer*				r	r		r										
288	Phasianidae	Crested Francolin	*Ortygornis sephaena*			r	pr	r		r	pr	r								
375	Anatidae	White-faced Whistling-duck	*Dendrocygna viduata*																	p
558	Columbidae	Speckled Pigeon	*Columba guinea*		p															
598	Columbidae	African Olive-pigeon F	*Columba arquatrix*																	p
616	Columbidae	Eastern Bronze-naped Pigeon FF ALT	*Columba delegorguei*																	p
620	Columbidae	Lemon Dove FF ALT	*Aplopelia larvata*																	pr
622	Columbidae	Dusky Turtle-dove ALT	*Streptopelia lugens*																	p
631	Columbidae	Mourning Collared-dove	*Streptopelia decipiens*																	r
632	Columbidae	Red-eyed Dove	*Streptopelia semitorquata*	p	p	p	pr	p	r	p	pr	r	r			r				
633	Columbidae	Ring-necked Dove	*Streptopelia capicola*	r	pr	r	r	r	r	r	r		r							
791	Columbidae	Emerald-spotted Wood-dove	*Turtur chalcospilos*	pr	p	pr	pr	p	pr	pr	pr	r	r	pr	pr	r	r			
793	Columbidae	Blue-spotted Wood-dove	*Turtur afer*					r												p

Continued

Continued

No. HBW	Family	English name	Species	0 m W	0 m D	200 m W	200 m D	400 m W	400 m D	≥600 m W	≥600 m D	Forest_ RS W	Forest_ RS D	Nets_ near SGR W	Nets_ near SGR D	Nets_ away from SGR W	Nets_ away from SGR D	Net_ forest_ RS W	Net_ forest_ RS D	Obs. & Lit. only
794	Columbidae	Tambourine Dove F	*Turtur tympanistria*			p		p	p	p	p		r							
796	Columbidae	Namaqua Dove	*Oena capensis*			r														
825	Columbidae	African Green-pigeon F	*Treron calvus*										r							p
1056	Caprimulgidae	European Nightjar *	*Caprimulgus europaeus*																	p
1072	Caprimulgidae	Fiery-necked Nightjar F	*Caprimulgus pectoralis*																	pr
1085	Caprimulgidae	Slender-tailed Nightjar	*Caprimulgus clarus*																	p
1086	Caprimulgidae	Mozambique Nightjar	*Caprimulgus fossii*											r		r				p
1127	Apodidae	Böhm's Spinetail F	*Neafrapus boehmi*										r							p
1164	Apodidae	White-rumped Swift	*Apus caffer*																	p
1179	Apodidae	African Palm-swift	*Cypsiurus parvus*	pr	p	p	pr	pr	p	pr	pr		r							
1188	Apodidae	Horus Swift	*Apus horus*							r										p
1190	Apodidae	Little Swift*	*Apus affinis*	p		p				p										
1619	Cuculidae	White-browed Coucal	*Centropus superciliosus*	pr		pr	p	pr	p	pr	pr	r	r							
1634	Cuculidae	Whistling Yellowbill F	*Ceuthmochares australis*		p	p		p		p	p	r								
1647	Cuculidae	Jacobin Cuckoo*	*Clamator jacobinus*		r		p				p									
1648	Cuculidae	Levaillant's Cuckoo [AFR]	*Clamator levaillantii*							p	p					r				
1672	Cuculidae	Thick-billed Cuckoo	*Pachycoccyx audeberti*																	pr
1689	Cuculidae	Klaas's Cuckoo[AFR]	*Chrysococcyx klaas*	r	p		r		r		p		r							
1690	Cuculidae	African Emerald Cuckoo F	*Chrysococcyx cupreus*																	p
1691	Cuculidae	Diederik Cuckoo[AFR]	*Chrysococcyx caprius*											r		r				p
1704	Cuculidae	Barred Long-tailed Cuckoo FF [ALT]	*Cercococcyx montanus*																	p
1718	Cuculidae	Red-chested Cuckoo F[AFR]	*Cuculus solitarius*								p		r							
1719	Cuculidae	Black Cuckoo FF[AFR]	*Cuculus clamosus*																	pr
1721	Cuculidae	Common Cuckoo*	*Cuculus canorus*							r	p									
1726	Cuculidae	Lesser Cuckoo*	*Cuculus poliocephalus*																	pr

Continued

Continued

No. HBW	Family	English name	Species	0 m W	0 m D	200 m W	200 m D	400 m W	400 m D	≥600 m W	≥600 m D	Forest_ RS W	Forest_ RS D	Nets_ near SGR W	Nets_ near SGR D	Nets_ away from SGR W	Nets_ away from SGR D	Net_ forest_ RS W	Net_ forest_ RS D	Obs. & Lit. only
1727	Cuculidae	Madagascar Cuckoo[AFR]	*Cuculus rochii*																	pr
1732	Sarothruridae	Buff-spotted Flufftail FF	*Sarothrura elegans*																	p
1926	Otididae	Black-bellied Bustard	*Lissotis melanogaster*												r					
1954	Musophagidae	Purple-crested Turaco	*Gallirex porphyreolophus*										r							
1958	Musophagidae	Livingstone's Turaco F	*Tauraco livingstonii*							p			r							
2149	Ciconiidae	African Openbill	*Anastomus lamelligerus*																	p
2182	Threskiornithidae	Hadada Ibis	*Bostrychia hagedash*					r												
2217	Ardeidae	Dwarf Bittern	*Ixobrychus sturmii*																	p
2226	Ardeidae	Black-crowned Night-heron	*Nycticorax nycticorax*																	p
2230	Ardeidae	Green-backed Heron	*Butorides striata*			p				p										
2231	Ardeidae	Squacco Heron	*Ardeola ralloides*																	p
2238	Ardeidae	Grey Heron	*Ardea cinerea*																	r
2243	Ardeidae	Black-headed Heron	*Ardea melanocephala*																	p
2247	Ardeidae	Goliath Heron	*Ardea goliath*				p													
2267	Scopidae	Hamerkop	*Scopus umbretta*		p			pr					r							
2328	Anhingidae	African Darter	*Anhinga rufa*																	p
2334	Burhinidae	Water Thick-knee	*Burhinus vermiculatus*																	p
2383	Charadriidae	African Three-banded Plover	*Charadrius tricollaris*			p	p				p									
2520	Scolopacidae	Great Snipe* NT	*Gallinago media*				p													
2533	Scolopacidae	Common Sandpiper*	*Actitis hypoleucos*	r																
2548	Turnicidae	Common Buttonquail	*Turnix sylvaticus*																	r
2779	Strigidae	Pearl-spotted Owlet	*Glaucidium perlatum*																	r
2786	Strigidae	African Barred Owlet F	*Glaucidium capense*										r							p
2929	Strigidae	African Wood-owl F	*Strix woodfordii*																	pr
2939	Strigidae	Spotted Eagle-owl	*Bubo africanus*																	pr

Continued

Continued

No. HBW	Family	English name	Species	0 m W	0 m D	200 m W	200 m D	400 m W	400 m D	≥600 m W	≥600 m D	Forest_ RS W	Forest_ RS D	Nets_ near SGR W	Nets_ near SGR D	Nets_ away from SGR W	Nets_ away from SGR D	Net_ forest_ RS W	Net_ forest_ RS D	Obs. & Lit. only
2943	Strigidae	Verreaux's Eagle-owl	*Bubo lacteus*																	p
2975	Elanidae	Black-winged Kite	*Elanus caeruleus*	r	p	r		r												
2981	Accipitridae	European Honey-buzzard* F	*Pernis apivorus*																	p
2989	Accipitridae	African Cuckoo-hawk F	*Aviceda cuculoides*																	p
2995	Accipitridae	African Harrier-hawk	*Polyboroides typus*																	p
2997	Accipitridae	Palm-nut Vulture	*Gypohierax angolensis*	p	p	pr	p				p	r								
3007	Accipitridae	Bateleur EN	*Terathopius ecaudatus*	p				r		p										
3011	Accipitridae	Black-chested Snake-eagle	*Circaetus pectoralis*	p				pr				r								
3012	Accipitridae	Brown Snake-eagle	*Circaetus cinereus*																	p
3013	Accipitridae	Southern Banded Snake-eagle NT F	*Circaetus fasciolatus*																	pr
3017	Accipitridae	Hooded Vulture CR	*Necrosyrtes monachus*																	p
3020	Accipitridae	White-backed Vulture CR	*Gyps africanus*	p																
3024	Accipitridae	Rüppell's Vulture CR	*Gyps rueppelli*																	p
3028	Accipitridae	Bat Hawk F	*Macheiramphus alcinus*																	p
3032	Accipitridae	Crowned Eagle NT FF	*Stephanoaetus coronatus*										r							p
3043	Accipitridae	Martial Eagle EN	*Polemaetus bellicosus*																	p
3049	Accipitridae	Long-crested Eagle	*Lophaetus occipitalis*																	p
3053	Accipitridae	Tawny Eagle VU	*Aquila rapax*																	p
3054	Accipitridae	Steppe Eagle* EN	*Aquila nipalensis*																	p
3062	Accipitridae	African Hawk-eagle	*Aquila spilogaster*																	p
3064	Accipitridae	Wahlberg's Eagle[AFR]	*Hieraaetus wahlbergi*		p					p										
3065	Accipitridae	Booted Eagle	*Hieraaetus pennatus*																	p
3069	Accipitridae	Lizard Buzzard	*Kaupifalco monogrammicus*							r		r								p
3070	Accipitridae	Dark Chanting-goshawk	*Melierax metabates*								r	r								

Continued

Continued

No. HBW	Family	English name	Species	0 m		200 m		400 m		≥600 m		Forest_ RS		Nets_ near SGR		Nets_ away from SGR		Net_ forest_ RS		Obs. & Lit. only
				W	D	W	D	W	D	W	D	W	D	W	D	W	D	W	D	
3072	Accipitridae	Eastern Chanting-goshawk	*Melierax poliopterus*	r		r		r												
3096	Accipitridae	African Goshawk F	*Accipiter tachiro*								p		r						r	
3098	Accipitridae	Shikra	*Accipiter badius*																	p
3118	Accipitridae	Little Sparrowhawk	*Accipiter minullus*								p	r								
3135	Accipitridae	Black Sparrowhawk F	*Accipiter melanoleucus*							p										
3146	Accipitridae	African Fish-eagle	*Haliaeetus vocifer*																	p
3209	Accipitridae	Augur Buzzard	*Buteo augur*																	p
3211	Accipitridae	Eurasian Buzzard*	*Buteo buteo*																	p
3220	Coliidae	Speckled Mousebird	*Colius striatus*		p		p		p		p									r
3224	Coliidae	Blue-naped Mousebird	*Urocolius macrourus*					r			r									p
3227	Trogonidae	Narina Trogon F	*Apaloderma narina*								p		r				p	r	r	
3229	Trogonidae	Bar-tailed Trogon FF	*Apaloderma vittatum*																	p
3270	Bucerotidae	Southern Ground-hornbill VU	*Bucorvus leadbeateri*									r	r							
3272	Bucerotidae	Pale-billed Hornbill	*Lophoceros pallidirostris*																	pr
3273	Bucerotidae	African Grey Hornbill	*Lophoceros nasutus*				r	r			r									
3278	Bucerotidae	Crowned Hornbill	*Lophoceros alboterminatus*		r				r	r	pr	r	r							
3298	Bucerotidae	Trumpeter Hornbill F	*Bycanistes bucinator*					r		p	p	r								
3332	Upupidae	Common Hoopoe (African)	*Upupa epops africana*																	p
3335	Phoeniculidae	Green Woodhoopoe	*Phoeniculus purpureus*									r	r					r		p
3341	Phoeniculidae	Common Scimitarbill	*Rhinopomastus cyanomelas*									r	r						r	p
3349	Meropidae	White-throated Bee-eater*	*Merops albicollis*														r			
3352	Meropidae	Northern Carmine Bee-eater[AFR]	*Merops nubicus*																	p
3362	Meropidae	Blue-cheeked Bee-eater*	*Merops persicus*				r		r	p										
3363	Meropidae	European Bee-eater*	*Merops apiaster*		pr	p	pr		pr	p	pr		r							

Continued

Continued

No. HBW	Family	English name	Species	0 m W	0 m D	200 m W	200 m D	400 m W	400 m D	≥600 m W	≥600 m D	Forest_ RS W	Forest_ RS D	Nets_ near SGR W	Nets_ near SGR D	Nets_ away from SGR W	Nets_ away from SGR D	Net_ forest_ RS W	Net_ forest_ RS D	Obs. & Lit. only
3365	Meropidae	Böhm's Bee-eater	*Merops boehmi*																	p
3366	Meropidae	Swallow-tailed Bee-eater	*Merops hirundineus*																	r
3370	Meropidae	Little Bee-eater	*Merops pusillus*	p	p	pr	pr	p	r	p	pr	r	r	pr	r					
3379	Coraciidae	Lilac-breasted Roller	*Coracias caudatus*	r	r															
3381	Coraciidae	European Roller*	*Coracias garrulus*	p																r
3384	Coraciidae	Broad-billed Roller[AFR]	*Eurystomus glaucurus*								p		r							
3412	Alcedinidae	African Pygmy-kingfisher[AFR]	*Ispidina picta*			p				p				pr		p		r	r	
3415	Alcedinidae	Malachite Kingfisher	*Corythornis cristatus*			p								r	p					
3448	Alcedinidae	Half-collared Kingfisher F	*Alcedo semitorquata*												p					
3455	Alcedinidae	Pied Kingfisher	*Ceryle rudis*																	p
3471	Alcedinidae	Grey-headed Kingfisher[AFR]	*Halcyon leucocephala*	p	r	p											r			
3472	Alcedinidae	Brown-hooded Kingfisher	*Halcyon albiventris*	p	pr		pr		p	p		r	r	p	p	p			r	
3473	Alcedinidae	Striped Kingfisher	*Halcyon chelicuti*	p				r				r	r		r	r				
3475	Alcedinidae	Woodland Kingfisher[AFR]	*Halcyon senegalensis*																	pr
3476	Alcedinidae	Mangrove Kingfisher	*Halcyon senegaloides*							p				p		p				
3700	Lybiidae	Green Barbet FF	*Cryptolybia olivacea*																	p
3707	Lybiidae	White-eared Barbet F	*Stactolaema leucotis*																	p
3712	Lybiidae	Eastern Green Tinkerbird FF	*Pogoniulus simplex*					r		p	p	r								
3713	Lybiidae	Moustached Green Tinkerbird FF	*Pogoniulus leucomystax*																	p
3717	Lybiidae	Yellow-rumped Tinkerbird F	*Pogoniulus bilineatus*	p	p	p	p	p	p	p	p	r	r	p	p	p				
3719	Lybiidae	Red-fronted Tinkerbird	*Pogoniulus pusillus*	pr		r	p				r				p					
3735	Lybiidae	Black-collared Barbet	*Lybius torquatus*	r									r	r						
3736	Lybiidae	Brown-breasted Barbet	*Pogonornis melanopterus*						r		p	r								

Continued

Continued

No. HBW	Family	English name	Species	0 m W	0 m D	200 m W	200 m D	400 m W	400 m D	≥600 m W	≥600 m D	Forest_ RS W	Forest_ RS D	Nets_ near SGR W	Nets_ near SGR D	Nets_ away from SGR W	Nets_ away from SGR D	Net_ forest_ RS W	Net_ forest_ RS D	Obs. & Lit. only
3752	Indicatoridae	Pallid Honeyguide	*Indicator meliphilus*								p			p						
3754	Indicatoridae	Lesser Honeyguide	*Indicator minor*				p				p	r								
3756	Indicatoridae	Scaly-throated Honeyguide	*Indicator variegatus*																	pr
3759	Indicatoridae	Greater Honeyguide	*Indicator indicator*									r	r							
3840	Picidae	Reichenow's Woodpecker	*Campethera scriptoricauda*																	r
3842	Picidae	Golden-tailed Woodpecker F	*Campethera abingoni*								p		r				r			
3845	Picidae	Little Spotted Woodpecker	*Campethera cailliautii*									r								p
3959	Picidae	Cardinal Woodpecker	*Dendropicos fuscescens*					r		p	pr	r						r		
4068	Falconidae	Eurasian Hobby*	*Falco subbuteo*					p												
4115	Psittacidae	Brown Parrot	*Poicephalus meyeri*				r				r		r							
4117	Psittacidae	Brown-headed Parrot F	*Poicephalus cryptoxanthus*																	p
4540	Pittidae	African Pitta FF [AFR]	*Pitta angolensis*							p				p		p				
4575	Calyptomenidae	African Broadbill FF	*Smithornis capensis*										r							p
6287	Oriolidae	Green-headed Oriole F	*Oriolus chlorocephalus*																	p
6289	Oriolidae	Eastern Black-headed Oriole	*Oriolus larvatus*		r		r	r			r	r	r		r				r	p
6294	Oriolidae	African Golden Oriole[AFR]	*Oriolus auratus*							p		r								
6482	Campephagidae	Grey Cuckooshrike F	*Ceblepyris caesius*																	p
6483	Campephagidae	Black Cuckooshrike	*Campephaga flava*																	p
6593	Vangidae	Retz's Helmetshrike	*Prionops retzii*									r	r					r		p
6595	Vangidae	Chestnut-fronted Helmetshrike F	*Prionops scopifrons*																	pr
6603	Vangidae	Black-and-white Shrike-flycatcher	*Bias musicus*										r							p
6609	Platysteiridae	Forest Batis FF	*Batis mixta*																	p

Continued

Continued

No. HBW	Family	English name	Species	0 m		200 m		400 m		≥600 m		Forest_ RS		Nets_ near SGR		Nets_ away from SGR		Net_ forest_ RS		Obs. & Lit. only
				W	D	W	D	W	D	W	D	W	D	W	D	W	D	W	D	
6615	Platysteiridae	Chinspot Batis	*Batis molitor*									r								
6618	Platysteiridae	East Coast Batis F	*Batis soror*															r		
6620	Platysteiridae	Eastern Black-headed Batis	*Batis minor*	r			r		r		r			r	r	r	r			p
6639	Platysteiridae	Black-throated Wattle-eye F	*Platysteira peltata*		pr	p	r		p	p	pr	r	r	p	p		p			
6646	Malaconotidae	Grey-headed Bush-shrike	*Malaconotus blanchoti*					p		p	pr		r							
6654	Malaconotidae	Black-backed Puffback F	*Dryoscopus cubla*	p	p		pr	pr	pr	p	pr	r	r	pr			pr			
6657	Malaconotidae	Marsh Tchagra	*Bocagia minuta*		p															
6658	Malaconotidae	Brown-crowned Tchagra	*Tchagra australis*		p					p	p	r		r	r	r		r		
6661	Malaconotidae	Black-crowned Tchagra	*Tchagra senegalus*	pr	p	p	pr	pr	r	pr	r		r		r	r	r			
6665	Malaconotidae	Black-fronted Bush-shrike FF	*Chlorophoneus nigrifrons*																	pr
6668	Malaconotidae	Orange-breasted Bush-shrike	*Chlorophoneus sulfureopectus*	p	p	pr	pr	pr	pr	pr	pr		r	pr		r	pr			
6678	Malaconotidae	Tropical Boubou	*Laniarius aethiopicus*																	p
6680	Malaconotidae	Zanzibar Boubou	*Laniarius sublacteus*	pr	pr	pr	pr	pr	pr	pr	pr	r	r	p			p			
6692	Malaconotidae	Gorgeous Bush-shrike F	*Telophorus viridis*		r	r		r		r	p	r	r							
6759	Dicruridae	Square-tailed Drongo F	*Dicrurus ludwigii*						p	p	p	r	r							
6761	Dicruridae	Fork-tailed Drongo	*Dicrurus adsimilis*				r				r	r	r				r			p
6870	Monarchidae	Eastern Crested-flycatcher FF	*Trochocercus bivittatus*						p		p		r			p	p			
6889	Monarchidae	African Paradise-flycatcher	*Terpsiphone viridis*						p		p					p	p			
6901	Laniidae	Red-backed Shrike*	*Lanius collurio*	p						p			r	r	r		r			
6911	Laniidae	Lesser Grey Shrike *	*Lanius minor*	p																
6995	Corvidae	Pied Crow	*Corvus albus*		p						p									
7018	Corvidae	House Crow	*Corvus splendens*	p	p	p	p	p												
7306	Alaudidae	Rufous-naped Lark	*Mirafra africana*																	r

Continued

Continued

No. HBW	Family	English name	Species	0 m W	0 m D	200 m W	200 m D	400 m W	400 m D	≥ 600 m W	≥ 600 m D	Forest_ RS W	Forest_ RS D	Nets_ near SGR W	Nets_ near SGR D	Nets_ away from SGR W	Nets_ away from SGR D	Net_ forest_ RS W	Net_ forest_ RS D	Obs. & Lit. only
7311	Alaudidae	Flappet Lark	*Mirafra rufocinnamomea*	r		r	r	r	r					r		r	r			
7365	Nicatoridae	Eastern Nicator F	*Nicator gularis*		p	pr	p	pr	pr	r	p		r			p	p		r	
7369	Macrosphenidae	Red-faced Crombec	*Sylvietta whytii*				p		r		r			r						
7381	Macrosphenidae	Moustached Grass-warbler	*Melocichla mentalis*	p	p	p		pr	p	p	p			r						
7386	Macrosphenidae	Kretschmer's Longbill FF	*Macrosphenus kretschmeri*					p		p	p	r	r							
7416	Cisticolidae	Bar-throated Apalis FF	*Apalis thoracica*							p			r							
7418	Cisticolidae	Yellow-breasted Apalis	*Apalis flavida*				p	r	pr	pr	p	r	r				r			
7425	Cisticolidae	Black-headed Apalis FF	*Apalis melanocephala*																	pr
7442	Cisticolidae	Miombo Wren-warbler	*Calamonastes undosus*													r	r	r		
7443	Cisticolidae	Stierling's Wren-warbler	*Calamonastes stierlingi*																	p
7446	Cisticolidae	Bleating Camaroptera	*Camaroptera brachyura*	r		pr	p	pr	pr	p	pr	r	r	p	pr	pr	pr	r	r	
7458	Cisticolidae	Red-faced Cisticola	*Cisticola erythrops*	p	p	p	p	p	p	p	p			p			p			
7459	Cisticolidae	Singing Cisticola	*Cisticola cantans*																	r
7470	Cisticolidae	Rattling Cisticola	*Cisticola chiniana*	pr	pr	pr	pr	pr	pr	pr	pr	r	r	pr	p	r				
7478	Cisticolidae	Winding Cisticola	*Cisticola marginatus*		p	p	p	p	p	pr	p									
7489	Cisticolidae	Croaking Cisticola	*Cisticola natalensis*							r										p
7493	Cisticolidae	Short-winged Cisticola	*Cisticola brachypterus*	p		pr	r	pr		r				p	r		r			
7499	Cisticolidae	Zitting Cisticola	*Cisticola juncidis*	r		r		r		r										
7528	Cisticolidae	Tawny-flanked Prinia	*Prinia subflava*	p	p	p	p	pr	pr	r	pr	r	r	r	r					
7536	Cisticolidae	Red-winged Prinia	*Prinia erythroptera*				r		r								r			p
7561	Acrocephalidae	Olivaceous Warbler*	*Iduna pallida*											p		r				
7573	Acrocephalidae	Sedge Warbler*	*Acrocephalus schoenobaenus*		p					p										
7577	Acrocephalidae	Common Reed-warbler*	*Acrocephalus scirpaceus*											p						
7581	Acrocephalidae	Basra Reed-warbler* EN	*Acrocephalus griseldis*												r					
7701	Hirundinidae	Black Saw-wing	*Psalidoprocne pristoptera*								p									

Continued

Continued

No. HBW	Family	English name	Species	0 m W	0 m D	200 m W	200 m D	400 m W	400 m D	≥600 m W	≥600 m D	Forest_ RS W	Forest_ RS D	Nets_ near SGR W	Nets_ near SGR D	Nets_ away from SGR W	Nets_ away from SGR D	Net_ forest_ RS W	Net_ forest_ RS D	Obs. & Lit. only
7718	Hirundinidae	Lesser Striped Swallow	*Cecropis abyssinica*	pr	p	p	p	p	p	pr	p									
7720	Hirundinidae	Mosque Swallow	*Cecropis senegalensis*								p									
7722	Hirundinidae	Red-rumped Swallow*	*Cecropis daurica*	p	p					p	p									
7730	Hirundinidae	Wire-tailed Swallow	*Hirundo smithii*																	r
7732	Hirundinidae	Barn Swallow*	*Hirundo rustica*	r	pr	pr	pr	pr	r	pr	pr		r							
7747	Hirundinidae	African Plain Martin	*Riparia paludicola*	p																
7785	Pycnonotidae	Sombre Greenbul	*Andropadus importunus*	pr	pr	pr	pr	pr	pr	pr	pr		r	pr	pr	pr	pr			
7800	Pycnonotidae	Yellow-bellied Greenbul F	*Chlorocichla flaviventris*			p	r	p	pr	pr	pr	r	r		p	p	p	r	r	
7803	Pycnonotidae	Olive-headed Bulbul FF ALT	*Arizelocichla striifacies*																	p
7813	Pycnonotidae	Little Greenbul F	*Eurillas virens*				p	p	p	p	p				p					
7822	Pycnonotidae	Lowland Tiny Greenbul FF	*Phyllastrephus debilis*							p			r							
7828	Pycnonotidae	Terrestrial Brownbul F	*Phyllastrephus terrestris*	p		pr	r	pr	pr	p	p		r	r		pr		r	r	
7829	Pycnonotidae	Fischer's Greenbul FF	*Phyllastrephus fischeri*				p		p	p	p	r	r				p	r		
7830	Pycnonotidae	Northern Brownbul	*Phyllastrephus strepitans*						r						pr		p		r	
7834	Pycnonotidae	Yellow-streaked Greenbul FF	*Phyllastrephus flavostriatus*										r					r	r	p
7915	Pycnonotidae	Common Bulbul	*Pycnonotus barbatus*	pr	pr	pr	pr	pr	pr	pr	pr	r	r	pr	pr	r	pr	r		
7957	Phylloscopidae	Willow Warbler *	*Phylloscopus trochilus*				r							r		r				
7986	Phylloscopidae	Yellow-throated Woodland-warbler F	*Phylloscopus ruficapilla*																	p
8021	Scotocercidae	Little Yellow Flycatcher FF	*Erythrocercus holochlorus*				p		p			r	r					r		
8073	Sylviidae	Garden Warbler*	*Sylvia borin*					p						r						
8213	Zosteropidae	African Yellow White-eye F	*Zosterops senegalensis*																	p
8356	Pellorneidae	Pale-breasted Illadopsis FF	*Illadopsis rufipennis*															r		p
8441	Leiothrichidae	Arrow-marked Babbler	*Turdoides jardineii*	p		p	p	p		p	p									
8752	Sturnidae	Kenrick's Starling FF	*Poeoptera kenricki*																	p

Continued

Continued

No. HBW	Family	English name	Species	0 m		200 m		400 m		≥600 m		Forest_ RS		Nets_ near SGR		Nets_ away from SGR		Net_ forest_ RS		Obs. & Lit. only
				W	D	W	D	W	D	W	D	W	D	W	D	W	D	W	D	
8773	Sturnidae	Lesser Blue-eared Starling	*Lamprotornis chloropterus*										r						r	
8780	Sturnidae	Violet-backed Starling[AFR]	*Cinnyricinclus leucogaster*			r		r					r	r						p
8783	Sturnidae	Black-bellied Starling F	*Notopholia corusca*	r	p							r	r							
8871	Turdidae	Red-tailed Ant-thrush FF	*Neocossyphus rufus*							p	p	r	r							
8938	Turdidae	Spotted Ground-thrush FF VU [AFR]	*Geokichla guttata*							p										
8951	Turdidae	Orange Ground-thrush FF [ALT]	*Geokichla gurneyi*																	p
8986	Turdidae	Kurrichane Thrush	*Turdus libonyana*									r	r							
9053	Muscicapidae	Bearded Scrub-robin	*Tychaedon quadrivirgata*	r			p		p	p	p	r	r		p	p	p	r	r	
9059	Muscicapidae	White-browed Scrub-robin	*Cercotrichas leucophrys*		pr		pr		r	r	pr		r	p	pr	r	r			
9081	Muscicapidae	African Dusky Flycatcher F	*Muscicapa adusta*																	pr
9084	Muscicapidae	Spotted Flycatcher*	*Muscicapa striata*						pr				r							
9093	Muscicapidae	Pale Flycatcher	*Agricola pallidus*				r										r			
9097	Muscicapidae	Grey Tit-flycatcher	*Fraseria plumbea*				p					r								
9099	Muscicapidae	Ashy Flycatcher F	*Fraseria caerulescens*						p	p				p						
9175	Muscicapidae	White-starred Robin F [ALT]	*Pogonocichla stellata*																	pr
9181	Muscicapidae	White-chested Alethe FF	*Chamaetylas fuelleborni*																	p
9185	Muscicapidae	White-browed Robin-chat	*Cossypha heuglini*	p		r	r			pr				pr	r	r				
9186	Muscicapidae	Red-capped Robin-chat F [AFR]	*Cossypha natalensis*				p	p	p	p	p		r		p	p	p	r	r	
9194	Muscicapidae	Collared Palm-thrush	*Cichladusa arguata*																	p
9196	Muscicapidae	Spotted Palm-thrush	*Cichladusa guttata*	r			r			r				r	r	r				p
9203	Muscicapidae	East Coast Akalat FF NT	*Sheppardia gunningi*		p		p		p		p		r					r	r	
9237	Muscicapidae	Common Nightingale *	*Luscinia megarhynchos*																	p
9513	Nectariniidae	Uluguru Violet-backed Sunbird F	*Anthreptes neglectus*																	pr

Continued

Continued

No. HBW	Family	English name	Species	0 m		200 m		400 m		≥600 m		Forest_ RS		Nets_ near SGR		Nets_ away from SGR		Net_ forest_ RS		Obs. & Lit. only
				W	D	W	D	W	D	W	D	W	D	W	D	W	D	W	D	
9519	Nectariniidae	Collared Sunbird F	*Hedydipna collaris*		p	p	pr	p	pr	pr	pr	r	r	p	p		p			
9533	Nectariniidae	Olive Sunbird FF	*Cyanomitra olivacea*		p	p	p	p	p	p	p	r	r	p	p	p	p	r	r	
9534	Nectariniidae	Mouse-coloured Sunbird	*Cyanomitra verreauxii*								p			p	p					r
9538	Nectariniidae	Amethyst Sunbird	*Chalcomitra amethystina*									r								p
9539	Nectariniidae	Scarlet-chested Sunbird	*Chalcomitra senegalensis*		pr	r	p	p							r					
9576	Nectariniidae	Mariqua Sunbird	Cinnyris mariquensis	p				p		p										
9581	Nectariniidae	Purple-banded Sunbird	*Cinnyris bifasciatus*		pr	r	pr		pr				r		r				r	
9595	Nectariniidae	Variable Sunbird	*Cinnyris venustus*		p															
9662	Ploceidae	Thick-billed Weaver	*Amblyospiza albifrons*	p	p	p	p		p	p	p									
9665	Ploceidae	Red-billed Quelea	*Quelea quelea*	p		r	p			p		r		r		r				
9670	Ploceidae	Black-winged Bishop	*Euplectes hordeaceus*	p		p	p		p	p	p									
9673	Ploceidae	Zanzibar Red Bishop	*Euplectes nigroventris*	pr		pr		p		pr		r		pr		r	r			
9674	Ploceidae	Red-collared Widowbird	*Euplectes ardens*	pr		pr	p	pr		pr										
9676	Ploceidae	Yellow Bishop	*Euplectes capensis*	pr	p	pr	p	pr	pr	pr	pr	r		pr		p	r			
9678	Ploceidae	Fan-tailed Widowbird	*Euplectes axillaris*			p		p												
9702	Ploceidae	Spectacled Weaver	*Ploceus ocularis*	pr	p	pr	p	p	pr	p	pr		r	pr	r	p	r			
9709	Ploceidae	African Golden Weaver	*Ploceus subaureus*	pr	p	pr		p		p										
9721	Ploceidae	Lesser Masked Weaver	*Ploceus intermedius*											r						
9730	Ploceidae	Village Weaver	*Ploceus cucullatus*	p			r			pr	p				r		r			
9738	Ploceidae	Golden-backed Weaver	*Ploceus jacksoni*											r						
9752	Ploceidae	Dark-backed Weaver F	*Ploceus bicolor*		p	p	pr	p		p	p	r	r							
9772	Ploceidae	Southern Red-headed Weaver	*Anaplectes rubriceps*									r								
9777	Estrildidae	Red-billed Firefinch	*Lagonosticta senegala*	pr	r			p	p	p				p		pr				
9780	Estrildidae	Jameson's Firefinch	*Lagonosticta rhodopareia*							p										

Continued

Continued

No. HBW	Family	English name	Species	0 m W	0 m D	200 m W	200 m D	400 m W	400 m D	≥600 m W	≥600 m D	Forest_ RS W	Forest_ RS D	Nets_ near SGR W	Nets_ near SGR D	Nets_ away from SGR W	Nets_ away from SGR D	Net_ forest_ RS W	Net_ forest_ RS D	Obs. & Lit. only
9782	Estrildidae	African Firefinch	*Lagonosticta rubricata*	p	p									p		p				
9791	Estrildidae	Green-winged Pytilia	*Pytilia melba*		r	p	pr	pr	pr	p				pr	r	pr	pr			
9793	Estrildidae	Red-throated Twinspot F	*Hypargos niveoguttatus*					p		p	p	r	r	p		p	p	r		
9798	Estrildidae	Blue-breasted Cordon-bleu	*Uraeginthus angolensis*			p	r			r	r			r	r	r				
9799	Estrildidae	Red-cheeked Cordon-bleu	*Uraeginthus bengalus*	p																
9806	Estrildidae	Lesser Seedcracker	*Pyrenestes minor*																	pr
9813	Estrildidae	Crimson-rumped Waxbill	*Estrilda rhodopyga*			p														
9816	Estrildidae	Common Waxbill	*Estrilda astrild*	p		p	p	p	p	p	p	r								
9823	Estrildidae	Green Twinspot FF	*Mandingoa nitidula*								p							r		
9830	Estrildidae	Yellow-bellied Waxbill	*Coccopygia quartinia*																	p
9842	Estrildidae	Cut-throat Finch	*Amadina fasciata*	r																
9846	Estrildidae	Zebra Waxbill	*Amandava subflava*			p				p										
9848	Estrildidae	Bronze Mannikin	*Spermestes cucullata*	pr	pr	p	pr	pr		p		r	r	pr						
9849	Estrildidae	Black-and-white Mannikin	*Spermestes bicolor*					p		p								r		
9915	Viduidae	Pin-tailed Whydah	*Vidua macroura*	pr		pr		pr		r	r						r			
9916	Viduidae	Long-tailed Paradise-whydah	*Vidua paradisaea*	p		r		pr		pr			r							
9932	Viduidae	Village Indigobird	*Vidua chalybeata*	pr		p				p				pr						
9936	Passeridae	House Sparrow	*Passer domesticus*	p	p		p				p									
9955	Passeridae	Northern Grey-headed Sparrow	*Passer griseus*																	p
9956	Passeridae	Southern Grey-headed Sparrow	*Passer diffusus*																	p
9965	Passeridae	Yellow-throated Bush-sparrow	*Gymnoris superciliaris*									r	r							
9979	Motacillidae	Sokoke Pipit FF EN	*Anthus sokokensis*																	pr
10,014	Motacillidae	African Pipit	*Anthus cinnamomeus*	pr	p		p			r										

Continued

Continued

No. HBW	Family	English name	Species	0 m		200 m		400 m		≥600 m		Forest_ RS		Nets_ near SGR		Nets_ away from SGR		Net_ forest_ RS		Obs. & Lit. only
				W	D	W	D	W	D	W	D	W	D	W	D	W	D	W	D	
10,024	Motacillidae	Yellow-throated Longclaw	*Macronyx croceus*	r	r	r		p			r									
10,035	Motacillidae	African Pied Wagtail	*Motacilla aguimp*																	p
10,200	Fringillidae	Reichenow's Seedeater	*Crithagra reichenowi*	p																
10,206	Fringillidae	Yellow-fronted Canary	*Crithagra mozambica*		p				p											
10,300	Emberizidae	Golden-breasted Bunting	*Emberiza flaviventris*																	p
10,308	Emberizidae	Cinnamon-breasted Bunting	*Emberiza tahapisi*		p				r			r		r		r	r	r	r	

The presence of some species in Pugu Hills still needs confirmation; for example, records of the Western Violet-backed Sunbird *Anthreptes longuemarei* (in Pugu and Ruvu South) may be a confusion with the Uluguru Violet-backed sunbird, whereas the records of Abyssinian Scimitarbil *Rhinopomastus minor* in Pugu may have been referring to the Common Scimitarbil. The Grey Parrot *Psittacus erithacus* is recently being recorded in the forest near Kisarawe Township, and is considered as an escapee of illegal trade or release from illegal husbandry. Black Crake Zaporni flavirostra and a pair of the Near-Threatened Plain-backed Sunbird Anthreptes reichenowi were recently (early 2024) recorded at Pugu Hills (Edigar Apolinary, pers. comm. 7 May 2024).

Key

- No. HBW refers to the taxonomic ordering (global identification number) as provided in the *Handbook of the Birds of the World* (HBW) and BirdLife International, 2023.
- Presence is noted by p or r, or both, where *p* = Pugu Hills, r = Ruvu South. W, wet season; D, dry season
- Distance 0–600 m refers to distance of the sampling in relation to SGR line.
- Nets near SGR = nets placed between 50 m and 100 m from SGR; Nets_away from SGR = nets placed 300–400 m from SGR.
- Forest_RS is a forest site within Ruvu South, south-west of the reserve.
- Obs. & lit. only refers to records that are known from literature and other records away from the systematic fieldwork.
- Forest dependency: FF, forest specialist species; F, forest generalist species
- ALT, altitudinal migrant
- AFR, Afrotropical/intra-African migrant; these include partial migrants and those whose movements in the area are not well understood.
- *, Palearctic migrant (long-distance movements to northern Africa, Europe and Asia).
- Threat categories: CR, Critically Endangered; EN, Endangered; VU, Vulnerable; NT, Near Threatened

9 Ecological Status of Galagos in Pugu Hills and Ruvu South Forest Reserves

Mohamed Julius Kibaja*, Fatina Athumani Mturi and Flora John Magige
College of Natural Resources and Applied Sciences, University of Dar es Salaam, Dar es Salaam, Tanzania

Abstract

This study aims at establishing the distribution, abundance, and activity budget of the galagos (bush babies) along the standard-gauge railway (SGR) route at Pugu Nature and Ruvu South forest reserves in Tanzania. Pugu Nature and Ruvu South forests are among the patches of coastal forests that harbour three sympatric galagos: the Rondo galago (*Galagoides rondoensis*), the Zanzibar galago (*Galagoides zanzibaricus*) and the Greater galago (*Otolemur garnettii*). The study revealed that galagos were mainly encountered in evergreen forest/thickets in both forest reserves, implying that conservation of intact evergreen thickets is crucial for securing all the species. There were also variations in density of the three galago species, the density of Rondo galago being the highest in both forest reserves. The activity budgets of all galagos were dominated by alarm and advertisement calls, indicating that these calls may have ecological and behavioural adaptive functions including those related to anti-predation and reproduction.

9.1 Introduction

Tanzania is home to diverse primate species and subspecies occurring in various habitats (De Jong *et al.*, 2011). Some of the species feature in the world's known biodiversity hotspot areas (Burgess *et al.*, 2003) and priority areas for primate conservation such as the eastern arc mountains and coastal forests of East Africa (Davenport *et al.*, 2014). In Tanzania, examples of endemic primate species that occur in the eastern arc mountains and coastal forests include Iringa red colobus (*Piliocolobus gordonorum*), Zanzibar red colobus (*Piliocolobus kirkii*), Zanzibar galago (*Galagoides zanzibatricus*) and Rondo galago (*Galagoides rondoensis*) (Burgess *et al.*, 2003). Pugu Nature Reserve and Ruvu South Forest Reserve are examples of forest patches of coastal forests that are home to sympatric galagos: *G. rondoensis*, *G. zanzibaricus* and the Greater galago (*Otolemur garnettii*) (Honess et al., 2008).

Galagos, or bush babies, are prosimians belonging to the order Primates, Family Galagonidae and Genus *Galagoides* (Masters *et al.*, 2017). They are nocturnal and communicate with conspecifics with vocalizations, vision and olfactory senses. The sympatric species of galagos inhabiting Pugu and Ruvu South forest Reserves can be distinctively identified through their vocalizations and body structures and size (Honess, 1996; Perkin, 2007; Honess *et al.*, 2008; Perkin, 2020). For example, the *G. rondoensis* can be distinguished from other dwarf galagos by its bottle-brush-shaped tail, its reproductive anatomy, and its distinctive

*Corresponding author: mkibaja@gmail.com

DOI: 10.1079/9781800626942.0009

'double unit rolling call' (Honess, 1996). While *G. rondoensis* is classified as Endangered by IUCN (Perkin, 2020), due to its extremely limited and fragmented range in a few remnant patches of coastal forests (Perkin, 2020), *G. zanzibaricus* and *O. garnettii* are not (Butynski *et al.*, 2008; Perkin, 2020). The *G. rondoensis* is sympatric with *G. zanzibaricus* in the northern part of its range and *G. granti* in the southern part of its range and *O. garnettii* throughout its range (Gwegime, 2016).

Primates play diverse roles in the ecosystem including seed dispersal (Bourlière, 1985). In East Africa, some species of galago, for instance *O. garnettii* and *G. zanzibaricus*, include fruits in their diet (Harcourt and Nash, 1986); thus, they can aid in seed dispersal of certain coastal forest plant species. Ecological services provided by primates depend, among other influences, on distribution, abundance and interactions with hetero-species in their habitats (Gutierrez-Arellano and Mulligan, 2018). For instance, it has been reported that defaunation of mammalian species across the neo-tropical region has been closely linked with dramatic losses of critical ecosystem services (Bogoni *et al.*, 2020). Abundance and distribution patterns of particular species can partly serve to indicate suitability of habitat for that species and the capacity of a species to provide ecosystem services.

Primate species richness, biomass and density can vary across seasons due to seasonal fluctuations of food resources or other parameters (Hanya *et al.*, 2011; Hending *et al.*, 2023). On the other hand, some primate species can exist in high abundance in particular vegetation types; for instance, arboreal Ashy red colobus (*P. tephrosceles*) has higher encounter rates in evergreen gallery forest than open woodland vegetation types in the Greater Mahale Ecosystem (Moyer *et al.*, 2006). Also, the *G. rondoensis* have been reported to occur most frequently in mid-canopies of evergreen thickets in comparison to other vegetation types (Gwegime, 2016). However, further systematic studies are required to examine the association of galago species with different vegetation types. It is therefore important to assess the galagos' abundances in different seasons and across vegetation types to predict how well primate species can adapt to changes in environmental conditions in their habitats in the face of ongoing climate change and anthropogenic activities. It is also critical to conduct general baseline surveys on distribution and population status for galago species as these can provide information to identify priority habitats for conservation and provide baseline population sizes that can be monitored in the future. The persistence of primate species in their habitats also depends on their behavioural flexibility to respond to variability of environmental conditions (McLennan *et al.*, 2017; Kalbitzer and Chapman, 2018). For instance, some primate species can adjust their activities by allocating more time moving in the dry season than wet seasons while others can respond otherwise (Kibaja *et al.*, 2023). It is critical to study primates' overall activity patterns and budgets to get some insights into their survival strategies across heterogeneous landscapes.

Escalation of anthropogenic activities such as deforestation, agriculture, logging, urbanization and infrastructure development of roads and railway lines also provide strong rationale for conducting primate surveys. Different forms of habitat alterations threaten most primates with extinction (Estrada *et al.*, 2017; Estrada and Garber, 2022). Most of those activities have been reported to affect *G. rondoensis* and its habitats across various parts of coastal forests including the Pugu Nature Forest Reserve (Gwegime, 2016). Even though primate species respond variously to different forms of habitat alterations, some species, for example, respond by declining after habitat modification (Estrada and Garber, 2022) while others become compressed in small forest fragments (Decker, 1994; Siex, 2003). Population compression results in large density for some primate species in such small fragments and this could have a profound effect on their behaviour patterns as well (Siex, 2003). Many studies have been carried out, describing taxonomy, morphology, distribution and ecology for nocturnal primates along coastal forests of East Africa (Perkin, 2007).

Very few studies, however, have been conducted to systematically estimate galago density in the coastal forests of Tanzania (Kiwia, 2006; Murphy, 2015; Gwegime, 2016). Little is known of their behavioural patterns and time budget. Furthermore, the Tanzanian government, through the Tanzania Railways Corporation (TRC), has embarked on a project to upgrade the central line to an electrified standard-gauge railway (SGR). Since this is an important project traversing through the Pugu and Ruvu Forest reserves, it is imperative to carry out a baseline study along its footprint at these sites to investigate the ecological status of galagos. Knowledge of ecological status of primates is crucial for planning effective conservation guides for future monitoring, and planning appropriate mitigation measures for galagos and their habitats. This study specifically determines the galagos' population size and density, their spatial distribution, and their behavioural patterns.

9.2 Materials and Methods

9.2.1 Study area and sites

Galago surveys were carried out in two study areas: Pugu Nature Forest Reserve and Ruvu South Forest Reserve, which consist of parts of remnant patches of coastal forest in Tanzania (see Chapter 1). The forest composition of most of these coastal forest patches varies considerably from one to another, largely due to the past removal of valued timber, but also reflecting differing soil and groundwater conditions. Pugu Nature Forest Reserve covers 2180 ha but less than 400 ha remain in reasonable condition. It consists of dry forest and thickets found mainly on ridge tops, and open grassland with poor tree cover (Clarke and Dickinson, 1995). It also has plantation forest of *Senna siamea*, *Eucalyptus* sp., *Grevillea robusta* and a few teak (*Tectona grandis*), which has been re-colonized by indigenous species (Gwegime *et al.*, 2014). The emergent forest tree species of 35m tree heights include *Antiaris toxicaria* and *Dialium holtzii* in the Msimbazi valley. Ruvu South Forest Reserve covers an area of 35,000 ha though only 10,000 ha in the western part of the reserve can be characterized as Zanzibar-Inhambane coastal forest/thicket, but extensive areas were illegally harvested for charcoal. Ruvu South Forest Reserve (FR) is also characterized by coastal *Brachystegia* woodland, stands of riparian forest, small wetlands and grassland.

Study sites were selected based on potentially suitable vegetation types for the Rondo galago. Transects were established at three sites: Kaolin Camp (Transects 1–5), Kinyanyiko Hill (T6 and T8) and Minaki Dam (T7) in the Pugu Nature Forest Reserve and the Mpiji site (T1–T4) in Ruvu South Forest Reserve Figs. 9.1 and 9.2. In Pugu Hills forest there were eight transects: five transects at the Kaolin Camp at the 24 km SGR point (≈ 50m distance from SGR); 2 transects in the Kinyanyiko site and only one transect in the Minaki Dam site (Fig. 9.1). In Ruvu South forest (Mpiji site) there were four transects positioned near the 38 km SGR point (≈ 80 m distance from SGR; Fig. 9.2).

9.2.2 Data collection

Spatial distribution, density and population size

Night surveys were carried out along pre-existing paths (herein referred to as line transects) in the Pugu Nature Forest Reserve while in the Ruvu South Forest Reserve pre-existing paths were not used; instead, trails were established by cutting and marking them using red-coloured boundary/flagging tape. Cutting was kept to the minimum required to make the transect line visible and passable at night. All transects were tracked using GPS to present them in a digital map of both forest reserves. The surveys were done during the first half of the night (18:30–23:00 (herein referred as night walks)) and the last third of the night (03:30–06:30) (herein referred as morning walks)). At least four hours per night of night walks and three hours per night of morning walks were conducted. A total of 28 days were spent for surveying the Pugu Nature Forest Reserve and nine days in Ruvu South Forest Reserve. Since at least five hours per night of night walks should be conducted (De Jong and Butynski, 2009), this study has fulfilled the requirement.

At the start of each systematic night walk, the location, time, date, weather conditions, moon phase and a description of the dominant vegetation type on the tape were recorded (De Jong and Butynski, 2009). In this study, vegetation types were evergreen forest, mixed evergreen forests/thickets, thickets, shrubland and degraded forests/woodlands. Walks were conducted slowly at ≤0.5 km/hr, pausing to observe any galagos and record vocalizations

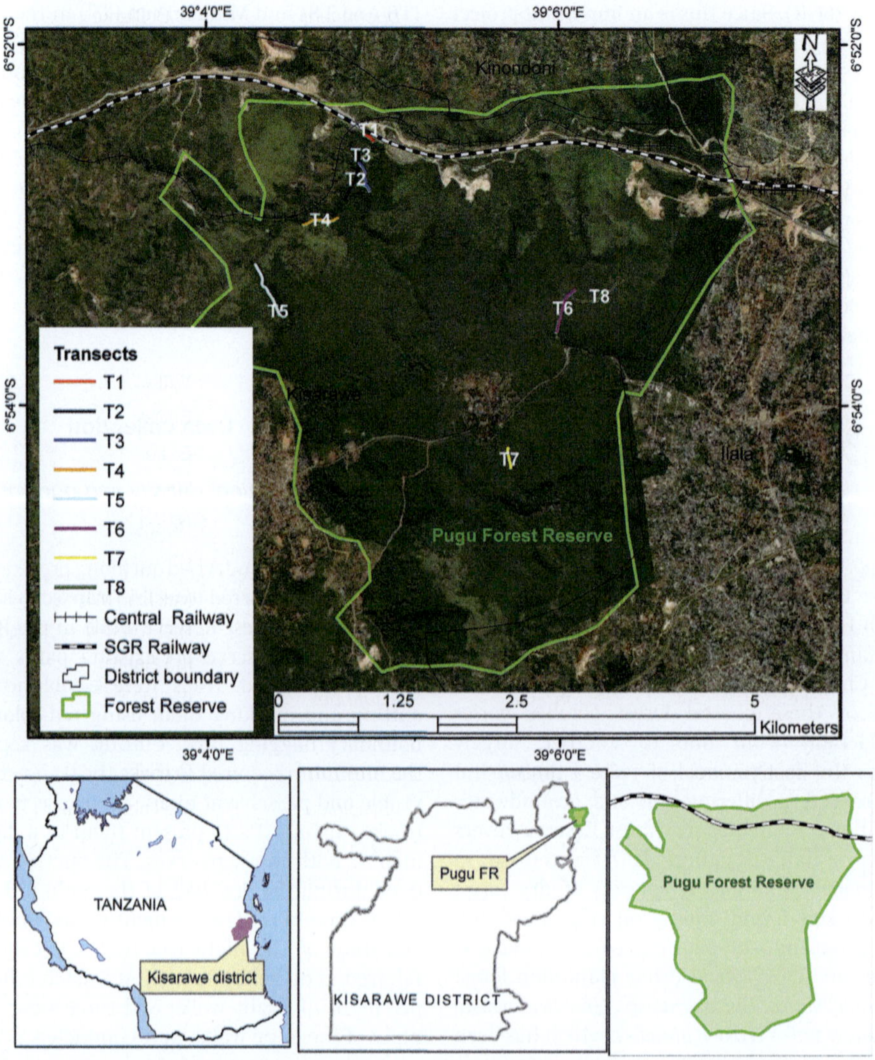

Fig. 9.1. Map showing transects in Pugu Nature Forest Reserve.

heard. During the night census walks, galago vocalizations were sound-recorded and these were used for species identification. Vocalizations were recorded using a cheap modern tape recorder (Sony pcm-A10, Zoom H1n). Start and finish times were noted as well as time taken to record and/or observe animals (De Jong and Butynski, 2009). Transect walks were digitally tracked using a GPS to determine their length, and thus distance walked. However, during night walks way points were recorded for each galago encounter (sightings and vocalizations) or upon encounter of any interesting event such as signs of human activities along the transects. It should be noted that with the constraints and difficulties of detecting galagos at night, estimating population densities can be subject to high degrees of error. Applying 'distance' methods has proved to be difficult for nocturnal work mainly due to over-counting where galagos can 'overtake' a recorder, which results in them being double-counted (Ambrose and Perkin, 2000). The double-counting problem was eliminated or highly reduced in this study by

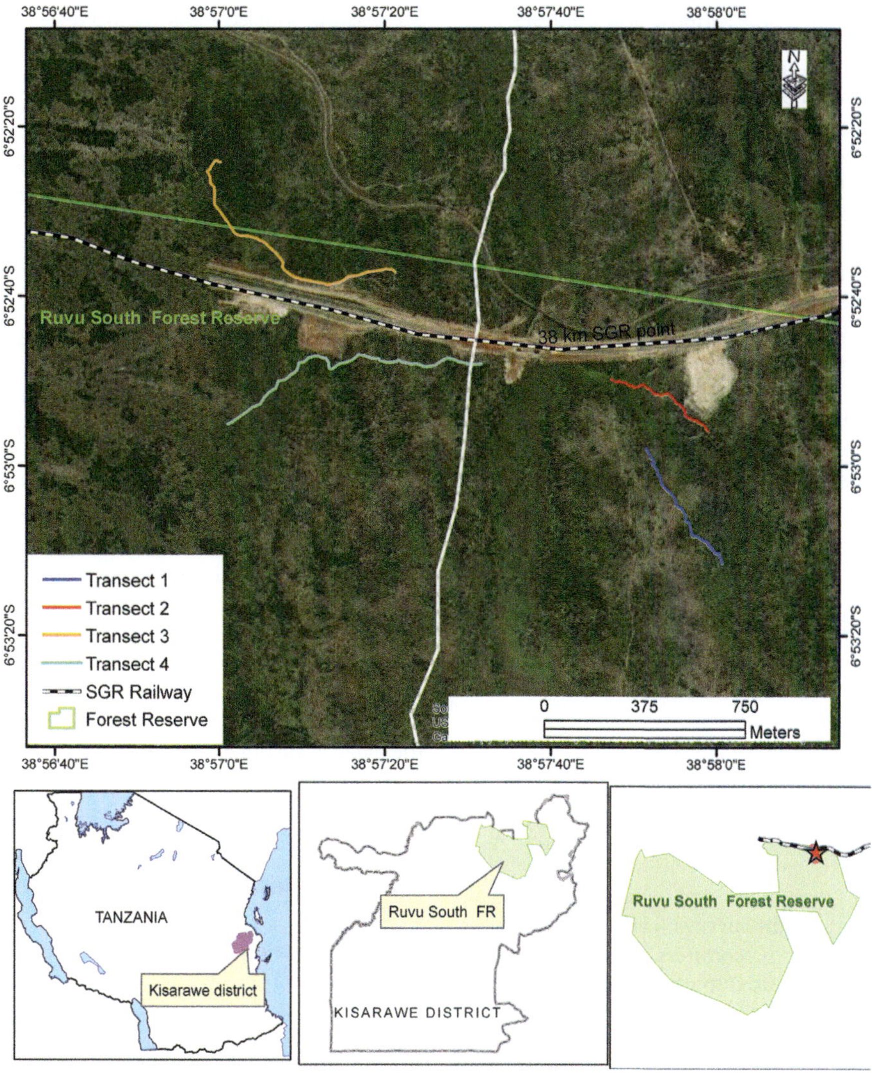

Fig. 9.2. Map showing transects in Ruvu South Forest Reserve.

adopting sweep night census walks in each site, whereby all transects at each site were walked at the same time by two or three people including a galago expert.

Recording galago vocalizations, particularly the 'species-specific advertising call', provides a reliable way of recognizing galago species in the field and is an important field survey tool, especially where it is difficult to trap individuals (Zimmermann, 1990; Bearder *et al.*, 2013). A reference database of galago vocalizations is held at Oxford Brookes University, UK. However, in the meantime, the galago calls have been identified in the field by the Rondo galago scientist (JG). Future extensive surveys, however, should make the necessary engagement with the reference database of galago vocalizations in UK. To obtain enough data for mapping the focal *G. rondoensis* species, live trapping using Sherman traps and opportunistic recordings of *G. rondoensis* were also made. When Rondo galago was trapped, it was transferred into a well-ventilated cage before taking measurements. It was later released back to the site where it was trapped.

Behavioural patterns and budgets

Behaviour sampling method and instantaneous recording rule (Bateson and Martin, 2007) were used to collect behavioural data. Follows were not possible for these cryptic, elusive and nocturnal species, thus the choice of this method enabled us to collect behaviours of animals when they were first sighted during transect walks (Croze, 1974; Kabigumila, 2001). Behaviours or activities recorded were calling (different types of calls), breeding, foraging/feeding, resting and moving. Breeding included behaviours such as mating (copulation); pregnant female; lactation; female with young; scrotal male; courtship – adult male and female in physical contact or close together in time and space. Moving included behaviours such as walking and leaping. Vocalizations were of four categories: alarm call, advertisement call, buzzing/puzzle and whistle call, and 'other calls' (uncategorized/unidentified calls). Foraging involved several behaviours, such as picking, holding, and ingesting and chewing food.

9.2.3 Data analysis

Spatial distribution of galagos and their population density estimates

Galago sightings and vocalizations from systematic night surveys (census night walks), opportunistic encounters and live galago trapping were used to obtain data on presence or absence of the Rondo galago. The GPS location for the presence of Rondo galago from the presence/absence survey, traps and night galago walks surveys (including sightings and vocalizations) were plotted in Google Earth maps in order to produce a distribution map for the Rondo galago. Google Earth maps were very informative as they clearly show vegetation types where galago species were recorded. The number of encounters and their percentages of all galago species (*G. rondoensis*, *G. zanzibaricus* and *O. garnettii*) were computed in each vegetation type and these were used as indicators of their habitat preferences. In this survey, population density was first computed in each transect by dividing the total number of encounters (including sightings and vocalizations) of galagos divided by the product of the total distances walked from pre-existing transects and the fixed strip width of 50 m (0.05 km) (25 m (0.025 km) either side of transect). The mean population density was obtained by averaging the total density between dry and wet seasons. Two sample tests were used to determine if there was a significant difference in the mean density of galagos between seasons.

Behaviour patterns and budgets

Behaviour (activity) patterns for all galago species were examined by calculating percentage of encounters of different behaviours (activities) in different hours of the night galago walk. Behaviour budget or activity budget was examined by calculating the percentage records or encounters of different behaviours/activities from the entire study period. This was done by dividing the number of behaviour records or encounters for each (behaviour) activity category with the total number of all activity records (Mekonnen *et al.*, 2017). A subdividing Chi-Square test was used to determine if there was a significant difference in the observed proportions of different behaviours exhibited by the galagos for the entire study period (Md-Zain and Ch'ng, 2011).

9.3 Results and Discussion

9.3.1 Spatial distribution of galagos

Rondo galagos were recorded to be present at seven transects except transect 4 in the Pugu Nature Forest Reserve (Fig. 9.3) while in the Ruvu South Forest Reserve, they were recorded at transects 2 and 3. Zanzibar galagos and Greater galagos were found in all transects except transect 1 in Pugu Nature Forest Reserve (Fig. 9.4) while in Ruvu South Forest Reserve they were only found in transect 2 (Fig. 9.5). The majority of sightings of all species, especially the *G. rondoensis* (Fig. 9.6) were recorded in evergreen forest and thicket vegetation types in both forest reserves, especially in those transects which traversed mainly evergreen vegetation types (Table 9.1). Our findings conform to those of previous reports that have documented the evergreen forest and thickets as preferable habitats for the *G. rondoensis* (Burgess and Clarke,

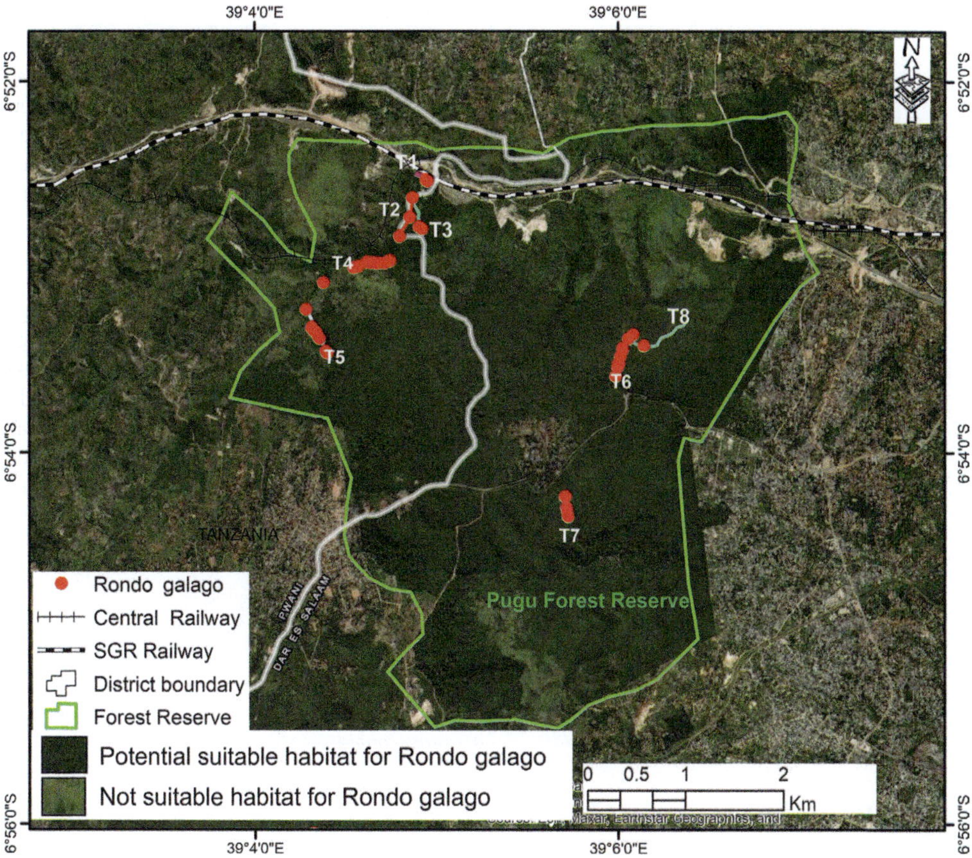

Fig. 9.3. Map showing the presence of Rondo galago in the Pugu Nature Forest Reserve.

2000; Gwegime, 2016; Perkin, 2020). It is important to note that some of these evergreen thickets are found along the SGR route and this provides a daunting future for galagos along the railway line. Studies have shown that lesser-bodied galagos prefer dense tangles of vines in thickets compared to large-bodied ones such as *O. garnettii*, which can inhabit open vegetation (Harcourt and Nash, 1986). In this study, we also observed that even the large-bodied galagos like *O. garnettii* were largely encountered in evergreen thickets (Table 9.1). Our observation does not provide the impression that *O. garnettii* also exhibits strong association with closed thickets but rather that it could be possible that during our night surveys the *O. garnettii* did not give out loud calls in patches of open vegetation to avoid detection by predators; hence it could be an under-estimation. On the other hand, the *G. rondoensis* is the smallest of all galago species (Perkin, 2019); therefore it is expected to have the shortest leaping distance. As such, the Rondo galago would avoid open habitats with large gaps which need long leaps to remain above ground. This observation indicates that the small-bodied galago is highly susceptible to habitat degradation, especially resulting from any form of forest clearing.

9.3.2 Population density of galagos

The overall population density for *G. rondoensis* in Pugu Nature Forest Reserve was 108.9 individuals per km^2 and this was large in comparison to densities of *G. zanzibaricus* with 34.1 animals per km^2 and *O. garnettii* with 39.2 animals per km^2.

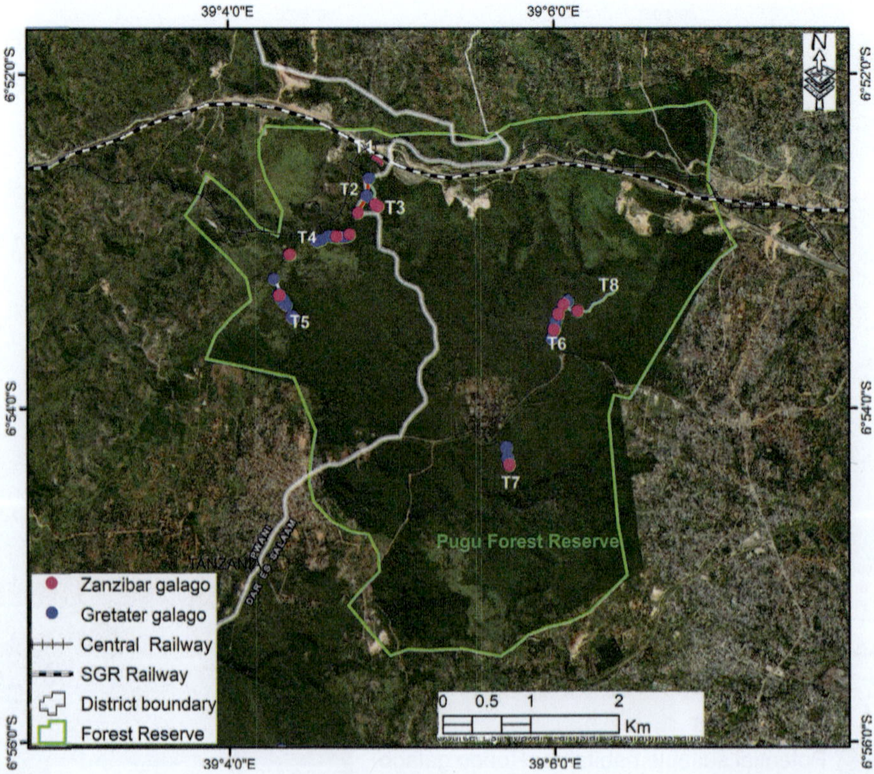

Fig. 9.4. Map showing the presence of Zanzibar and Greater galago in the Pugu Nature Forest Reserve.

Similarly, the overall density for *G. rondoensis* in Ruvu South Forest Reserve (close to the SGR line) was 25.4 animals per km^2, which was larger than that of *G. zanzibaricus* and *O. garnettii* in that forest (Table 9.2; Appendix 9.1–9.3). While the density of *O. garnettii* was significantly higher in the wet than the dry season in Pugu Nature Forest Reserve ($t = 2.692$, $df = 10$, $p = 0.023$), the densities for the other galagos were not statistically different between seasons (*G. rondoensis*: $t = 0.115$, $df = 10$, $p = 0.911$; *G. zanzibaricus*: $t = 1.800$, $df = 10$, $p = 0.101$; Table 9.2). In the Ruvu South Forest Reserve, the densities of *G. zanzibaricus* and *O. garnettii* were higher in the wet than the dry season while that of *G. rondoensis* was higher in dry season (Table 9.2; Appendix 9.1–9.3). The variations of density among species in both sites can be partly due to species-specific behaviours. During surveys along transects, it was noted that the *G. rondoensis* appeared to be disturbed by spotlights that made them respond by making frequent alarm calls. Production of alarm calls along transects influenced other conspecifics to resurface from their hideouts, hence leading to their highest detection by observers. This could partly explain its highest density in comparison to other syntopic galago species. Conversely, the other galago species were very indifferent to our presence or flash lights and sometimes they did not even move or vocalize when they were spotted.

The overall density estimates for *G. rondoensis* in Pugu Nature and Ruvu South forests are lower than that reported in earlier studies (8 animals/ha ☒ 800 animals per km^2) (Gwegime, 2016; Perkin, 2020). The variation in the overall and seasonal densities of the three galago species can be largely due to differences in size of surveyed area and sampling effort. For instance, the density of eight animals per ha (i.e. 800 animals per km^2) was only recorded from a 0.5 ha forest fragment (Perkin, 2020) while the present density was computed from eight transects (total cumulative distance 32.6 km) across the large expanse of Pugu Nature Forest Reserve and four

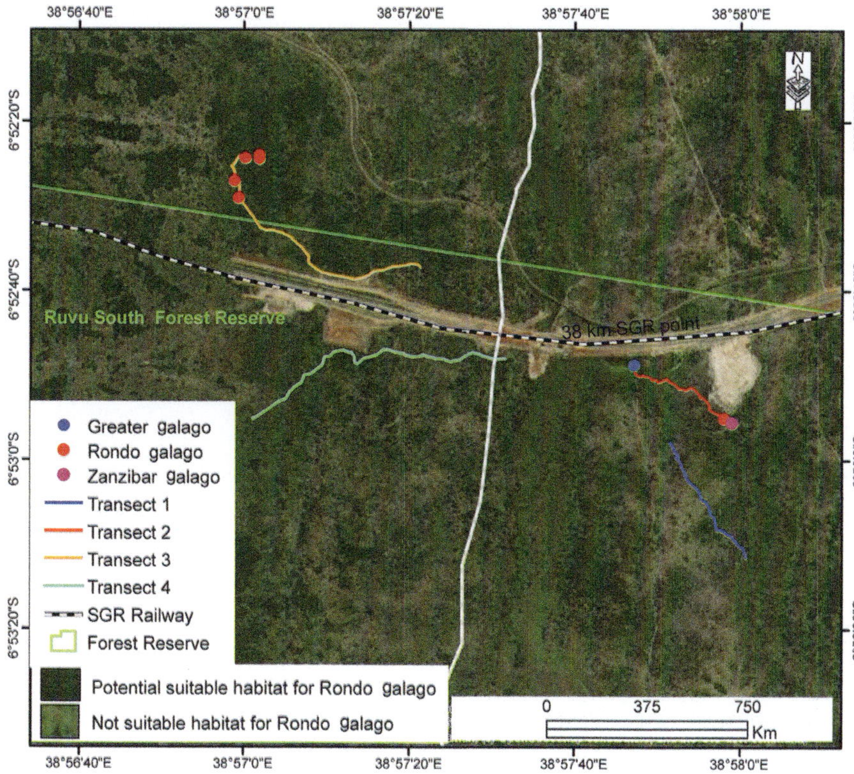

Fig. 9.5. Map showing the presence of Rondo galago, Greater galago and Zanzibar galago along the transects in the Ruvu South Forest Reserve.

transects (7.4 km) in Ruvu South Forest Reserve along the SGR footprint. In our survey, only four transects (5.8 km) were also surveyed during the wet season while eight transects (25.9 km) were traversed during the dry season in the Pugu Nature Forest Reserve. Small sampling effort could overestimate densities if many individuals were crowded in such small areas. Thus, the discrepancy in sampling efforts could explain the variations of observed density estimates for three galago species across sites and seasons. Otherwise, because the population trend of *G. rondoensis* is decreasing (Perkin, 2020), thus the lower than reported density in this study may be an indication of its declining population in the study areas because of habitat perturbations emanating from anthropogenic activities.

The higher density of *G. rondoensis* in the dry season for Ruvu South could be attributed to the fact that of the two additional transects added during the dry season, one passed through an evergreen forest fragment (about 500 m outside the forest boundary) with 45.0 animals per km^2. The observation of several individuals in such forest fragment in Ruvu South Forest Reserve provide a strong rationale to conduct detailed surveys by targeting the small remaining forest fragments in such a vast landscape. Based on density estimates, it can be preliminarily concluded that seasonality can have a profound effect on the vocal activity of *O. garnettii* and *G. zanzibaricus* but slightly on *G. rondoensis*, provided the latter occurs in the homogeneous vegetation type.

9.3.3 Activity/time budgets and patterns for galagos

In Pugu Nature Forest Reserve, galagos were recorded to exhibit a variety of activities, with calling being the most frequently observed

Fig. 9.6. Rondo galago trapped in a cage at Pugu Nature Forest Reserve in 2021.

Table 9.1. Encounters of Rondo galago and other sympatric species of galago in different vegetation types during the entire study period.

	Pugu Nature FR			Ruvu South FR		
Vegetation types	RG	ZG	GG	RG	ZG	GG
Dry evergreen forest*	24 (14.6)	1 (3.2)	3 (9.4)	-	-	-
Dry evergreen forest/thicket	108 (65.9)	29 (77.4)	29 (68.8)	18 (94.7)	0 (0.0)	0 (0.0)
Riverine forest*	13 (7.9)	4 (12.9)	0 (0.0)	-	-	-
Thicket	17 (10.4)	6 (3.2)	13 (21.8)	1 (5.3)	1(100)	1(100)
Thicket/shrubland	2 (1.2)	1 (3.2)	0 (0.0)	0 (0.0)	0 (0.0)	0 (0.0)
Regenerating woodland	0 (0.0)	0 (0.0)	0 (0.0)	0 (0.0)	0 (0.0)	0 (0.0)
Shrubland/degraded forest patches	0 (0.0)	0 (0.0)	0 (0.0)	0 (0.0)	0 (0.0)	0 (0.0)

The numbers without brackets are frequencies and numbers inside brackets are percentages of encounters.
RG, Rondo galago; ZG, Zanzibar galago; GG, Greater galago; * = vegetation type was not present in the surveyed sites in Ruvu South Forest Reserve.

behaviour for all species. However, in Ruvu South Forest Reserve, all the three galago species were observed to engage in one type of behaviour (i.e. calling activity). Their calling activities were mainly dominated by alarm and advertisement calls (Tables 9.3 and 9.4). Such observed activity patterns and budgets could be seriously affected by the sampling and recording rule used to collect such data. For reliable activity data, several individuals should have been followed systematically for a specified duration, but this was not possible due to elusiveness of the study animals. Thus, the researchers were compelled to use such a method that enabled them to collect data when animals were first seen (Croze, 1974; Kabigumila, 2001) but this method could be biased to most conspicuous or common behaviours (Bateson and Martin, 2007), especially the vocal activities (i.e. alarm and advertisement calls) for bushbabies. However,

Table 9.2. Overall and seasonal density estimates (density (mean density ±SE) per km² of Rondo galago, Zanzibar galago and Greater galago).

Site	Species	Wet season	Dry season	Overall density
Pugu NFR	Rondo galago	112.2±47.6	105.5±33.2	108.9±3.4
	Zanzibar galago	46.0±16.2	22.2±5.2	34.1±11.9
	Greater galago	56.7±6.6	21.7±8.4	39.2±17.5
Ruvu South FR	Rondo galago	5.7±5.7	45.0±45.0	25.4±19.7
	Zanzibar galago	5.7±5.7	0±0.0	2.9±2.9
	Greater galago	5.7±5.7	0±0.0	2.9±2.9

despite such methodological issues, the results are worth attention as they can provide useful interpretations.

The observation of highest encounters of calling activity could be associated with various roles including reduction of risks from predators and other forms of dangers (Bearder, 2007) and for social cohesion such as attraction of companions for mating or other functions (Schneiderová *et al.*, 2020). During the surveys, there were opportunistic records of the presence of several potential predators such as owls, big snakes (e.g. pythons), eagles, African Palm civet and genets. Even though there was no direct or indirect observation of predation of any galago species by any of these predators during the study, frequent production of alarm calls can be associated with response to predators and/or our presence.

Of all the species, the *G. rondoensis* was observed to produce most of advertisement calls compared to other sympatric galagos. These loud calls provide the impression that *G. rondoensis* mostly depends on advertisement calls for intra-specific communication. During the night walks, there was an encounter of an aggregation of at least five individuals; thus, this observation provides the impression that social communication is an integral part of their social cohesion. The less frequently observed behaviours in this study should be regarded as rare, but the method used to study behaviour appeared to under-estimate them. For instance, mating and feeding are very important behaviours but were not most frequently recorded behaviours. Nevertheless, mating activity was not observed in *G. rondoensis*; their advertisement calls provide evidence of their reproductive activities. Feeding was only observed in *G. rondoensis* in the Pugu Nature Forest Reserve. During sightings there, encounters of feeding in sites mainly with flower-laden plants (tree, shrub, lianas, climbers forbs), which either could be potential foods or attract various nocturnal arthropods that fell prey to *G. rondonensis*, were recorded.

The three species of galagos were observed to be very active at dusk, especially from 19:00 to 21:00 and at dawn from 04:00 to 06:00 for both Pugu Nature Forest Reserve and Ruvu South Forest Reserve (Tables 9.3 and 9.4). These findings conform well to reports of galago surveys conducted elsewhere (De Jong and Butynski, 2009), surveys conducted during times when the galagos are most active (i.e. dusk and dawn) (De Jong and Butynski, 2009). However, to get comprehensive data to allow reliable interpretation, future studies can be carried out from dusk to dawn in their natural environments along transects, rather than surveying them for four hours after dusk or dawn.

9.4 Conclusion

The widespread presence of *G. rondoensis* and other sympatric galagos in surveyed transects in Pugu Nature Forest Reserve and Ruvu South Forest Reserve provides a strong rationale to strengthen conservation efforts for the entire Pugu Nature and Ruvu South reserves. The observation of many encounters of all three species of sympatric galagos in evergreen forest fragments in both study forests is indication that such evergreen vegetation types are suitable for the species compared to open habitats. This observation implies that conservation of intact forest fragments is key to the survival of the

Table 9.3. Activity budget for the three species of galagos in Pugu Nature Forest Reserve.

	Other calls	Alarm calls	Advertisement calls	Buzzing/PW calls	All calls	Moving	Resting	Mating	Feeding	Total
Rondo galago										
19–20 hours	3 (18.8)	8 (50.0)	2 (12.5)		13 (81.3)	2 (12.5)			1(6.3)	16 (9.6)
20–21 hours	10 (15.4)	17 (26.2)	17 (26.2)	1 (1.5)	45 (69.2)	8 (12.3)	9 (13.8)		3 (4.6)	65 (38.9)
21–22 hours		1 (100.0)			1 (100.0)					1 (0.6)
4–5 hours	2 (10.5)	10 (52.6)	2 (10.5)		14 (73.7)	3 (15.8)	2 (10.5)			19 (11.4)
5–6 hours	6 (9.1)	36 (54.5)	8 (12.1)	6 (9.1)	56 (84.8)	8 (12.1)	1 (1.5)		1 (1.5)	66 (39.5)
Activity budget	21 (12.6)	72 (43.1)	29 (17.4)	7 (4.1)	129 (77.2)*	21 (12.6)	12 (7.2)		5 (2.9)	
Zanzibar galago										
19–20 hours	2 (33.3)	4 (66.7)			6 (100)					6 (12.8)
20–21 hours	2 (18.2)	7 (63.6)			9 (81.8)	2 (18.2)				11(23.4)
21–22 hours										0 (0.0)
3–4 hours						3 (60.0)	1 (20.0)	1 (20.0)		5 (10.6)
4–5 hours		3 (37.5)			3 (37.5)	1 (12.5)	4 (50.0)			8 (17.0)
5–6 hours		8 (47.1)	5 (29.4)		13 (76.5)	1 (5.9)	3 (17.6)			17 (36.2)
Activity budget	4 (8.5)	22 (46.8)	5 (10.6)		31 (65.9)*	7 (14.9)	8 (17.0)	1 (2.1)		
Greater galago										
18–19 hours		1 (50.0)			1 (50.0)	1 (50.0)				2 (3.6)
19–20 hours	2 (50.0)				2(50.0)	2 (50.0)				4 (7.3)
20–21 hours	3 (21.4)	8 (57.1)	2 (14.3)		13 (92.9)		1 (7.1)			14 (25.5)
21–22 hours		2 (100.0)			2 (100.0)					2 (3.6)
3–4 hours										
4–5 hours		13 (72.2)	2 (11.1)		15 (83.3)	1 (5.6)	2 (11.1)			18 (32.7)
5–6 hours	3 (20.0)	6 (40.0)	4 (26.7)		13 (86.7)	0 (0.0)	2 (13.3)			15 (27.3)
Activity budget	8 (14.5)	30 (54.5)	8 (14.5)		46 (83.6)*	4 (7.2)	5 (9.1)			

The numbers without brackets are frequencies and numbers inside brackets are percentages of encounters. PW, Puzz/whistle, *, the frequency of calling activity was significantly highest to the rest activities. Calling activity was significant different from the rest activities for Rondo galago: subdividing $\chi 2=37.371$, df = 1, $p<0.0001$: *G. zanzibaricus:* $\chi 22=4.787$, df = 5, $p<0.0001$ and *O. garnettii:* $\chi 22=24.891$, df = 1, $p<0.0001$)

Table 9.4. Activity budget for the three species of galagos in Ruvu South Forest Reserve.

Rondo galago	Other calls	Alarm call	Advertisement call	Buzzing/PW calls	All calls	Moving	Resting	Mating	Feeding	Total
19–20 hours			14(100.0)		14(100.0)					14(70.0)
20–21 hours		6(100.0)			6(100.0)					6 (30.0)
Activity budget		6(30.0)	14 (70.0)		20(100.0)					
Zanzibar galago										
19–20 hours										
20–21 hours		1(100.0)			1(100.0)					1(100.0)
Activity budget		1(100.0)			1(100.0)					
Greater galago										
19–20 hours										
20–21 hours		1(100.0)								1(100.0)
Activity budget		1(100.0)								

NOTE: The numbers without brackets are frequencies and numbers inside brackets are percentages of encounters, PW = Puzz/whistle.

galagos, especially the endangered *G. rondoensis*. The overall densities of *G. rondoensis* in Pugu Nature and Ruvu South forest reserves are lower than those reported previously. Some potential reasons for the low density have been attributed to possible variations in sampling efforts/intensity, spatial coverage of the landscape and other causes related to habitat degradation and destruction. This observation suggests that further intensive and extensive surveys are needed to assess their comprehensive spatial-temporal distribution across the vast landscape of these forests. Despite of the behavioural diversity of galagos, in these surveys it was observed that the activity budget of most galagos was dominated by calls, especially alarm and advertisement calls. These calls are attributed to behavioural adaptation to reduce risks of predation and other sources of danger and other cohesive roles.

9.5 Recommendation

Based on the findings of this study, further monitoring surveys on galagos in Pugu Nature and Ruvu South forest reserves and elsewhere in their geographic range are recommended before it is too late. The study recommends monitoring be carried out twice annually (wet and dry seasons) every two years to be enlightened on the impact after operation of the SGR and the progress and impact of the habitat restoration by TRC. Besides transect surveys, camera traps are advised to be used in future monitoring. Current surveys in the Ruvu South FR have concentrated in places close to the railways. Because Ruvu South FR is a large landscape of about 35,000 ha, we recommend extension of the scope of sampling to approximately 50% of the entire landscape by focusing on the suitable evergreen forest patches. The study recommends stronger, more effective and sustainable law enforcement and conservation efforts through having qualified, sufficient and committed human resource that is provisioned with equipment and facilities for effecting their responsibilities. The study calls for combined conservation efforts involving all stakeholders such as the government, through Tanzania Forest Services (TFS), Tanzania Wildlife Authority (TAWA), Tanzania National Parks (TANAPA), the National Environmental Management Council (NEMC) and non-governmental Institutions (NGOs) at local, national, regional and international level. The wildlife conservation authorities in the country should implement constructive and sound results, suggestions and recommendations from scientific research to achieve sustainable conservation and development in the country.

Acknowledgements

We are very grateful to Almighty God for enabling us to accomplish the field study and writing the manuscript. Special thanks to our employers for giving us the permission to conduct the study and provision of some research equipment. We are grateful to the Tanzania Railways Corporation (TRC) for funding the galago surveys and the write-up of this section. We are very indebted to Professor Andrew Perkin for his advice regarding the methodology and handling of galagos, especially the endangered *G. rondoensis*. Thanks to the galago scientist Justine Gwegima, who identified the galagos in the field. Thanks also go to Magdalena Kitila (TRC) for logistics and co-ordinating the surveys from the begining to the end. We finally thank Bennedict K. Minja, Swalehe Mweyo, Mataba Bulugu Lisesi, Arafat F. Napunda, Alloyce Akwilini Mayombo and Bashiru Salum for helping in data collection and other field-related activities.

References

Ambrose, L. and Perkin, A.W. (2000) A survey of nocturnal prosimians at Moca on Bioko Island, Equatorial guinea. *African Primates* 4, 4–10.

Bateson, M. and Martin, P. (2007) *Measuring Behaviour: An Introductory Guide*. Cambridge University Press, Cambridge, UK.

Bearder, S.K. (2007) A comparison of calling patterns in two nocturnal primates, *Otolemur crassicaudatus* and *Galago moholi*, as a guide to predation risk. In: Gursky, S.L. and Nekaris, K.A.I. (eds) *Primate Anti-predator Strategies*. Springer, Boston, Massachusetts, pp. 206–221.

Bearder, S.K., Butynski, T.M. and De Jong, Y.A. (2013) Vocal profiles for the Galagos: A tool for identification. *Primate Conservation* 75.

Bogoni, J.A., Peres, C.A. and Ferraz, K.M. (2020) Effects of mammal defaunation on natural ecosystem services and human well being throughout the entire Neotropical realm. *Ecosystem Services* 45, 101173.

Bourlière, F. (1985) Primate communities: their structure and role in tropical Ecosystems. *International Journal of Primatology* 6, 1–26.

Burgess, N., Butynski, T., Gordon, I., Luke, Q., Sumbi, P, *et al*. (2003) Eastern Arc Mountains and coastal forests of Tanzania and Kenya: biodiversity hotspots. Ecosystem profile report prepared for Critical Ecosystem Partnership Fund. Unpublished.

Burgess, N.D. and Clarke, G.P. (2000) *Coastal Forests of Eastern Africa*. IUCN Gland, Switzerland.

Butynski, T.M., De Jong, Y., Perkin, A., Bearder, S. and Honess, P. (2008) *Galagoides zanzibaricus*. IUCN Red List of Threatened Species 2008. Available at: https://www.researchgate.net/publication/336144904_Galagoides_zanzibaricus_The_IUCN_Red_List_of_Threatened_Species_-_2008 (accessed 6 August 2023).

Clarke, G. and Dickinson, A. (1995) "*Status Reports for 11 Coastal Forests in Coast Region, Tanzania*" (Frontier-Tanzania Technical Report no.17). UK/University of Dar es Salaam, Tanzania. Unpublished: Society for Environmental Exploration.

Croze, H. (1974) The Seronera bull problem: the elephants. *African Journal of Ecology* 12, 1–27.

De Jong, Y.A. and Butynski, T. (2009) Primate Biogeography, Diversity, Taxonomy and Conservation of the Coastal Forests of Kenya. In: *Report to the Critical Ecosystem Partnership Fund*. Eastern Africa Primate Diversity and Conservation Programme, Nanyuki, Kenya.

De Jong, Y.A., Butynski, T.M. and Perkin, A.W. (2011) List of the Primates of Tanzania. Unpublished Report. Available at: www.WildSolutions.nl (accessed 6 August 2023).

Davenport, T.R., Nowak, K. and Perkin, A. (2014) Priority primate areas in Tanzania. *Oryx* 48, 39–51.

Decker, B.S. (1994) Effects of habitat disturbance on the behavioral ecology and demographics of the Tana river red colobus (*colobus badius rufomitratus*). *International Journal of Primatology* 15, 703–737.

Estrada, A. and Garber, P.A. (2022) Principal drivers and conservation solutions to the impending primate extinction crisis: introduction to the special issue. *International Journal of Primatology* 43, 1–14.

Estrada, A., Garber, P.A., Rylands, A.B., Roos, C., Fernandez-Duque, E, *et al*. (2017) Impending extinction crisis of the world's primates: why primates matter. *Science Advances* 3, e1600946.

Gutierrez-Arellano, C. and Mulligan, M. (2018) A review of regulation ecosystem services and disservices from faunal populations and potential impacts of agriculturalisation on their provision, globally. *Nature Conservation* 30, 1–39.

Gwegime, J., Mwangoka, M., Said, H., Mulungu, E., Nowak, K, *et al*. (2014) *Two Surveys of the Biodiversity and Forest Condition of Rondo Proposed Nature Reserve in 2012 and 2013*. TFCG Technical Paper 39, TFCG, Dar es Salaam.

Gwegime, J. (2016) *A Survival Blueprint for the Rondo galago* (*Galagoides rondoensis*). Tanzania Forest Conservation Group (TFCG), Dar es Salaam.

Hanya, G., Stevenson, P., Van Noordwijk, M., Te Wong, S., Kanamori, T, *et al*. (2011) Seasonality in fruit availability affects frugivorous primate biomass and species richness. *Ecography* 34, 1009–1017.

Harcourt, C.S. and Nash, L.T. (1986) Species differences in substrate use and diet between sympatric galagos in two Kenyan coastal forests. *Primates* 27, 41–52.

Hending, D., Randrianarison, H., Andriamavosoloarisoa, N.N.M., Ranohatra-Hending, C., Solofondranohatra, J.S, *et al*. (2023) Seasonal differences in the encounter rate of the fat-tailed dwarf lemur (*Cheirogaleus medius*) in the transitional forests of Northwest Madagascar: Implications for reliable population density assessment. *International Journal of Primatology* 44, 482–498.

Honess, P.E. (1996) Descriptions of the dwarf galago species of Tanzania. *African Primates* 2, 75–79.

Honess, P.E., Perkin, A. and Bearder, S.K. (2008) Rondo dwarf galago. *Folia Primatologica* 74, 337–354.

Kabigumila, J. (2001) Occurrence and activity budget of the leopard Tortoise, Geochelone Pardalis, in northern Tanzania. *Tanzania Journal of Science* 27, 87–99.

Kalbitzer, U. and Chapman, C.A. (2018) Primate responses to changing environments in the Anthropocene. In: Kalbitzer, U. and Jack, K. (eds) *Primate Life Histories, Sex Roles, and Adaptability*. Developments in Primatology: Progress and Prospects series, Springer, Cham, Switzerland, pp. 283–310.

Kibaja, M.J., Mekonnen, A., Reitan, T., Nahonyo, C.L., Levi, M. *et al.* (2023) On the move: Activity budget and ranging ecology of endangered ashy red colobus monkeys (*Piliocolobus tephrosceles*) in a savanna woodland habitat. *Global Ecology and Conservation* 43, e02440.

Kiwia, H. (2006) Species richness and abundance of large mammals in Zaraninge forest, coast region, Tanzania. *Tanzania Journal of Science* 32, 105–116.

Masters, J.C., Génin, F., Couette, S., Groves, C.P., Nash, S.D. *et al.* (2017) A new genus for the eastern dwarf galagos (primates: Galagidae). *Zoological Journal of the Linnean Society* 181, 229–241.

McLennan, M.R., Spagnoletti, N. and Hockings, K.J. (2017) The implications of primate behavioral flexibility for sustainable human–primate coexistence in anthropogenic habitats. *International Journal of Primatology* 38, 105–121.

Md-Zain, B. and Ch'ng, C. (2011) The activity patterns of a group of Cantor's dusky leaf monkeys (*Trachypithecus obscurus halonifer*). *International Journal of Zoological Research* 7, 59–67.

Mekonnen, A., Fashing, P.J., Bekele, A., Hernandez-Aguilar, R.A., Rueness, E.K. *et al.* (2017) Impacts of habitat loss and fragmentation on the activity budget, ranging ecology and habitat use of bale monkeys (*Chlorocebus djamdjamensis*) in the southern Ethiopian Highlands. *American Journal of Primatology* 79, e22644.

Moyer, D., Plumptre, A., Pintea, L., Hernandez-Aguilar, A., Moore, J, *et al.* (2006) "Surveys of chimpanzees and other biodiversity in Western Tanzania" Unpublished Report to United States Fish and Wildlife Service.

Murphy, A. (2015) *Anthropogenic Disturbance in Nocturnal Primates and Conservation Perception in Zaraninge Forest, Tanzania*. Lund University, Sweden.

Perkin, A.W, *et al.* (2019) Rondo Dwarf Galago *Paragalago rondoensis* Honess in Kingdon, 1997. In: Schwitzer, C., Mittermeier, R.A., Rylands, A.B., Chiozza, F. and Williamson, E.A. (eds) *Primates in Peril: The World's 25 Most Endangered Primates 2018–2020*. IUCN SSC Primate Specialist Group/ International Primatological Society/Global Wildlife Conservation/Bristol Zoological Society.

Perkin, A.W. (2007) Comparative penile morphology of East African Galagos of the genus Galagoides (family Galagidae): implications for taxonomy. *American Journal of Primatology: Official Journal of the American Society of Primatologists* 69, 16–26.

Perkin, A.W. (2020) *Paragalago rondoensis*. The IUCN Red List of Threatened Species 2020: e.T40652A17962115.

Schneiderová, I., Singh, N.J., Baklová, A., Smetanová, M., Gomis, N.B. *et al.* (2020) Northern lesser galagos (*Galago Senegalensis*) increase the production of loud calls before and at dawn. *Primates* 61, 331–338.

Siex, K.S. (2003) Effects of population compression on the demography, ecology, and behavior of the Zanzibar Red Colobus Monkey *(Procolobus kirkii)*. Duke University, Durahm, North Carolina.

Zimmermann, E. (1990) Differentiation of vocalizations in bushbabies (Galaginae, Prosimiae, primates) and the significance for assessing Phylogenetic relationships. *Journal of Zoological Systematics and Evolutionary Research* 28, 217–239.

Appendix 9.1a. Individual density of Rondo galago along transects in Pugu Nature Forest Reserve and Ruvu South Forest Reserve during the wet season, April 2021.

Pugu Nature Forest Reserve							
Transect	Days	Transect length (km)	Cumulative transect length (km)	Number of individuals	Strip width (2L) (km)	Area surveyed (km2)	Density /km2 per transect
T1	4	0.173	0.692	8	0.05	0.035	231.2
T2	5	0.637	3.185	15	0.05	0.159	94.2

Continued

Continued

Transect	Days	Transect length (km)	Cumulative transect length (km)	Number of individuals	Strip width (2L) (km)	Area surveyed (km2)	Density /km2 per transect
Pugu Nature Forest Reserve							
T3	4	0.487	1.948	12	0.05	0.097	123.2
T4	3	0.176	0.528	0	0.05	0.026	0
Overall density (Average)							112.2 ± SE47.6
Ruvu South Forest Reserve							
T1	3	0.588	1.764	1	0.05	0.088	11.3
T2	4	1	4	0	0.05	0.2	0
Overall density (Average)							5.7 ± SE5.7

Appendix 9.1b. Individual density of Rondo galago along transects in Pugu Nature Forest Reserve and Ruvu South Forest Reserve during the dry season, December 2021.

Transect	Days	Transect length (km)	Cumulative transect length (km)	Number of individuals	Strip width (2L) (km)	Area surveyed (km^2)	Density /km^2 per transect
Pugu Nature Forest Reserve							
T1	2	0.7	2.8	0	0.05	0.14	0.0
T2	4	0.7	4.9	12	0.05	0.245	49.0
T3	2	0.7	2.8	3	0.05	0.14	21.4
T4	3	0.7	4.2	23	0.05	0.21	109.5
T5	2	0.7	2.8	39	0.05	0.14	278.6
T6	2	0.7	2.8	26	0.05	0.14	185.7
T7	2	0.7	2.8	20	0.05	0.14	142.9
T8	2	0.7	2.8	8	0.05	0.14	57.1
Overall density (Average)							105.5 ± SE 33.2
Ruvu South Forest Reserve							
T1	1	0.7	1.4	0	0.05	0.07	0.0
T2	1	0.7	1.4	0	0.05	0.07	0.0
T3	2	1	2	18	0.05	0.1	180.0
T4	1	1	1	0	0.05	0.05	0.0
Overall density (Average)							45.0 ± SE 45.0

Appendix 9.2a. Individual density of Zanzibar galago along transects in Pugu Nature Forest and Ruvu South Forest Reserve during the wet season, April 2021.

Pugu Nature Forest Reserve							
Transect	Days	Transect length (km)	Cumulative transect length (km)	Number of individuals	Strip width (2L) (km)	Area surveyed (km^2)	Density /km^2 per transect
T1	4	0.173	0.692	2	0.05	0.035	57.1
T2	5	0.637	3.185	12	0.05	0.159	75.5
T3	4	0.487	1.948	5	0.05	0.097	51.5
T4	3	0.176	0.528	0	0.05	0.026	0
Overall density (Average)							46.0 ± SE16.2
Ruvu South Forest Reserve							
T1	3	0.588	1.764	1	0.05	0.088	11.4
T2	4	1	4	0	0.05	0.2	0
Overall density (Average)							5.7 ± SE5.7

Appendix 9.2b. Individual density of Zanzibar galago along transects in Pugu Nature Forest Reserve and Ruvu South Forest Reserve during the dry season, December 2021.

Pugu Nature Forest Reserve							
Transects	Days	Transect length	Cumulative transect length	Number of individuals	Strip width (2L)	Area surveyed	Density per transect
T1	2	0.7	2.8	0	0.05	0.14	0
T2	4	0.7	4.9	9	0.05	0.245	36.7
T3	2	0.7	2.8	6	0.05	0.14	42.9
T4	3	0.7	4.2	4	0.05	0.21	19.0
T5	2	0.7	2.8	4	0.05	0.14	28.6
T6	2	0.7	2.8	4	0.05	0.14	28.6
T7	2	0.7	2.8	1	0.05	0.14	7.1
T8	2	0.7	2.8	2	0.05	0.14	14.3
Overall density (Average)							22.2 ± SE5.2
Ruvu South Forest Reserve							
T1	1	0.7	1.4	0	0.05	0.07	0
T2	1	0.7	1.4	0	0.05	0.07	0
T3	2	1	2	0	0.05	0.1	0
T4	1	1	2	0	0.05	0.1	0

Continued

Continued

Transects	Days	Transect length	Cumulative transect length	Number of individuals	Strip width (2L)	Area surveyed	Density per transect
Pugu Nature Forest Reserve							
Overall density (Average)							Mean = 0 ± SE 0.0

Appendix 9.3a. Individual density of Greater galago along transects in Pugu Nature Forest Reserve and Ruvu South Forest Reserve during the wet season, April 2021.

Transect	Days	Transect length (km)	Cumulative transect length (km)	Number of individuals	Strip width (2L) (km)	Area surveyed (km^2)	Density /km^2 per transect
Pugu Nature Forest Reserve							
T1	4	0.173	0.692	2	0.05	0.035	57.1
T2	5	0.637	3.185	11	0.05	0.159	69.2
T3	4	0.487	1.948	6	0.05	0.097	61.9
T4	3	0.176	0.528	1	0.05	0.026	38.5
Overall density (Average)							56.7 ± SE6.6
Ruvu South Forest Reserve							
T1	3	0.588	1.764	1	0.05	0.088	11.36364
T2	4	1	4		0.05	0.2	0
Overall density (Average)							5.7 ± SE5.7

Appendix 9.3b. Individual density of Greater galago along transects in Pugu Nature Forest Reserve and Ruvu South Forest Reserve during the dry season, December 2021.

Transect	Days	Transect length	Cumulative transect length	Number of individuals	Strip width (2L)	Area surveyed	Density per transect
Pugu Nature Forest Reserve							
T1	2	0.7	2.8	0	0.05	0.14	0
T2	4	0.7	4.9	14	0.05	0.245	57.1
T3	2	0.7	2.8	8	0.05	0.14	57.1
T4	3	0.7	4.2	5	0.05	0.21	23.8
T5	2	0.7	2.8	2	0.05	0.14	14.3
T6	2	0.7	2.8	0	0.05	0.14	0

Continued

Continued

Transect	Days	Transect length	Cumulative transect length	Number of individuals	Strip width (2L)	Area surveyed	Density per transect
Pugu Nature Forest Reserve							
T7	2	0.7	2.8	3	0.05	0.14	21.4
T8	2	0.7	2.8	0	0.05	0.14	0
Overall density (Average)							21.7 ± SE8.4
Ruvu South Forest Reserve							
T1	1	0.7	1.4	0	0.05	0.07	0
T2	1	0.7	1.4	0	0.05	0.07	0
T3	2	1	2	0	0.05	0.1	0
T4	1	1	2	0	0.05	0.1	0
Overall density (Average)							0 ± SE0.0

10 Bats of Pugu-Kazimzumbwi and Ruvu South Forests, Tanzania

Chacha Werema*
College of Natural and Applied Sciences, University of Dar es Salaam, Tanzania

Abstract

Bats are the second most speciose mammalian order with over 1400 species known worldwide. Bats are more poorly studied than other mammalian taxa in Tanzania. Among the areas where they have been little studied are Pugu-Kazimzumbwi (PKNFR) and Ruvu South (RSFR) forests. These forests represent an important ecosystem of the coastal forests with high and unique biological resources. Using mist nets, nine bat species, belonging to seven genera and seven families, were captured and, in combination with the previous studies in the study area, PKNFR and RSFR have 17 and eight species of bats, respectively. This means PKNFR has the highest number of bat species in comparison to the other coastal forests whose bat fauna have been investigated. Therefore, these forests are of considerable importance in conservation of bats and continuing to conserve them is recommended.

10.1 Introduction

Bats make up more than 20% of mammals (Simmons, 2005), forming the second-most speciose mammalian order, with over 1400 species known worldwide (Burgin *et al.*, 2018). Out of 221 bat species known from Africa, Tanzania has 105 species which have been described (Patterson and Webala, 2012). These constitute a major part of Tanzania's mammal diversity. Bats perform important roles in agriculture (e.g. pollination and seed dispersal) (Kunz *et al.*, 2011; Seltzer *et al.*, 2013); forestry (Muscarella and Fleming, 2007; Seltzer *et al.*, 2013), and may be a concern for public health (Wibbelt *et al.*, 2010). They are also important indicators of ecosystem health and providers of a variety of ecosystem services (Jones *et al.*, 2009) making them an essential group to conserve to maintain functioning habitats. This makes bats important in fragmented and intensively used habitats where seed dispersal by animals plays an important role for the persistence and reforestation of different plant species (Muscarella and Fleming, 2007). For example, fruit bats can function as dispersal agents for several plant genera and species as they can travel some distances from the parent plant (Seltzer *et al.*, 2013). In dispersing seeds, the fruit bats may connect fragmented habitats which might be more difficult for other dispersal agents, for example rodents (Shilton *et al.*, 1999).

Bats utilize diverse roosting structures and a wide range of diets (Kunz and Pierson, 1994; Meyer *et al.*, 2008) and they are sensitive to anthropogenic alterations such as changes in habitat quality, which make them valuable indicators of habitat disruption (Jones *et al.*, 2009). Any form of habitat disturbance such as forest fragmentation and selective logging can induce major changes in the distribution and abundance of bat species (Meyer *et al.*, 2016). At the landscape scale, remaining patches of forest may become

*cwerema@udsm.ac.tz

 Biodiversity and Conservation along an East African Railway (eds C.A. Masao *et al.*)
DOI: 10.1079/9781800626942.0010

too small, too isolated and too influenced by edge effects to maintain viable populations of some bat species (Meyer *et al.*, 2008).

While the knowledge about bat species' richness and diversity is crucial for conservation and management of biodiversity (Margules and Pressey, 2000), the detailed studies on these aspects are rare because bats have been more poorly studied (Monadjem *et al.*, 2020) than other mammalian groups, partly because of the complexity involved in their trapping and identification (Trentin and Rovero, 2011). This is also the case in Tanzania where only few studies exist on bat fauna (see Katunzi *et al.*, 2021), making the taxonomy and species distribution of this taxon little known about, let alone the responses of different bats species to habitat loss.

Studies of bats in Pugu-Kazimzumbwi and Ruvu South forests exist, however, because of various human activities which have led to habitat degradation and fragmentation (Burgess *et al.*, 1992), wildfires (Mligo, 2019), exotic plant species (Rocky and Mligo, 2012) and the construction of the standard-gauge railway (SGR), the distribution and diversity of bats might have been affected. As such, in view of the existing threats, this study is an attempt to assess the diversity of bats in Pugu-Kazimzumbwi and Ruvu South forest reserves. This assessment is based on studies which were conducted between 2019 and 2021. Additionally, literature was sought to come up with a comprehensive list of bats which have been recorded in the study area.

10.2 Materials and Methods

10.2.1 Study area

The study was conducted in Pugu-Kazumzumbwi Nature Forest Reserve (PKNFR) and Ruvu South Forest Reserve (RSFR), in Kisarawe district, Pwani region (see Chapter 8). The vegetation found in the study area falls under Northern Zanzibar-Inhambane Coastal Forest Mosaic eco-region. There were two study sites in PKNFR and one site in RSFR. There were two sampling sites in PKNFR to include the existing kaolin mines in the reserve as well as the forested habitat. The kaolin mines are known to have congregations of bats of three bat species: Persian trident bat *Triaenops persicus*, Lander's horseshoe bat *Rhinolophus landeri* and Noack's leaf-nosed bat *Hipposideros ruber* (Howell, 1976).

10.2.2 Methods

Bat surveys were conducted in three periods: 15–22 November 2019; 29 November–4 December 2019; and 1–20 April 2021 (Table 10.1). Within PKNFR, bats were surveyed in the two kaolin mines, in the vicinity of the kaolin mines and surrounding environment (forest habitat). Within RSFR, bats were surveyed in a woodland. To be able to catch and identify bats, standard mist nets (each 12 m long, 2.6 m high; Castro-Luna *et al.*, 2007, Castro-Arellano *et al.*, 2009, 2010) and scoop nets were used. The latter method was used in the kaolin mines only. In the kaolin mines, in addition to observations, the bats were sampled during the day and night hours. Mist nets were set at each site and for logistical reasons sampling efforts varied (Table 10.1). The bats mist-netted were released at the site of capture after they were identified to species. Species identification was aided by field guides and relevant published papers and books on the subject, particularly those by Patterson and Webala (2012), Happold and Happold (2013) and Monadjem *et al.* (2020).

Table 10.1. Mist-netting effort in each sampling period and forest reserve.

Study area	Mist-netting period	Mist-netting effort (metre-net-hours)
PKNFR	15–22 November 2019	672
	29 November–4 December 2019	540
	1–20 April 2021	5106
RSFR	1–20 April 2021	3264

10.3 Results

10.3.1 Overall species richness

Nine bat species belonging to seven genera and seven families were mist-netted in the study area (Table 10.2). Of these, all nine species were mist-netted in PKNFR. These include Natal long-fingered bat *M. natalensis*, which was mist-netted for the first time in PKNFR (Table 10.2). Only four species were caught in RSFR (Table 10.2). In combination with the previous studies in these forests, PKNFR and RSFR have 17 and 8 species of bats, respectively (Table 10.2; Howell, 1976; Cockle *et al.*, 1998; Mbao, 2022). The two forest reserves share six species (Table 10.2).

10.3.2 Bats of PKNFR

Nine species were mist-netted in PKNFR (Table 10.2). Only four species, the Lander's horseshoe bat *R. landeri*, Persian trident bat *T. persicus*, Noack's leaf-nosed bat *H. ruber* and Natal long-fingered bat *M. natalensis* were found roosting in the existing two mines. However, only a few individuals of the Natal long-fingered bat *M. natalensis* were observed in the kaolin mines. The numbers varied daily (1–8 individuals). Thousands of the Persian trident bat *T. persicus*, Noack's leaf-nosed bat *H. ruber* and Lander's horseshoe bat *R. landeri* were observed in the mines. In the mines there was coexistence such that in one of the mines there were Lander's horseshoe bat *R. landeri* and Persian trident bat *T. persicus*, and in the other mine the Noack's leaf-nosed bat *H. ruber* only

Table 10.2. Bat species mist-netted in PKNFR and RSFR.

		Study site			
		PKNFR		RSFR	
Family	Species	CS	PS	CS	PS
Pteropodidae	Little epauletted fruit bat *Epomophorus labiatus*	√	x	√	
	Wahlberg's epauletted fruit bat *E. wahlbergi*	√	x		x
	Egyptian rousette *Roesettus aegyptiacus*		x		
	Angolan soft-furred fruit bat *Lissonycteris angolensis*		x		
	Bergman's collared bat *Myonycteris relicta*				x
Hipposideridae	Noack's leaf-nosed bat *Hipposideros ruber*	√	x		x
	Striped leaf-nosed bat *Macronycteris vittatus*	√	x		
Rhinonycteridae	Persian trident bat *Triaenops persicus*	√	x	√	x
Rhinolophidae	Lander's horseshoe bat *Rhinolophus landeri*	√	x	√	x
Megadermatidae	Yellow-winged bat *Lavia frons*	√	x	√	
Nycteridae	Andersen's slit-faced bat *Nycteris aurita*				x
	Hairy slit-faced bat *Nycteris hispida*		x		
	Egyptian slit-faced bat *Nycteris thebaica*	√	x		
Miniopteridae	Natal long-fingered bat *Miniopterus natalensis*	√			
Verspertillionidae	Temmink's myotis *Myotis tricolor*		x		
	Neoromicia capensis		x		
	Pipistrellus		x		
	Yellow-bellied house bat *Scotophilus dinganii*		x		
	White-bellied house bat *Scotophilus hirundo*		x		

√ = species mist-netted and those which have been recorded in the study area (x = species which have been recorded by other researchers). CS = current study; PS= previous studies coexisted with Persian trident bat *T. persicus*. The latter occupied the innermost and anoxic parts of both mines.

Near the kaolin mines, seven species were mist-netted (Table 10.3). The most abundant, also with the highest catch rate, was the Persian trident bat *T. persicus*. Six species of bats were captured in the forested sites within PKNFR (Table 10.4). Most of these species had low abundances, hence low catch rates (Table 10.4).

10.3.3 Species mist-netted in Ruvu South Forest Reserve

Four bat species were mist-netted in Ruvu South Forest Reserve (Table 10.5). All four species had low capture rates.

10.4 Discussion

In comparison with previous studies, the number of species mist-netted in the two forests in the current study was relatively low, and the number of species captured in RSFR was a subset of those caught in PKNFR. This paucity was probably because the use of mist nets to sample bats biases collections since only bats which fly within 0–3m can be trapped. This has been witnessed by several authors including Cranbrook *et al.* (1965), and Kunz and Brook (1975), who have discussed the problems and biases of using mist nets. It is also possible that there are some bat species that avoid mist nets (Laval, 1970; Kunz, 1973; Laval and Fitch, 1977) especially micro-bats, which depend on echolocation in orientating and foraging. Capture biases may also reflect flight behaviour, as it has been presented by Kingdon (1974) on the heights at which bats fly. Nevertheless, mist nets are still used to catch bats because there is evidence that shows high-flying species getting caught in nets at 3m or lower (e.g. *Epomops*, which fly at 10m, and *Tadarida* at 3000m; Kingdon, 1974). Also, proportionately higher numbers of mega-bats are recovered from nets compared to micro-bats.

In a recent study in PKNFR by Mbao (2022), which covered different habitats

Table 10.3. Abundance of bats mist-netted in the vicinity of kaolin mines in PKNFR.

Species	Mist-netting effort (mnh)	Number of individuals mist-netted	Catch rate (per 1000 mnh)
Natal long-fingered bat (*M. natalensis*)	3018	269	89.13
Striped leaf-nosed bat (*M. vittatus*)	3018	2	0.66
Lander's horseshoe bat (*R. landeri*)	3018	3	0.99
Persian trident bat (*T. persicus*)	3018	1192	394.96
Egyptian slit-faced bat (*N. thebaica*)	3018	2	0.66
Wahlberg's epauletted fruit bat (*E. wahlbergi*)	3018	4	1.33

Table 10.4. Abundance of bats mist-netted at forested sites within PKNFR.

Species	Mist-netting effort (mnh)	Number of individuals mist-netted	Catch rate (per 1000 mnh)
Lander's horseshoe bat (*R. landeri*)	3408	2	0.59
Persian trident bat (*T. persicus*)	3408	42	12.32
Egyptian slit-faced bat (*N. thebaica*)	3408	1	0.29
Wahlberg's epauletted fruit bat (*E. wahlbergi*)	3408	12	3.52
Little epauletted fruit bat (*E. labiatus*)	3408	1	0.29
Yellow-winged bat (*L. frons*)	3408	1	0.29

Table 10.5. Bat species mist-netted in Ruvu South Forest Reserve.

Species	Mist-netting effort (mnh)	Number of individuals mist-netted	Catch rate (per 1000 mnh)
Lander's horseshoe bat (*R. landeri*)	3264	1	0.306
Persian trident bat (*T. persicus*)	3264	3	0.918
Little epauletted fruit bat (*E. labiatus*)	3264	1	0.306
Yellow-winged bat (*L. frons*)	3264	3	0.918

including forest, grassland, riverine forest and built environment, ten species were captured in mist nets. Of the species that were mist-netted by Mbao (2022), seven species, including Wahlberg's epauletted fruit bat *E. wahlbergi*, Little epauletted fruit bat *E. labiatus*, Noack's leaf-nosed bat *H. ruber*, Striped leaf-nosed bat *H. vittatus*, Persian trident bat *T. persicus*, Lander's horseshoe bat *R. landeri* and Egyptian slit-faced bat *Nycteris thebaica*, were caught. Like the findings of this study, the most abundant species in Mbao's (2022) study were also abundant in this study. By mist-netting ten bat species, the study by Mbao (2022) supports the fact that with mist-netting bats only a few species can be captured, as does the study by Cockle *et al.* (1998) in the coastal forests of Tanzania.

Despite the paucity of bat species mist-netted in the current study in the two forests, in combination with the previous studies, PKNFR and RSFR have 17 and eight species of bats, respectively (see Howell, 1976; Cockle *et al.*, 1998; Mbao, 2022). This makes PKNFR to have the highest number of bat species in comparison with the other coastal forests whose bat fauna have been investigated, including Genda Genda (15 species), Mkulumuzi (10), Tongwe (10), Zaraninge (9), Tong'omba (8), Mafia Island (7), Kiwengoma (7), Somanga (4), Kisiju (3), Kama (2), Ruvu North (2) and Msubugwe (1) (Cockle *et al.*, 1998). The list of bats in PKNFR includes the Natal long-fingered bat *M. natalensis*, which was mist-netted for the first time in PKNFR and has not been recorded in the other coastal forests where bat research has been conducted (see Cockle *et al.*, 1998).

In Pande Game Reserve, another coastal forest, some 16 km from the study area, Hayes (2003) mist-netted 12 confirmed species, a number which is about 70% of the species that have been recorded in PKNFR. He mist-netted four species (*E. labiatus, E. wahlbergi, H. ruber* and *T. persicus*), which were also mist-netted in this study. Elsewhere in Tanzania, in relatively large areas of sampling, the number of species of bats mist-netted have been relatively lower than those which have been recorded in PKNFR. For example, in the Uzungwa Scarp Forest Reserve, Trentin and Rovero (2011) mist-netted 16 species of bats, including three fruit bat species (*E. wahlbergi, E. labiatus* and *R. aegyptiacus*) and one insectivoruous bat, Noack's leaf-nosed bat *H. ruber*), which have been recorded in PKNFR. In and around Chome Nature Forest Reserve, with bats being sampled in five different land uses, Katunzi *et al.* (2021) captured ten species using mist nets and harp traps. Four out of 11 species (*E. wahlbergi, R. aegyptiacs, M. vittatus* and *M. natalensis*) recorded by Katunzi *et al.* (2021) have been mist-netted in PKNFR.

Elsewhere in East Africa, in and around Lake Bogoria National Park, within Kenya's Great Rift Valley, Wechuli *et al.* (2016) mist-netted 11 species. In eastern Kenya, Webala *et al.* (2004) mist-netted 15 species in three vegetation types found in Meru National Park. However, in larger forests, and with increased sampling, the number of bat species could be higher. This has been well demonstrated in south-western Kenya by Webala *et al.* (2019) who mist-netted 26 bat species, four of which were captured in mist nets in the present study (*E. labiatus, E. wahlbergi, L. frons* and *N. thebaica*).

Generally, few bat species have been mist-netted by different authors in different areas in Tanzania (e.g. Cockle *et al.*, 1998; Hayes, 2003; Trentin and Rovero, 2011; Katunzi *et al.*, 2021; Mbao, 2022) and Kenya (Webala *et al.*, 2004; Wechuli *et al.*, 2016). The differences in number of species found in the study area and those recorded in other areas in Tanzania (e.g. Cockle *et al.*, 1998; Trentin and Rovero, 2011; Katunzi *et al.*, 2021) are mainly due to variations in size of the forest areas and varying sampling efforts. Larger patches of habitat tend to have more bat species because they offer a range of habitats which account for the numbers and individuals (Cockle *et al.*, 1998; Webala *et al.*, 2019). There is a significant correlation between the number of net-nights (i.e. sampling effort) and the number of individuals as well as the number of species caught in mist nets (Cockle *et al.*, 1998). In this case, Cockle *et al.* (1998) mist-netted more species in forests where they spent more mist-netting nights. For example, they captured 15 species in Genda Genda and 14 species in Kazimzumbwi because these forests received more mist-netting nights than the rest of the remaining 15 coastal forests.

Biogeographical aspects could also explain the differences in species mist-netted in this study and those from other studies in Tanzania. For example, there were more species and higher catch rates of bats mist-netted in the PKNFR than RSFR due to the presence of kaolin mines in the former. Similarly, there were more bat species mist-netted in the vicinity of these mines. These mines are known to have thousands of Noack's leaf-nosed bat and Lander's horseshoe bat in two separate mines, each coexisting with Persian trident bat (Howell, 1976).

10.5 Conclusion

With the present study and previous studies, 17 and 8 species of bats have been recorded in PKNR and RSFR, respectively, making a total of 19 species in the two forests. With 17 species, PKNFR has the highest bat species among the coastal forests of Tanzania where bat studies have been conducted (see Cockle *et al.*, 1998; Hayes, 2003). The kaolin mines found in the PKNFR are of conservation importance as thousands of Lander's horseshoe bat *R. landeri*, Persian trident bat *T. persicus* and Noack's leaf-nosed bat *H. ruber* inhabit them. Also, of interest is the Natal long-fingered bat *M. natalensis*, which was recorded for the first time in PKNFR in the current study. As such, PKNFR is important for conservation of a number of bat species and there is a need to continue conserving it with the kaolin mines included.

References

Burgess, N.D., Mwasumbi, L.B., Hawthorne, W.J., Dickinson, A. and Dogget, R.A. (1992) Preliminary assessment of the distribution, status and biological importance of coastal forests in Tanzania. *Biological Conservation* 62(3), 205–218. DOI: 10.1016/0006-3207(92)91048-W.

Burgin, C.J., Colella, J.P., Kahn, P.L. and Upham, N.S. (2018) How many species of mammals are there? *Journal of Mammalogy* 99, 1–14. DOI: 10.1093/jmammal/gyx147.

Castro-Arellano, I., Lacher, T.E., Willig, M.R. and Range, T.F. (2010) Assessment of assemblage-wide temporal niche segregation using null models. *Methods in Ecology and Evolution* 1(3), 311–318. DOI: 10.1111/%20j.2041-210X.2010%2000031.x.

Castro-Arellano, I., Presley, S.J., Willig, M.R., Wunderle, J.M. and Saldanha, L.N. (2009) Reduced impact logging and temporal activity of understory bats in lowland Amazonia. *Biological Conservation* 142(10), 2131–2139. DOI: 10.1016/j.biocon.2009.04.013.

Castro-Luna, A.A., Sosa, V.J. and Castillos-Campos, G. (2007) Bat diversity and abundance associated with the degree of secondary succession in a tropical forest mosaic in South-Eastern Mexico. *Animal Conservation* 10(2), 219–228. DOI: 10.1111/j.1469-%201795.2007.00097.x.

Cockle, A., Kock, D., Stubblefield, L., Howell, K.M. and Burgess, N.D. (1998) Bat assemblages in Tanzanian coastal forests. *Mammalia* 62(1), 53–68.

Cranbrook, The Earl of, Barrett, H.G. and Yates, F. (1965) Observations on noctule bats (*noctalus noctula*) captured while feeding. *Proceedings of the Zoological Society of London* 144, 1–24. DOI: 10.1111/j.1469-7998.1965.tb05163.x.

Happold, M. and Happold, D.C. (2013) *Mammals of Africa. Volume IV: Hedgehogs, Shrews, and Bats* Bloomsbury, London.

Hayes, B. (2003) Bats. In: Doggart, N. (ed.) *Pande Game Reserve: A Biodiversity Survey*. TFCG Technical Paper 7. Tanzania Forest Conservation Group, Dar es Salaam, pp. 40–44.

Howell, K.M. (1976) An ecological study of three species of insectivorous bats near Kisarawe, Tanzania. Phd thesis, University of Dar es Salaam.

Jones, G., Jacobs, D.S., Kunz, T.H., Willig, M.R. and Racey, P.A. (2009) Carpe noctem: The importance of bats as bioindicators. *Endangered Species Research* 8, 93–115. DOI: 10.3354/esr00182.

Katunzi, T., Soisook, P., Webala, P.W., Armstrong, K.N. and Bumrungsri, S. (2021) Bat activity and species richness in different land-use types in and around Chome nature forest reserve, Tanzania. *African Journal of Ecology* 59, 117–131. DOI: 10.1111/aje.12783.

Kingdon, J. (1974) In: *East African Mammals. An Atlas of Evolution in Africa,* Volume 2. Part A. Academic Press, London and New York.

Kunz, T.H. (1973) Resource utilisation: Temporal and spatial components of bat activity in central Iowa. *Journal of Mammalogy* 54, 14–32. DOI: 10.2307/1378869.

Kunz, T.H. and Brook, C.E. (1975) A comparison of mist nets and ultrasonic detectors for monitoring flight activity of bats. *Journal of Mammalogy* 56, 907–911. DOI: 10.2307/1379662.

Kunz, T.H. and Pierson, E.D. (1994) Bats of the world: an introduction. In: Nowak, R.W. (ed.) *Walker's Bats of the World*. Johns Hopkins University Press, Baltimore, Maryland, pp. 1–46.

Kunz, T.H., Braun de, E., Bauer, D., Lobova, T. and Fleming, T.H. (2011) Ecosystem services provided by bats. *Annals of the New York Academy of Sciences* 1223(1), 1–38. DOI: 10.1111/j.1749-6632.2011.06004.x.

Laval, R.K. (1970) Banding returns and activity periods of some Costa Rican bats. *Southwestern Naturalist* 15, 1–10.

Laval, R.K. and Fitch, H.S. (1977) Structure, movements and reproduction in three Costa Rican bat communities. *Occasional Papers Museum of Natural History, University of Kansas* 69, 1–28.

Margules, C.R. and Pressey, R.L. (2000) Systematic conservation planning. *Nature* 405, 243–253. DOI: 10.1038/35012251.

Mbao, J. (2022) Assessment of plants and bats species diversity and distribution in Pugu-Kazimzumbwi Nature forest reserve, Kisarawe, Tanzania. MSc dissertation, University of Dar es Salaam.

Meyer, C.F.J., Fründ, J., Lizano, W.P. and Kalko, E.K.V. (2008) Ecological correlates of vulnerability to fragmentation in Neotropical bats. *Journal of Applied Ecology* 45(1), 381–391. DOI: 10.1111/j.1365-2664.2007.01389.x.

Meyer, C.F.J., Struebig, M. and Willing, M.R. (2016) Responses of tropical bats to habitat fragmentation, logging, and deforestation. In: Voigt, C.C. and Kingston, T. (eds) *Bats in the Anthropocene: Conservation of Bats in a Changing World*. Springer, Heidelberg, pp. 63–103.

Mligo, C. (2019) Post fire regeneration of indigenous plant species in the Pugu forest reserve, Tanzania. *Global Ecology and Conservation* 18(2019), e00611.

Monadjem, A., Taylor, P.J., Cotterill, F.P.D. and Schoeman, M.C. (2020) *Bats of Southern and Central Africa: A Biogeographic and Taxonomic Synthesis*, 2nd edn. Wits University Press, Johannesburg, South Africa.

Muscarella, R. and Fleming, T.H. (2007) The role of frugivorous bats in tropical forest succession. *Biological Reviews* 82, 573–590. DOI: 10.1111/j.1469-185X.2007.00026.x.

Patterson, B.D. and Webala, P.W. (2012) Keys to the bats (Mammalia: Chiroptera) of East Africa. *Fieldiana: Life and Earth Sciences* 6, 1–60. DOI: 10.3158/2158-5520-12.6.1.

Rocky, J. and Mligo, C. (2012) Regeneration pattern and size-class distribution of indigenous woody species in exotic plantation in Pugu forest reserve, Tanzania. *International Journal of Biodiversity and Conservation* 4(1), 1–14. DOI: 10.5897/IJBC11.198.

Seltzer, C.E., Ndangalasi, H.J. and Cordeiro, N.J. (2013) Seed dispersal in the dark: Shedding light on the role of fruit bats in Africa. *Biotropica* 45, 450–456. DOI: 10.1111/btp.12029.

Shilton, L.A., Altringham, J.D., Compton, S.G. and Whittaker, R.J. (1999) Old world fruit bats can be long-distance seed dispersers through extended retention of viable seeds in the gut. *Proceedings of the Royal Society of London B. Biological Sciences* 266, 219–223. DOI: 10.1098/rspb.1999.0625.

Simmons, N.B. (2005) Order Chiroptera. In: Wilson, D.E. and Reeder, D.M. (eds) *Mammal Species of the World: A Taxonomic and Geographic Reference*, Vol. 1. Johns Hopkins University Press, Baltimore, Maryland.

Trentin, L. and Rovero, F. (2011) Results of an inventory of bats in the Uzungwa Scarp forest reserve, Tanzania. *Journal of East African Natural History* 100, 45–57. DOI: 10.2982/028.100.0103.

Webala, P.W., Mwaura, J., Mware, J.M., Ndiritu, G.G. and Patterson, B.D. (2019) Effects of habitat fragmentation on the bats of Kakamega forest, Western Kenya. *Journal of Tropical Ecology* 35, 260–269. DOI: 10.1017/S0266467419000221.

Webala, P.W., Oguge, N.O. and Bekele, A. (2004) Bat species diversity and distribution in three vegetation communities of Meru National Park Kenya. *African Journal of Ecology* 42, 171–179. DOI: 10.1111/j.1365-2028.2004.00505.x.

Wechuli, D.B., Webala, P.W., Patterson, B.D. and Ochieng, R.S. (2016) Bat species diversity and distribution in a disturbed regime at the lake Bogoria national reserve, Kenya. *African Journal of Ecology* 55, 465–476. DOI: 10.1111/aje.12376.

Wibbelt, G., Moore, M.S., Schountz, T. and Voigt, C.C. (2010) Emerging diseases in Chiroptera: Why bats? *Biology Letters* 6, 438–440. DOI: 10.1098/rsbl.2010.0267.

11 Wildlife Conservation Implications of Large Mammals along the Standard-gauge Railway from Dar es Salaam to Isaka, Kahama, Tanzania

Deusdedith F. Bwenge[1*], Catherine A. Masao[2], Felix A. Shayo[1] and Baraka N. Mbwambo[1]

[1]Tanzania Wildlife Research Institute, Njiro, Arusha, Tanzania; [2]Institute of Resource Assessment, University of Dar es Salaam, Tanzania

Abstract

Sustainable development strives for harmonization of economic development while protecting environmental resources for present and future generations. The government of Tanzania is implementing the construction of a standard-gauge railway (SGR) from Dar es Salaam to Isaka, Mwanza. This study examines the conservation implications in the pursuit of implementation of the SGR and protection of wildlife resources. A combination of methods was used to collect data including ground-walking transects, camera traps and key informant interviews. A total of 26 species of large mammals were observed along the proposed SGR alignment that are within eight orders and 12 families. Out of 26 observed species, African elephant, giraffe, leopard, lion, zebra and wild dog are listed by IUCN and CITES as species of conservation concern. Installation of electrified and high-tensile steel-wire fences and provision of crossings to facilitate free movements of wildlife along existing corridors are recommended.

11.1 Introduction

11.1.1 Background information

The sustainability component of the Sustainable Development Goals strikes a balance between present and future human well-being while limiting the levels of interruption of environmental resources by human activities, by living within the boundaries of ecological equilibrium for present and future generations (Arrow *et al.*, 1995; Alberti, 1996; Millennium Ecosystem Assessment, 2005). Infrastructural development is a very important investment for national development but in most cases it is done at the cost of environmental destruction. Therefore, during infrastructural development, planning needs to consider environmental health and natural resource concerns to enable managing and reducing possible impacts on ecosystems (Chen *et al.*, 2014; Kopperoinen *et al.*, 2014).

Sustainable development has begun to reflect on traditional development thinking while striving to find a development strategy that harmonizes industrial economic development and resource protection, and wildlife protection

*Corresponding author: deusdedith.fidelis@tawiri.or.tz

DOI: 10.1079/9781800626942.0011

gradually attracted people's attention through protecting wild animals and maintaining ecological balance while at the same time promoting economic development. The approach has recently attracted conservation concerns, working harder to balance natural resource protection in terms of long-term viability of species and ecosystems on the one hand, and socio-economic development on the other (Newmark and Hough, 2000; Du *et al.*, 2023). This is supported by many wildlife conservation efforts in Africa which focus on providing local communities with economic incentives to utilize wildlife as a form of land use in order to achieve the twin goals of ecological preservation as well as sustainable human economic development (Nelson, 2000; Jones *et al.*, 2012; Taruvinga, 2019).

The government of the United Republic of Tanzania through the Tanzania Railways Corporation (TRC) is currently implementing the construction of the standard-gauge railway (SGR) project from Dar es Salaam to Isaka under four lots covering a total of 1224 km for the whole project. This project passes across different areas with different land uses including ecologically sensitive areas like wildlife corridors and migration routes, and forests which are fundamental habitats for wildlife species.

Wildlife corridors and migration routes play a vital role to ensure wildlife protected areas host wildlife species that are important for the national economy and local communities for ecosystem services (Davenport and Davenport, 2006; TAWIRI (2009). Therefore, all proposed infrastructural developments must consider the welfare of wildlife and natural resources found within these areas. This is critically important because the SGR will be a high-speed railway and will definitely need to be fenced to keep people, livestock and wildlife away from the line. This has raised concerns over wildlife migration route blockage, habitat loss and fragmentation affecting wildlife species populations in the project area (Yapi Merkezi, 2019).

The proposed standard-gauge railway is an important catalyst for economic growth but it is expected to result in the loss of biological diversity including blockage of wildlife corridors and/or migratory routes and, consequently, fragmentation of habitats from changes in land use and land cover (Davenport and Davenport, 2006; Peres, 2010). Biodiversity loss offsets must be considered in the context of the mitigation hierarchy, namely: avoidance, minimization, restoration and offset. Avoidance measures are taken to prevent adverse effects on biological diversity, while minimization refers to reducing the duration, intensity or spatial extent of effects that cannot be avoided. Moreover, restoration refers to the rehabilitation of ecosystems adversely affected by infrastructure development, and offsets are measures taken to compensate for any adverse effects on biological diversity that cannot be avoided, minimized or mitigated (BBOP, 2009; Quintero and Mathur, 2011).

The objective of this chapter is to document the baseline information of large mammal species of wildlife and interpret the ecological consequences of spatial use for the SGR project and the conservation implications along the proposed railway from Dar es Salaam to Isaka in lots 1–4, by documenting the species composition and distribution. This study further investigates the permeability of the SGR to develop an opinion on the severity of wildlife fragmentation impacts. This document will, further, inform the development of a Biodiversity Action Plan (BAP) for the SGR project.

11.2 Methodology

11.2.1 Study area

This chapter focuses on larger mammals observed along the SGR project from Dar es Salaam to Isaka, Kahama, implemented under four different phases, namely: Lot 1 from Dar es Salaam to Morogoro (300 km) crosses different regions including Dar es Salaam in Ilala Municipal; Pwani in Kisarawe and Kibaha districts; and Morogoro region through Morogoro and Mvomero districts. Lot 2 from Morogoro to Makutopora, Dodoma (426 km) crossing Morogoro, Dodoma and Singida regions passing through Mvomero, Kilosa, Mpwapwa, Chamwino, Dodoma City Council, Bahi and Kintinku in Manyoni; Lot 3 from Makutopora to Tabora (368 km); and Lot 4 from Tabora to Isaka, Kahama (130 km), in total covering 1224 km for the whole length of the project (Fig. 11.1).

The SGR project traverses different land uses ranging from small- to large-scale farming,

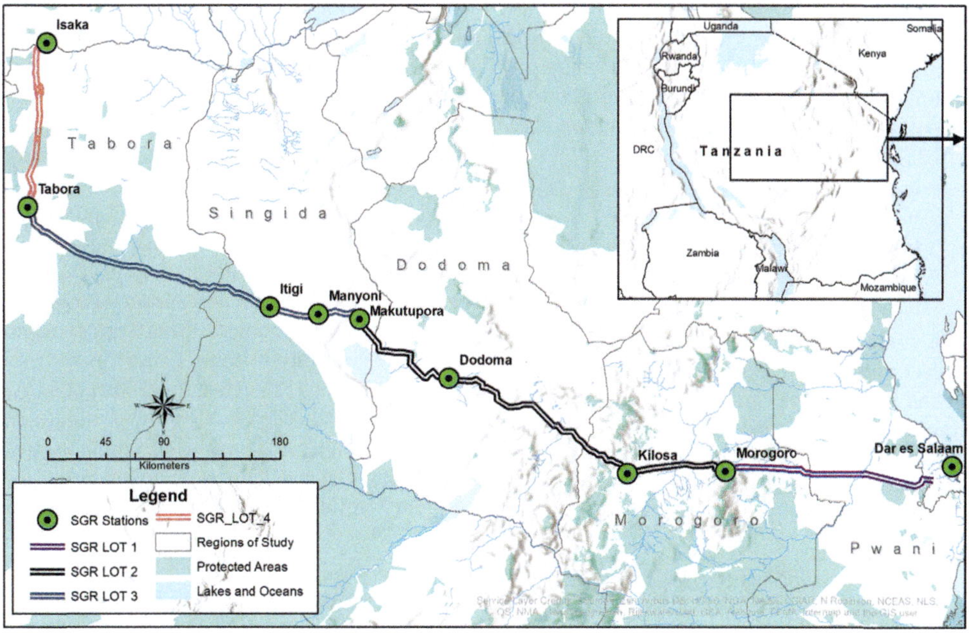

Fig. 11.1. Map of Tanzania showing the SGR line project area from Dar es Salaam to Isaka, Kahama.

residential areas, townships and conservation areas, including forest reserves and wildlife corridors, which are important habitats for wildlife species including large mammals. The forests include Pugu Forest Reserve; Kazimzumbwi Forest Reserve; Ruvu South Forest Reserve; Mkulazi and Ngerengere forest reserves, which form part of Nyerere-Wamimbiki Wildlife Corridor; Nyerere National Park; Wamimbiki Game Reserve; Mikumi-Wamimbiki Wildlife Corridor, which connects Mikumi National Park to Wamimbiki Game Reserve; Kizigo–Swagaswaga Wildlife Corridor in Manyoni and Itigi; Chaya Game Controlled Area in Itigi and Tabora; and the associated thickets which form important habitats for wildlife species in the study area extending from Rungwa, Kigosi, Kizigo and Ugalla river game reserves (TAWIRI (2009).

11.2.2 Study design

The study area was divided into two sections, namely: proposed Project Impact Zone (which is comprised of railway line buffer zone); and Control Zones. The Project Impact Zone comprised the SGR line from 0 to 10m within the rail corridor on both sides, where all observations and assessments were conducted. The control zones/lines were conducted 100m away from the impact zone on the right and left side of the proposed SGR line. Walking transects of 1 km length were designed within the proposed SGR line in both project impact zones and control zones, where a total of 56 transects (14 in Lot 1, 18 in Lot 2, 16 in Lot 3, and 8 in Lot 4) were conducted in the project impact lines while 112 transects (28 in Lot 1, 36 in Lot 2, 32 in Lot 3 and 16 in Lot 4) were conducted in the control zones/lines. The assessment was conducted on specific sampling sites that were selected during the preliminary surveys and were verified from maps; the selection considered areas with healthy forests and wildlife corridors with possibilities of inhabiting wildlife species.

The ground-walking transects for the assessment of wildlife species were established using a hand-held global position system (GPS) receiver (Garmin 64s), where transect length was set to 1 km through available trails in the project areas, and every transect was set to start at each assessment zone of the selected sampling site(s) throughout the study area as in Norton-Griffiths (1978). Also, in this study, wildlife species were categorized to be large mammals based on their body size where any animal species above 5 kg was categorized to be a large mammal as adopted in Emmons and Feer (1997) and Happold (2013).

11.2.3 Data-collection methods and equipment

The study used a combination of methods to collect data for large mammal wildlife species in the project area, including ground-walking transects, camera traps and key informant interviews to highlight wildlife species compositions, abundance and habitat uses in the project area, as described below.

Ground-walking transect surveys

The ground-walking transects of 1 km were conducted for the assessment of wildlife species (large mammals) at each of the selected samples of project areas. Data were collected by an observation team composed of five observers (two on the left and two on the right, and one person as data recorder, as adopted in Norton-Griffiths (1978) and Drapier *et al.* (2006). The observation team walked through established transects and split apart to a distance of up to 10m to maximize observations in all directions and increase sighting chances for wildlife species for direct and/or indirect signs observations.

Information collected along the walking transects included date, time, species name, group size, coordinates/GPS waypoint (position), habitat and project section (lot). Further evidence/signs of wildlife observed along the walking transects were photographed (Appendix 11.2).

Camera trapping

During the survey, camera traps were opportunistically deployed in the project area, specifically in areas which were selected during ground-walking transects. Opportunistic deployment of camera traps was done by selecting areas with high potential for the presence of wildlife species of large mammals based on what was observed during walking transects. The cameras were set on trails for walking animals and/or closer to water points for animals that would be coming to drink. However, cameras were only deployed after ascertaining security to avoid camera vandalism, because all the areas were in the village lands.

The camera traps used were the Bushnell (Browning trail cameras BTC-SHDP Strick force HD Pro), which were set with intervals of 5 seconds between photos to allow photo-bursts (rapid fire with no delay) to ensure all species passing the traps were captured, as in O'Connell *et al.* (2011) and Swann and Perkins (2014).

A total of 36 camera traps were deployed along the entire study area (4 cameras were set in Lot 1, 6 in Lot 2, 20 in Lot 3 and 5 in Lot 4). All cameras were checked every 7 days for batteries and functionality and were left for up to 20 trap nights; after that, photos were downloaded and species were identified for reporting (Appendix 11.3).

Key informant questionnaire interviews

Key informant questionnaire interviews were designed for community members with knowledge of wildlife by asking them if they had any idea of the presence of wildlife species and if they could show any sightings or signs of the presence of wildlife in the area. The species mentioned by respondents were cross-checked in the field guides (Kingdon 2015; Stuart and Stuart 2000) to ascertain the wildlife species mentioned by the respondents in the project area and to be sure of the exact species mentioned.

11.3 Results and Discussion

11.3.1 Large mammal species composition along the SGR

Results show a total of 26 species of large mammals were recorded along the proposed standard-gauge railway from Dar es Salaam to Isaka based on transects, camera trapping and interviews. The 26 species recorded fell within eight orders and 12 families. The eight orders observed are: Artiodactyla, Carnivora, Primate, Proboscidea, Rodentia, Lagomorpha, Perissodactyla and Tubulidentata, while the 12 families are: Bovidae, Canidae, Cercopithecidae, Felidae, Viverridae, Elephantidae, Equidae,

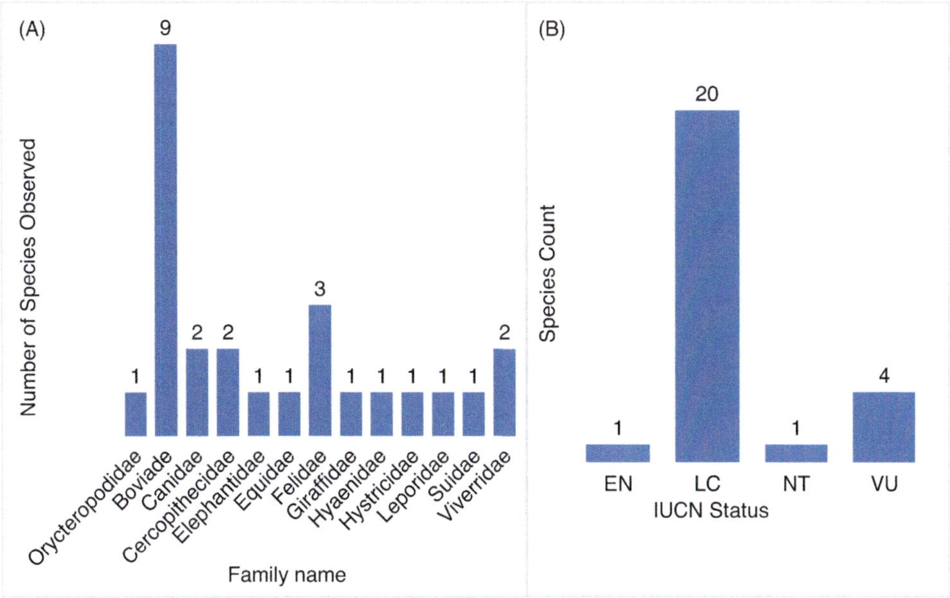

Fig. 11.2. Species count by family and IUCN status along the standard-gauge railway from Dar es Salaam to Isaka, Kahama (2019–2022). (VU, Vulnerable; EN, Endangered; NT, Near Threatened; LC, Least Concern). (A) Families and species observed, (B) IUCN species count.

Giraffidae, Hyaenidae, Hystricidae, Leporidae and Suidae (Appendices 11.1 and 11.3; Fig. 11.2).

Out of 26 species recorded, four (African elephant *Loxodonta africana*, giraffe *Giraffa camelopardalis*, leopard *Panthera pardus* and lion *Panthera leo* are listed as Vulnerable (VU) according to the IUCN Red List). Wild dog *Lycaon pictus* is listed as Endangered (EN), Zebra *Equus burchellii* as Near Threatened (NT), while the other remaining 20 species are listed as Least Concern (LC) (IUCN, 2024) (Table 11.1).

The IUCN Red List monitors and measures progress towards the biodiversity targets that were adopted by governments at the Conference of the Parties to the Convention on Biological Diversity (CBD) 2010, which aimed to prevent the extinction of known threatened species by 2020. Tanzania signed the CBD in 1992 and ratified the Convention in March 1996, which commits the parties/states to protect the biodiversity resources under the IUCN Red List of threatened species (IUCN, 2024).

Furthermore, the African elephant *Loxodonta africana*, leopard *Panthera pardus* and lion *Panthera leo* are listed in Appendix 1 of the Convention on International Trade in Endangered Species of Wild Fauna and Flora (CITES), which includes species threatened with extinction and provides the greatest level of protection, including a prohibition on commercial trade. Also, the giraffe, *Giraffa camelopardalis*, is listed in Appendix 2 of CITES, which includes species that are not currently threatened with extinction but may become so without trade controls. CITES, as a global agreement among state parties to regulate or ban international trade in species under threat, ensures that international trade in animals and plants does not threaten their survival in the wild. Tanzania ratified CITES on 29 November 1979, and the Convention entered into force on 27 February 1980. Therefore, as part of implementing this multi-lateral agreement, the country is obliged to regulate and maintain these biological resources for the present and the future.

The presence of 26 species recorded within eight orders and 12 families is also supported by Kingdon and Largen (1997) and Wilson and Reeder (2005). Worldwide, there are known 29 orders and 153 species, while in Africa there are 13 orders and 50 families. Therefore, the presence of the recorded species contributes to

Table 11.1. Species count by family and IUCN status along the standard-gauge railway from Dar es salaam to Isaka, Kahama, 2019–2022.

	Family	Order	Species	Common name	IUCN status
1	Orycteropodidae	Tubulidentata	Orycteropus afer	Aardvark	LC
2	Boviade	Artiodactyla	*Syncerus caffer*	Buffalo	LC
		Artiodactyla	*Tragelaphus scriptus*	Bushbuck	LC
		Artiodactyla	*Sylvicapra grimmia*	Common duiker	LC
		Artiodactyla	*Madoqua kirkii*	Kirk's dikdik	LC
		Artiodactyla	*Taurotragus oryx*	Eland	LC
		Artiodactyla	*Alcelaphus buselaphus*	Hartebeest	LC
		Artiodactyla	*Aepyceros melampus*	Impala	LC
		Artiodactyla	*Tragelaphus imberbis*	Lesser kudu	LC
		Artiodactyla	*Raphicerus campestris*	Steenbok	LC
3	Canidae	Carnivora	*Canis mesomelas*	Black-backed jackal	LC
		Carnivora	*Lycaon pictus*	Wild dog	EN
4	Cercopithecidae	Primate	*Cercopithecus aethiops*	Vervet monkey	LC
		Primate	*Papio cynocephalus*	Yellow baboon	LC
5	Elephantidae	Proboscidea	*Loxodonta africana*	Elephant	VU
6	Equidae	Perissodactyla	*Equus burchellii*	Zebra	NT
7	Felidae	Carnivora	*Panthera pardus*	Leopard	VU
		Carnivora	*Panthera leo*	Lion	VU
		Carnivora	*Leptailurus serval*	Serval cat	LC
8	Giraffidae	Artiodactyla	*Giraffa camelopardalis*	Giraffe	VU
9	Hyaenidae	Carnivora	*Crocuta crocuta*	Spotted hyena	LC
10	Hystricidae	Rodentia	*Hystrix cristata*	Porcupine	LC
11	Leporidae	Lagomorpha	*Lepuss axatilis*	Scrub hare	LC
12	Suidae	Artiodactyla	*Potamochoerus larvatus*	Wild pig	LC
13	Viverridae	Carnivora	*Civettictis civetta*	African civet	LC
		Carnivora	*Genetta genetta*	Small spotted genet	LC

both the global and African regional population of mammalian wildlife species. Forming the global and African regional wildlife species is the fundamental reason for conserving the species alongside infrastructural development.

The recorded species seem comparatively few from a global and African perspective, probably because the study was conducted while construction activities were ongoing throughout the project area; this might have caused fear and disturbance to animals because construction activities had removed vegetation cover in the area, forcing animals to move away from the project area. Such wildlife species composition and relative abundance is demonstrated in areas with space, cover, water and food resources (Mekonnen *et al.*, 2011; Bantihun and Bekele, 2015).

11.3.2 Large mammal species distribution along SGR

The habitat uses and distribution of large mammals along the SGR is shown in Table 11.1 and Appendix 11.1, where the species that were recorded in lots 1–4 include African civet, buffalo, black-backed jackal, common duiker, leopard, lion, porcupine, hare, serval cat, small spotted genet, spotted hyena, yellow baboon and vervet monkey. Elephant and impala were only recorded in lots 1, 2 and 3. Wild dogs were recorded in lots 2 and 3. Furthermore, lesser kudu, common eland and giraffe were only recorded in lots 3 and 4, while aardvark was only recorded in Lot 4.

The reported species of large mammals along the study area are also recorded in other studies adjacent to the study area in Zaraninge, Pugu, Ruvu and Kazimzumbwi forest reserves, Nyerere-Wamimbiki and Mikumi-Wamimbiki wildlife corridors (Kiwia, 2006; Azeria *et al.*, 2007; Bukombe *et al.*, 2022). This shows that the SGR project area has biological resources with the potential to be conserved and for taking precautions with infrastructure construction along the areas. The species distribution depends on the position of the area in relation to areas with high vegetation cover close to healthy ecosystems which provide enough food, cover and space resources, which attract numerous mammalian species in the area (Bakala and Mekonen, 2020). The species recorded in the study area are closer to natural wildlife ecosystems including Nyerere and Mikumi national parks, Wamimbiki and Rungwa, Kigosi, Kizigo and Ugalla river game reserves, and the associated wildlife corridors: Nyerere-Wamimbiki, Mikumi-Wamimbiki, Kizigo-Swagaswaga, Manyoni and Itigi thickets and, Chaya game controlled area (TAWIRI (2009). These natural ecosystems are interconnected and cross the SGR project area forming important habitats for wildlife species.

11.3.3 Conservation implications of large mammals along the standard-gauge railway

The documentation and compilation of this chapter learned that the Biodiversity Action Plan of the SGR project states that the whole SGR line project (from Dar es Salaam to Isaka) might need to be fenced with high-tensile steel wire as a barrier to prevent direct contact between the railway line and other objects including wild mammals, livestock, humans and other animals. The fencing will involve all wildlife corridors and elephant sites along the SGR being installed with electrified elephant-proof fences with high-tensile steel wire along the SGR line. Reliable pre-construction data on the patterns of wildlife migration routes are significant as an optimal effective mitigation measure (Ament *et al.*, 2023).

These findings are important to identify target species and priority areas for implementing mitigation measures. Since one or more target species at the project area can be selected as a focus for planning the mitigation measures, which will also serve other species in the area (Brennan *et al.*, 2022), the optimal consideration should be given to all wildlife species found at the area taking into account their ecological requirements, but with a focus on African elephant *Loxodonta africana* and giraffe *Giraffa camelopadalis* as target species of interest given their size, which increases their ecological requirements.

The fully grown African elephant ranges between 2.5 m and 4 m from shoulder to toe and weighs up to 7 tonnes, while the giraffe's average height ranges between 4.8 m and 5.5 m, and weighs about 1900 kg (Kingdon, 2015). The designed bridges, culverts and underpasses should be able to accommodate such height, and because they are the largest animals in the area, their dimensions automatically accommodate all animals in the area, thereby allowing free movement. These designed fences are intended to divert elephants towards appropriate animal-crossing structures to avoid risks in crossing the railway (Yapi Merkezi, 2019).

Since the SGR will be a high-speed railway, and to ensure safety, it will need to be fenced to keep people, livestock and wildlife away from the line. This raises concerns that wildlife migration routes could be blocked, leading to habitat fragmentation of the affected wildlife populations in the area, hence jeopardizing their conservation fate. Further to this, the placement of bridges, culverts and underpasses for elephants and other wildlife species is predicted to avoid fragmentation, allowing movements for genetic connectivity.

However, the actual response of elephants and other wildlife to new structures remains uncertain because the evidence of animals and their movements using the designated structures has not been proved. The presence of new structures along the wildlife corridors and/or migratory routes is also new to wildlife species; one cannot guarantee if wildlife will certainly use the designated structures (bridges, culverts, underpasses), which raises the question of conservation implications amid the infrastructural development.

The practical wildlife conservation implications may further include escalated human wildlife conflicts and also the possibility of limiting gene flow within wildlife species populations within the area as a result of blockage of wildlife corridors and/or migratory routes. This suggests wildlife movements and behavioural monitoring studies should be conducted after the construction of the mitigation structures to continue monitoring the effectiveness of the mitigations put in place against identified impacts and to ensure wildlife habitat connectivity among wildlife species populations between core protected areas.

11.4 Conclusion and Recommendations

11.4.1 Conclusion

This study illuminates the intricate interplay between infrastructure development and wildlife conservation, focusing on the construction of the SGR in Tanzania, from Dar es Salaam to Isaka. The imperative of addressing the impact of this substantial infrastructural expansion on biodiversity becomes strikingly evident given the intersection of the SGR with numerous wildlife corridors in the project area.

The results uncover a total of 26 large mammal species, each with its own conservation implications, including, but not limited to, Near Threatened, Vulnerable and Endangered species, according to the IUCN Red List. The documenting of species distribution across different lots accentuates the necessity for meticulously considering the SGR's permeability to mitigate the impact of wildlife fragmentation. Documenting baseline information regarding large mammals along the SGR and exploring the project's permeability form an important foundation for developing a comprehensive Biodiversity Action Plan.

11.4.2 Recommendations

The study recommends:

- in light of these compelling findings, and the Biodiversity Action Plan for the SGR project, the installation of high-tensile steel-wire fences along the entire project line, giving crossings at specific wildlife corridors and elephant sites (at Ngerengere, Mkata, Manyoni and Chala). The proposed crossing locations (bridges, culverts and underpasses) aim to allow free movement of wildlife while facilitating habitat connectivity and furthering species' genetic connectivity and assuage the potential fragmentation of wildlife populations. Viadacts are regarded as the most recommended solution for such areas with large numbers of wildlife, although the designs for phases 1 and 2 were already being implemented when this study was conducted;
- measures to deter livestock and wildlife, with specific crossing locations prompting concerns about potential impediments to wildlife migration routes and habitat fragmentation; and
- comprehensive and close monitoring of proposed mitigation measures, which are delineated in the mitigation hierarchy: avoidance, minimization, restoration and offset.

References

Alberti, M. (1996) Measuring urban sustainability. *Environmental Impact Assessment Review* 16, 381–424.

Ament, R., Clevenger, A. and van der Ree, R. (2023) "*Addressing Ecological Connectivity in the Development of Roads, Railways, and Canals*" (IUCN WCPA Technical Report Series).

Arrow, K., Bolin, B., Costanza, R., Dasgupta, P., Folke, C. *et al.* (1995) Economic growth, carrying capacity, and the environment. *Ecological Economics* 15, 91–95.

Azeria, E.T., Sanmartín, I., Ås, S., Carlson, A. and Burgess, N. (2007) Biogeographic patterns of the East African Coastal forest vertebrate fauna. *Biodiversity and Conservation* 16, 883–912.

Bakala, F. and Mekonen, G. (2020) Species diversity and relative abundance of medium and large-sized wild mammals: a study from Adaba community forest, West Arsi zone, Southeast Ethiopia. *African Journal of Ecology* 59(8), 1–6. DOI: 10.1111/aje.12827.

Bantihun, G. and Bekele, A. (2015) Population structure of small mammals with different seasons and Habitats in Arditsy forest, Awi zone, Ethiopia. *International Journal of Biodiversity and Conservation* 7(8), 378–387.

BBOP (Business and Biodiversity Offsets Programme) (2009) *Biodiversity Offset Design Handbook*. BBOP, Washington, DC.

Brennan, L., Chow, E. and Lamb, C. (2022) Wildlife overpass structure size, distribution, effectiveness, and adherence to expert design recommendations. *Peer Journal* 10, e14371.

Bukombe, J., Marealle, W., Kimaro, J., Kija, H., Kavana, P, *et al.* (2022) Viability assessment of the Wami-Mbiki game reserve to Nyerere National Park wildlife corridor in Southern Tanzania. *Global Ecology and Conservation* 39, 13. DOI: 10.1016/j.gecco.2022.e02259.

Chen, J., Sun, B.M., Chen, D., Wu, X., Guo, L.Z, *et al.* (2014) Land use changes and their effects on the value of ecosystem services in the small Sanjiang plain in China. *The Scientific World Journal* 752846. DOI: 10.1155/2014/752846.

Davenport, J. and Davenport, J.L. (eds) (2006) *The Ecology of Transportation: Managing Mobility for the Environment*. Springer, Dordrecht, The Netherlands.

Drapier, N.G., Fritz, H., Bourgarel, M., Renaud, P.C., Poilecot, P. *et al.* (2006) Cost and efficiency of large mammal census techniques; comparison of methods for a Participatory approach in a communal area, Zimbabwe. *Biodiversity and Conservation* 15, 735–754.

Du, N., Fathollahi-Fard, A.M. and Wong, K.Y. (2023) Wildlife resource conservation and utilization for achieving sustainable development in China: Main barriers and problem identification. *Environmental Science and Pollution Research*.

Emmons, L. and Feer, F. (1997) *Neotropical Rainforest Mammals: A Field Guide*. University of Chicago Press, Chicago, Illinois.

Happold, D.C.D. (2013) *Mammals of Africa. Volume III: Rodents, Hares, and Rabbits*. Bloomsbury, London. Available at: https://www.bloom sbury.com/us/mamma ls-of-africa-volume-iii-978140818 9924/ (accessed 6 May 2024).

IUCN (2024) *The IUCN Red List of Threatened Species*. Version 2023-1. Available at: https://www.iucnre dlist.org/ (accessed 6 May 2024).

Jones, T., Bamford, A.J., Ferrol-Schulte, D., McWilliam, N. and Rovero, F. (2012) Vanishing wildlife corridors and options for restoration: a case study from Tanzania. *Tropical Conservation Science* 5(4), 463–474. Available at: www.tropicalconservationscience.org

Kingdon, J. and Largen, M.J. (1997) The kingdon field guide to African mammals. *Zoological Journal of the Linnean Society* 120(4), 479.

Kingdon, J. (2015) *The Kingdon Field Guide to African Mammals*, 2nd edn. Bloomsbury Publishing, London.

Kiwia, H.Y.D. (2006) Species richness and abundance of large mammals in Zaraninge forest, coast region, Tanzania. *Tanzania Journal of Science* 32(2).

Kopperoinen, L., Itkonen, P. and Niemelä, J. (2014) Using expert knowledge in combining green infrastructure and Ecosystem services in land use planning: an insight into a new place-based methodology. *Landscape Ecology* 29(8), 1361–1375. DOI: 10.1007/s10980-014-0014-2.

Mekonnen, T., Yaba, M., Bekele, A. and Malcolm, J. (2011) Food selection and habitat association of Starck's hare (Lepus Starcki Petter, 1963) in the bale mountains National Park, Ethiopia. *Asian Journal of Applied Sciences* 4, 728–734. DOI: 10.3923/ajaps.2011.728.734.

Millennium Ecosystem Assessment (2005) *Ecosystems and Human Wellbeing: Synthesis*. Island Press, Washington, DC.

Nelson, F. (2000) Sustainable development and wildlife conservation in Tanzanian Maasailand. *Environment, Development and Sustainability* 2(2), 107–117.

Newmark, D.W. and Hough, J.L. (2000) Conserving wildlife in Africa: integrated conservation and development projects and beyond. *BioScience* 50(7), 585–592.

Norton-Griffiths, M. (1978) *Counting Animals: Handbook no.1 in a Series of Handbooks on Techniques in African Wildlife Ecology*. In: Grimsdell, J.J.R. (ed.) African Wildlife Leadership Foundation. Nairobi.

O'Connell, A.F., Nichols, D.J. and Karanth, U.K. (eds) (2011) *Camera Traps in Animal Ecology: Methods and Analyses*. Springer. DOI: 10.1007/978-4-431-99495-4_10.

Peres, C. (2010) Overexploitation. In: Sodhi, N.S. and Ehrlich, P.R. (eds) *Conservation Biology for All*. Oxford University Press, Oxford, United Kingdom.

Quintero, D.J. and Mathur, A. (2011) Biodiversity offsets and infrastructure. *Conservation Biology* 25(6), 1121–1123. DOI: 10.1111/j.1523-1739.2011.01769.x.

Stuart, C. and Stuart, T. (eds) (2000) *A Field Guide to the Tracks and Signs of Southern and East African Wildlife*, 3rd edn. Struik Publishers, South Africa.

Swann, D.E. and Perkins, N. (2014) Camera trapping for animal monitoring and management: a review of applications. *Paper*.

Taruvinga, P. (2019) *Stakeholders, conservation and socio-economic development: the case of Mapungubwe Cultural Landscape World Heritage Site, South Africa*. Dissertation, University of Cape Town.

TAWIRI (2009) Wildlife Corridors in Tanzania. (T. Jones, T. Caro, & T. R. B. Davenport, Eds). Tanzania Wildlife Research Institute. Available at: chrome-extension://oemmndcbldboiebfnladdacbdfmadadm/https://www.stzelephants.org/download/wildlife-corridors/Wildlife%20Corridors%20in%20Tanzania.pdf

Wilson, D.E. and Reeder, D.M. (eds) (2005) *Mammal Species of the World: A Taxonomic and Geographic Reference vol. 1*. Johns Hopkins University Press, Baltimore, Maryland. Available at: https://books.google.co.uk/books?id=JgAMbNSt8ikC&printsec=frontcover#v=onepage&q&f=false

Yapi Merkezi (2019) Standard Gauge Railway – Dar es Salaam to Makupopora, Tanzania: Biodiversity Action Plan. ERM Project no.0462176. Available at: chrome-extension://oemmndcbldboiebfnladdacbdfmadadm/https://www.trc.co.tz/uploads/publications/sw1572527026-BAP%20SGR%20latest.pdf (accessed 7 May 2024).

Appendix 11.1 Checklist of Large Mammal Species of Wildlife Recorded during Environmental and Social Impact Assessment Study Conducted along the Standard-gauge Railway from Dar es Salaam, 2019–2022 (VU, Vulnerable; EN, Endangered, NT, Near Threatened; LC, Least Concern)

S/N	Common name	Scientific name	Order	Family	IUCN Status	CITES Appendix	Method for data collection	Project location
1	Aardvark	*Orycteropus afer*	Tubulidentata	Orycteropodidae	LC	Not listed	Camera & Walking transects	Lot 3
2	Buffalo	*Syncerus caffer*	Artiodactyla	Boviade	LC	Not listed	Walking transects & Questionnaires	Lot 1, 2, 3 & 4
3	Black-backed Jackal	*Canis mesomelas*	Carnivora	Canidae	LC	Not listed	Questionnaires	Lot 1, 2, 3 & 4
4	Bushbuck	*Tragelaphus scriptus*	Artiodactyla	Bovidae	LC	Not listed	Walking transects	Lot 3
5	African civet	*Civettictis civetta*	Carnivora	Viverridae	LC	Not listed	Camera & Walking transects	Lot 1, 2, 3 & 4
6	Common duiker	*Sylvicapra grimmia*	Artiodactyla	Bovidae	LC	Not listed	Camera, Walking transects & Questionnaires	Lot 1, 2 & 3
7	Kirk's dikdik	*Madoqua kirkii*	Artiodactyla	Bovidae	LC	Not listed	Walking transects	Lot 3
8	Common Eland	*Taurotragus oryx*	Artiodactyla	Bovidae	LC	Not listed	Walking transects & Questionnaires	Lot 3 & 4
9	African Elephant	*Loxodonta africana*	Proboscidea	Elephantidae	VU	I	Walking transects & Questionnaires	Lot 1, 2, 3
10	Giraffe	*Giraffa camelopardalis*	Artiodactyla	Giraffidae	VU	II	Questionnaires	Lot 3 & 4

Continued

Continued

S/N	Common name	Scientific name	Order	Family	IUCN Status	CITES Appendix	Method for data collection	Project location
11	Hartebeest	*Alcelaphus buselaphus*	Artiodactyla	Bovidae	LC	Not listed	Walking transects	Lot 3
12	Impala	*Aepyceros melampus*	Artiodactyla	Bovidae	LC	Not listed	Walking transects	Lot 1, 2 & 3
13	Leopard	*Panthera pardus*	Carnivora	Felidae	VU	I	Walking transects & Questionnaires	Lot 1, 2, 3 & 4
14	Lesser Kudu	*Tragelaphus imberbis*	Artiodactyla	Bovidae	LC	Not listed	Camera, Walking transects & Questionnaires	Lot 3 & 4
15	Lion	*Panthera leo*	Carnivora	Felidae	VU	I	Questionnaires	Lot 1, 2, 3 & 4
16	Porcupine	*Hystrix cristata*	Rodentia	Hystricidae	LC	Not listed	Walking transects	Lot 1, 2, 3 & 4
17	Small spotted genet	*Genetta genetta*	Carnivora	Viverridae	LC	Not listed	Camera traps	Lot 1, 2, 3 & 4
18	Spotted hyena	*Crocuta crocuta*	Carnivora	Hyaenidae	LC	Not listed	Walking transects &Questionnaires	Lot 1, 2, 3 & 4
19	Steenbok	*Raphicerus campestris*	Artiodactyla	Bovidae	LC	Not listed	Camera traps	Lot 3
20	Vervet monkey	*Cercopithecus aethiops*	Primates	Cercopithecidae	LC	Not listed	Walking transects	Lot 1, 2, 3 & 4
21	Wild dog	*Lycaon pictus*	Carnivora	Canidae	EN	Not listed	Questionnaires	Lot 2 & 3
22	Wild pig	*Potamochoerus larvatus*	Artiodactyla	Suidae	LC	Not listed	Walking transects, Questionnaires	Lot 1, 2, 3 & 4
23	Yellow baboon	*Papio cynocephalus*	Primate	Cercopithecidae	LC	Not listed	Walking transects & Questionnaires	Lot 1, 2, 3 & 4
24	Zebra	*Equus burchellii*	Perissodactyla	Equidae	NT	Not listed	Questionnaires	Lot 2, 3 & 4

Appendix 11.2 Evidence/signs of Large Mammal Species of Wildlife Recorded during Environmental and Social Impact Assessment Study Conducted along the Standard-gauge Railway from Dar es Salaam, 2019–2022

SN	Photographic evidence/signs	Species name (sign)
1		African buffalo (dung)

Continued

Continued

SN	Photographic evidence/signs	Species name (sign)
2		African civet (dung)
3		African elephant (dung)
4		Common duiker (pellets)

Continued

Continued

SN	Photographic evidence/signs	Species name (sign)
5		Lesser kudu (pellets)
6		Kirk's dikdik (pellets)
7		Spotted hyena (dung)

Continued

Continued

SN	Photographic evidence/signs	Species name (sign)
8		Leopard (dung)

Appendix 11.3 Photos of Large Mammal Species of Wildlife Captured through Camera Traps during Environmental and Social Impact Assessment Study Conducted along the Standard-gauge Railway from Dar es Salaam, 2019–2022

SN	Photo	Species name
1		African civet
2		Common duiker

Continued

Continued

SN	Photo	Species name
3		Lesser kudu
4		Steenbok
5		Small spotted genet

Continued

Continued

SN	Photo	Species name
6		African wildcat

12 Prospects on Biodiversity along the Standard-gauge Railway from Dar es Salaam to Makutupora, Singida, Tanzania

Catherine A. Masao[1*], Joel Nobert[1] and Flora John Magige[2]
[1]Institute of Resource Assessment, University of Dar es Salaam, Mlimani, Dar es Salaam, Tanzania; [2]College of Natural and Applied Sciences, University of Dar es Salaam, Dar es Salaam, Tanzania

Abstract
The construction of the standard-gauge railway (SGR) from Dar es Salaam to Makutupora, Singida, represents one of major infrastructural advancements with significant economic potentials for Tanzania. The infrastructure traverses the Pugu-Kazimzumbwi Nature Forest Reserve, Pugu Hills Forest Reserve and Ruvu South Forest Reserve that already are experiencing adverse effects on their biodiversity following unchecked human activities in this Anthropocene era. Developmental activities, on the other hand, pose similar problems for the biodiversity, with some species on the verge of extinction. The standard-gauge railway construction is one of the developmental activities in Tanzania that has contributed to the clearance of the forests. Studies conducted on land-use and land-cover changes and various taxa (flora and fauna), including threatened, rare, and bio-indicator species, have expressed conservation concerns. Amongst important conservation concerns that already have impact on the biodiversity include habitat fragmentation and/or loss, edge effects, direct wildlife displacement and mortality, and alterations of ecosystems. Habitat fragmentation threatens species survival by isolating populations and reducing genetic diversity. Understanding these impacts, well planned mitigation measures must be considered to safeguard the biodiversity including establishing proper land-use management plans, forest restoration, control of invasive species, habitat conservation, and continuous monitoring of the biodiversity, for a successful outcome to be realized.

12.1 Introduction

Biodiversity is in crisis in this Anthropocene era where humans are to blame for degrading the natural habitats, polluting pristine environments, and fuelling climate change. Development activities such as road and railway constructions are among the activities that, if not checked, will have a negative impact on biodiversity, including species extinction.

The Standard-Gauge Railway (SGR) infrastructure developments represent one of mega-projects in East Africa in recent years. While the railways promise to boost economic growth by enhancing transportation efficiency, it also poses potential threats to the region's rich biodiversity. The Tanzania SGR (Dar es Salaam – Mwanza via Isaka Kahama) (TRC, 2021) is one of the good case examples in the East African region that traverses a diverse array of

*Corresponding author: masao.catherine@udsm.ac.tz; ndeutz@yahoo.com

 Biodiversity and Conservation along an East African Railway (eds C.A. Masao *et al.*)
DOI: 10.1079/9781800626942.0012

ecosystems, including coastal forests, miombo woodland, wetlands, savannas, and highland areas, each of which supports unique and often endemic species. The SGR alignment that runs from Dar es Salaam to Makutupora is a typical example of areas where human footprint and biodiversity degradation can be seen. Recent studies conducted along the alignment have indicated impacts on the environment emerging from the SGR construction and other economic activities (TRC, 2021).

The studies have provided baseline status of the areas which also highlights some clear patten of impacts of the project especially those resulting from cleared vegetation and blockage of wildlife movement corridors. Largely, the areas have rich biodiversity and require necessary measures to ensure no net loss and net gain according to IFC-PS6 (IFC, 2012). The data in the book serves as an input to the monitoring and other necessary mitigation measures that are proposed under the ESIA studies conducted prior to commencement of the project. The chapter is a narrative review of the existing literature of the studies conducted in study sites and beyond to establish the SGR impacts and prospects on biodiversity. The chapter benchmarks the current state of the art of the areas and set the scen to future feedback on the various mitigation measures already been implemented in the project areas. The same data and updates on success and failure stories may be tapped from similar studies in East African and beyond, thus, to increase the data on the biodiversity along the line projects.

12.2 Potential impacts and prospects of SGR on Biodiversity

12.2.1 Impacts of SGR on Biodiversity

Studies have indicated that, line projects especially SGR have both prospects and negative impacts on the biodiversity. Some key negative impacts include i) habitat fragmentation and loss (Caro *et al.*, 2017; Fahrig, 2003; Laurance *et al.*, 2014); species displacement and mortality (Francis *et al.*, 2009; and iii) edge effects (Harper *et al.*, 2005) as explained hereunder:

i. Habitat fragmentation and loss:
The SGR construction has led to habitat fragmentation, which can disrupt wildlife movement and gene flow, potentially leading to population declines and reduced biodiversity (Fahrig, 2003). The linear nature of the railway creates barriers for species that require large, contiguous habitats, particularly large mammals, and birds (Caro *et al.*, 2017; Laurance *et al.*, 2014).

ii. Edge effects:
The introduction of a railway line creates new habitat edges, which can alter microclimatic conditions and increase vulnerability to invasive species, predators, and human encroachment (Harper et al., 2005). These edge effects often result in reduced species richness and altered community compositions along the railway corridor, particularly in forested and woodland habitats.

iii. Species displacement and mortality:
Railway operations can lead to direct mortality of species through collisions, especially for slow-moving animals such as reptiles and amphibians. Also, the noise, vibrations, and light pollution associated with railway traffic can displace sensitive species, further impacting local biodiversity (Francis *et al.*, 2009).

12.2.2 Prospects of SGR on Biodiversity

Despite the negative impacts that are associated by SGR infrastructure, the implementation of effective mitigation strategies such as biodiversity offsetting, acceptable wildlife movement corridors, and continuous monitoring, there is potential to minimize effects and enhance biodiversity conservation in the target areas. Biodiversity offsetting seeks to compensate the losses caused by infrastructure projects by protecting or restoring habitats elsewhere (IFC, 2012). The biodiversity offset actions therefore provide opportunities to conserve areas beyond the project sites that may not necessarily have been degraded by the specific SGR project but will add a positive value to ecosystems in a particular area. Furthermore, construction of the wildlife movement corridors may enhance wildlife movements and channel their directions towards areas most potential for the resources they need. In a long run, conservation may be improved, and habitats would be restored and net gain in biodiversity be realized.

12.3 Impacts and Prospects of SGR on Biodiversity along the Dar es Salaam – Makutupora alignment - Tanzania

From the understanding that, SGR alignment representing other line infrastructure present significant impacts on the biodiversity, we present some of the findings from the studies to confirm this fact. Analyses of land-use cover change of the areas indicated that the SGR alignment from Dar es Salaam to Makutupora has contributed to the clearance of about 614 ha of forest to pave the way for construction activities in the Pugu-Ruvu ecosystem alone. Prior to the SGR activities, about 581 ha in the Ruvu-Pugu ecosystem were converted from forest to non-forest due to other disturbances that are not directly related to the SGR but have a contributory effect on the ecosystem (TRC, 2021). Such activities include settlement, kaolin mining and extraction of wood for timber and charcoal. Although various mitigation measures have been proposed in the Environmental and Social Impact Assessment (ESIA) reports, there was no sufficient evidence to conclude that the ecosystem could recover on its own through a process of natural succession. Earlier studies along the line projects had also indicated that habitat fragmentation and loss (Caro *et al.*, 2017; Laurance *et al.*, 2014) is among negative impacts of SGR projects thus necessitating the need for careful implementation of relevant mitigation measures to reduce or offset the impacts. Mitigation measures that are been used include restoration, offset and rehabilitation of the habitats.

Various chapters in this book have indicated that some areas along the SGR alignment still partly harbour key biodiversity species and important wildlife areas of biological significance. Therefore, the future of the biodiversity along the SGR alignment from Dar es Salaam to Makutupora (Chapters 2–10) and to Isaka (Chapter 11) largely depends on the effective implementation of mitigation measures and ongoing conservation efforts which already exist in these areas.

In Chapter 2, an analysis of the land-use and land-cover change of the areas has indicated a loss of about 3% of the forest cover that is within the Pugu-Kazimzumwi Nature Forest Reserve (PKNR) and the Ruvu South Forest Reserve (RSFR). The two forests are cited as the last vestiges of lowland coastal forest in Tanzania and contain a unique flora with 12 endemic species and one endemic genus containing the East African coastal bird the Sokoke Pipit, and a few bird species with unusual or restricted distributions are found in the forest, an important site for the Rondo dwarf galago, amongst others. The areas are also unique in terms of plants and contain a high plant diversity index (Shannon-Weiner Diversity Index) ranging between 3.4 and 4.1 (Chapter 4). Furthermore, 14 plant species that are either threatened or near-threatened according to the IUCN Red List have been recorded, indicating high conservation importance of the areas, especially those within PKNR and RSFR. This indicates the necessity to ensure that special measures are continually implemented to avoid net loss of biodiversity.

Despite clearings made through project activities, the traditional species composition that had previously been recorded was recorded and identified, which is an indication that species can be rescued. It was established that there are more intense anthropogenic activities in the Ruvu South reserve than in the Pugu-Kazimzumbwi reserve, probably because the latter has been uplifted to a nature reserve. The ongoing anthropogenic pressure includes logging, pole cutting and charcoal making, as evidenced by the existence of several charcoal kilns in the reserve. The most-logged tree species were *Pterocarpus angolensis*, *Baphia kirkii*, *Afzelia quanzensis*, *Milicia excelsa* and *Hymenaea verrucosa*. However, to a large extent, the observed anthropogenic pressure could not be directly associated with the SGR project activities, except in some areas where land excavation and stripped topsoil resulted in gully erosion and vegetation cover removal. Nevertheless, biodiversity loss emanating from the combined effect of various types of disturbance may be detrimental, considering the ecological significance of both PKNFR and RSFR.

PKNFR and RSFR host a significant number of plant species that belong to different conservation statuses, as per the IUCN and CITES categories, and a large part of the reserve can be regarded as a sensitive habitat. However, based on the overall population structure obtained from both PKNFR and RSFR, it is evident that most species, including the species of conservation concern, are regenerating. Several invasive species were identified within the two reserves amongst which *Lantana camara* and *Xanthium strumarium* were dominant. It is recommended that these species be controlled before they spread further into the reserve. *Leucaena* sp. is also spreading quite fast, and deliberate efforts should be made to curb it.

This book has further successfully documented 16 wetlands within the project footprint. It is interesting that some wetlands were recorded as emerging, with well-established habitats for wetlands species. Efforts to conserve these habitats are key to biodiversity net gain according to IFC (2012). Wetland species identified during wet and dry seasons are of least concern as regards the IUCN Red List. However, some species, such as the Ruvu spiny reed frog *Hyperolius ruvuensis*, which was recorded in 1990, was not recorded during this survey, hence calling for further studies to establish whether the species is extinct in the field (Chapters 3 and 8).

Arthropods are good bio-indicators of ecosystem status. An existing dataset of insect species within sites that cross the Pugu-Kazimzumbwi and Ruvu forest reserves has been provided. In both sites, bio-indicator species for disturbed habitats were identified. There was high relative abundance of some species towards alignment, whereas high diversity was recorded in less-disturbed habitats. This is a typical indication that the SGR may have some impact on the environment including decreasing species diversity, which may be recovered through post-construction restoration process. The findings are well matched with other authors who had earlier indicated that anthropogenic activities can positively and/or negatively influence the distribution of insect species and subsequently species richness (Chowdhury *et al.*, 2017; Boussougou *et al.*, 2018). Further, dry-season sampling yielded a high number of collections as compared to the wet season data in both sites. However, the average sample abundance was low in the dry season as compared to the wet season.

Species with known ecological functionality in the areas are also identified, indicating the necessity for a continued vigilant conservation of the habitats in the areas. Therefore, continuous monitoring of the diversity, abundance, and distribution of arthropods during construction and operation phases is recommended to establish the anthropoid species' trends, as a habitat restoration measure and for assurance of no net loss of some of the key species.

Macroinvertebrates and fish were used to find out if species could serve as indicators of predictable changes and modifications due to construction activities. This is because the two taxa groups have often been used for bioassessment as they are known to respond to habitat modification relatively easily and are quick to sample (Chapter 6). Selected wetland sites within a buffer of 60m along the SGR alignment were sampled both in wet and dry season for this purpose. Overall, 69% of the wetland assessed ranged from fair to very poor condition, while 31% ranged from good to very good condition, in terms of macroinvertebrates and fish indices. The loss and change in structure of macroinvertebrates and fish communities in the study area reflect a landscape modification. The modification altered species distributions, caused a simplification of aquatic community structure and composition, and increased the likelihood of non-native species. This situation, if unmediated, may lead to decreased capacity of wetlands along the SGR alignment to provide ecosystem services to human populations and the surrounding environment. It is confirmed that macroinvertebrates and fish are reliable bio-indicators for assessment of wetland status.

Despite the importance of herpetofauna as bio-indicators, there is a lack of studies on these taxa groups along the SGR alignment from Dar es Salaam to Makutupora. The assessment of species richness and composition revealed 51 amphibian and 110 reptilian species that occur within the project area. Only one species was recorded as endemic in the areas while one amphibian (reed frog *Hyperolius ruvuensis*) and two reptilian species were categorized as Critically Endangered, based on the IUCN Red List criteria. Although the reed frog *Hyperolius ruvuensis* had been previously recorded in the areas, the recent surveys showed no records of the same, indicating a need for further surveys to confirm the existence of the species in terms of its conservation status. The status confirms the previous reported findings that amphibians and reptiles are among threatened vertebrate groups resulting from habitat alteration caused by anthropogenic activities (Heatwole and Wilkinson, 2009; Menegon *et al.*, 2020). The state of the SGR alignment from Dar es Salaam to Makutupora could confirm the reality that the amphibian and reptile species within the areas could be threatened. This can be supported by the fact that the SGR already crosses the habitats known as home to two endemic and critically endangered herpetofauna species, including the Ruvu spiny reed frog *Hyperolius ruvuensis* and the blue gecko *Lygodactylus williamsi*. The railways may affect the herpetofauna in different ways including barrier effects, species invasions and environmental disturbances through noise and pollution (Borda-de-Água *et al.*, 2017). So far, studies to indicate the mortality rate of herpetofauna through collisions with trains are very few to negligible in many areas. The TRC has already embarked on monitoring studies as part of mitigation measures; however, it is recommended that herpetofauna passages in the form of tunnels, underpasses, and drainage culverts during construction, to prevent blockage of the movements of animals between the two sides of the railway, be provided. This may not have been a key issue during construction of the Tanzanian SGR alignment from Dar es Salaam to Makutupora.

Chapter 8 provides a compiled list of 267 (29% forest-dependent) and 201 (26% forest-dependent) bird species known in Pugu and Ruvu South reserves, respectively. This is an accumulation of the biodiversity data that have been collected since the 1990s. Interestingly, the bird list has been updated from 80 species in 1980 and 1990 to 300 species to date. The analysis has indicated no differences in species richness between wet and dry seasons, although the areas of the SGR alignment near the Msimbazi river seem to attract more bird species during the dry season. Further analysis has indicated a significant seasonal difference in bird diversity in the areas, which can be attributed to local movements, behavioural differences during the breeding season, and breeding plumage, pairing, and singing, which may influence detection. The areas are also harbouring endemic and threatened species that are listed in the IUCN Red List, including Southern Ground-hornbill, Bateleur, Great Snipe, Hooded Vulture, White-backed Vulture, Rüppell's Vulture, Martial Eagle, Tawny Eagle, and Steppe Eagle. Migratory bird species are also important species that have been recorded in the areas, including the African Pitta, an intra-African migrant.

This unique accumulation of bird data from the 1990s to the present is a significant contribution, especially amid the SGR construction activities that are underway and will have significant implications for the bird biodiversity in the reserves and along the alignment. The SGR adds a cumulative effect to the human footprint in the areas, since even before the SGR, the Pugu and Ruvu South reserves were facing serious threats, mainly from anthropogenic activities, leading to habitat degradation and fragmentation (Burgess *et al.*, 1992; Rocky and Mligo, 2012; Mligo, 2019). Their proximity to Dar es Salaam, the most populated region in Tanzania, poses threats due to the high demands for forest products such as charcoal, poles, and timber.

The increasing threat to the forests has implications for bird species that are already included in the IUCN Red List as Threatened. Some of the previous species that were not encountered in this survey include the Sokoke Pipit, known to have a very narrow range and was last recorded in 2002. Furthermore, only one individual of Spotted Ground-thrush (VU) was recorded, on 16 April 2021, in Pugu Forest Reserve, while previous surveys encountered 16 recorded sightings, in 1984 and 1991. This trend could be a good indication of the threat levels on birds in the areas.

The rich data of birds at the proximity and within the SGR alignment calls for special emphasis on habitat conservation and monitoring, principally due to the existence of many species that have significant conservation value, especially due to their threat levels globally. However, Pugu and Ruvu South forests remain among the best birding sites near Dar es Salaam and continue to offer opportunities for bird research and training for the universities and colleges. Continued restoration, monitoring and evaluation activities are among important mitigation measures that are necessary to ensure no net loss of biodiversity. On the other hand, the effect of the erected fence on both sides of the SGR alignment cannot be quantified, although it appears to be a potential barrier to movement of understorey bird species within the reserves. The effect of the fence needs to be studied using experimental capture-mark-recapture techniques and telemetry application to inform future SGR alignments within and beyond Tanzania.

Galago surveys were carried out both in Pugu Hills and Ruvu South Forest reserves, which consisted of remnant patches of coastal forest in Tanzania. Three species of galagos were encountered in which *G. rondoensis* was among the most recorded species in the evergreen forest and thicket vegetation in both reserves. Perkin (2020) indicates that the Rondo galagos are endangered; however, the current records have indicated a higher density of these species as compared to the other two galagos that coexist (Chapter 9). A quick reasoning would be that the other two galagos are more threatened than the Rondo galagos, but the reality could be that the other two are more shy and more elusive than the Rondo galago. Within the Pugu-Ruvu ecosystem, there are patchy forests that serve as habitats for these species, although some of the galagos were recorded outside the protected areas, indicating the need for efforts to conserve these forest remnants within the areas. It has been noted that project Environmental and Social Impact Assessment management plans had recommended a need for habitat restoration as part of compensation measures that could rescue the already threatened habitats for these species.

Bats form the second most speciose mammalian order, with over 1400 species known worldwide (Chapter 10). There are known to be 17 species that occur in the PKNFR, although the assessments along the SGR alignment within the reserve only recorded nine species. This is unique data as there have been no similar studies since 1998. Natal long-fingered bat *Miniopterus natalensis* was mist-netted for the first time among the nine species encountered. Despite the smaller number of species that we are reporting, since 1998 the number in the RSFR is comparatively low. Use of mist nets could contribute to observed differences in encounter rates because only bats, which fly within 3 m, can potentially be trapped (Cranbrook *et al.*, 1965). Despite some of these limitations, it can be concluded that the PKNFR has the highest number of bat species among the coastal forests of Tanzania. Kaolin mines found in the PKNFR are of special importance since thousands of Lander's horseshoe bat *R. landeri*, Persian trident bat *T. persicus* and Noack's leaf-nosed bat *H. ruber* roost in them. Therefore, it is recommended that the sites be restored and conserved to ensure existence of the species within the mines and their surrounding areas.

Despite the inevitable degradation of the two protected forests within the SGR alignment that are of significant conservation value, efforts are underway to ensure that various mitigation measures proposed through Environmental and Social Impact Assessment (ESIA) studies conducted in the areas indicate some positive change towards restoration of the sites soon. Tree planting and compensation measures have so far been implemented in part through collaboration by Tanzania Forest Service Agency (TFS) and other important stakeholders. So far, TFS (as one key stakeholder along the alignment) is participating in restoring their areas under close supervision of local and international experts who monitor the project after every three months, and TRC follow up the tasks on a routine basis. As such, it is hoped that, despite the degradation that had resulted from the project activities, the ongoing measures that are put in place will, to a large extent, rescue most of the biodiversity within these protected areas.

A total of 26 species of large mammals were recorded along the SGR alignment from Dar es Salaam to Isaka (Mwanza), Tanzania (Chapter 11). Among the recorded species, four species, namely, African elephant (*Loxodonta africana*), giraffe (*Giraffa Camelopardalis*), leopard (*Panthera pardus*) and lion (*Panthera leo*) were listed as Vulnerable (VU) according to the IUCN Red List, whereas wild dog (*Lycaon pictus*) is listed as Endangered (EN) and zebra (*Equus burchellii*) is listed as Near Threatened (NT). Some of the species that were recorded are also listed in Appendix I of the Convention on International Trade in Endangered Species of Wild Fauna and Flora (CITES), thus indicating that the species are threatened by extinction and require regulated trade mechanisms.

It is also important to mention that the SGR alignment from Dar es Salaam to Isaka has several wildlife corridors that are internationally recognized as sensitive habitats that require special attention amid the ongoing development activities and increased human footprint. The key aspects that have had conservation implications for large mammals' crossings, among important mitigation measures that were previously proposed, include the mitigation measures that involve construction of elephant-proof fences.

There has been valuable learning from Kenya SGR implications to elephant crossings and from the fact that the existence of this infrastructure contributed to change in behaviour where a group of elephants stopped to cross the underpasses that were created to allow continuation of their movements (Okita-Ouma *et al.*, 2016). If necessary, measures are not taken, such forced change in animal behaviour, and daily activity budget, may result in long-term changes in terms of population viability especially when the resources which were accrued from one side of the infrastructure cannot be accessed because of the inevitable changes. It is through continued monitoring and evaluation actions that the Kenya Wildlife Service (KWS) authorities managed to pick the changes and tried to start rectifying the problem, although, seemingly, the solution has worked out. According to news through KWS, it was reported that, 'it took elephants in Kenya's Tsavo Conservation Area just three months to learn where to find underpasses built beneath a controversial new railway line that splits their territory in two'. Monitoring studies always provide both negative and positive feedback that is key to decision making in biodiversity management, both inside and outside protected areas, and at proximity to such infrastructure like SGR. All chapters indicate the need for active monitoring and research activities that will need to inform this initial data along the alignment.

12.4 Conclusion and Recommendations

The Tanzania SGR alignment present a good case study where both negative impacts and prospects for biodiversity conservation have been demonstrated. Habitat fragmentation and loss, edge effects and species displacement and mortality are some of the negative impacts also presented by the project along the SGR alignment from Dar es Salaam to Makutupora. Despite the negatives, it is certain that, careful planning, and implementation of effective mitigation measures, it is possible to minimize the impacts and enhance biodiversity through habitat restoration and/or offset strategies. The Tanzania's SGR project offers an opportunity to demonstrate how large-scale infrastructure development can be integrated with conservation goals, ensuring that economic growth does not come at the expense of the country's rich natural heritage. The existing monitoring and evaluation efforts undertaken in the areas have provided rich data that is useful in ensuring the no net loss of the species with conservation value including the IUCN red listed species and other range restricted species. The continuation of research in the areas is promising and acts as positive consequence of the project. Future books may need to detail the updates on the performance of mitigation measures implemented along the SGR projects along high biodiversity areas in East Africa and beyond.

References

Borda-de-Água, L., Barrientos, R., Beja, P. and Pereira, H.M. (2017) Railway ecology. In: Borda-de-Água, L., Barrientos, R., Beja, P. and Pereira, H.M. (eds) *Railway Ecology*. Springer Open, Switzerland, pp. 29–35.

Boussougou, G.B., Brou, Y.T. and Valimba, P. (2018) Vulnerability of Pugu and Kazimzumbwi forest reserves under anthropogenic pressure in southeast Tanzania. In: Gervasi, O. (ed.) *Computational Science and Its Applications – ICCSA 2018*. Lecture Notes in Computer Science, vol. 10961, Springer, Cham, Switzerland, pp. 231–240.

Burgess, N.D., Mwasumbi, L.B., Hawthorne, W.J., Dickinson, A. and Doggett, R.A. (1992) Preliminary assessment of the distribution, status, and biological importance of Coastal forests in Tanzania. *Biological Conservation* 62, 205–218.

Caro, T., Darwin, J., Forrester, T., Ledoux-Bloom, C. and Wells, C. (2017) Conservation in the Selous Game Reserve, Tanzania. *Conservation Biology* 27(1), 46–54.

Chowdhury, S., Hesselberg, T., Böhm, M., Islam, M.R. and Aich, U. (2017) Butterfly diversity in a tropical urban habitat (lepidoptera: Papilionoidea). *Oriental Insects* 51(4), 417–430.

Cranbrook, The Earl of, Burrett, H.G. and Yates, F. (eds) (1965) Observations on noctule bats (Noctalus noctula) captured while feeding. *Proceedings of the Zoological Society of London* 144, 1–24. Available at: https://doi.org/10.1111/j.1469-7998.1965.tb05163.x

Fahrig, L. (2003) Effects of Habitat Fragmentation on Biodiversity. *Annu Rev Ecol Evol Syst* 34, 487–515.

Francis, C.D., Ortega, C.P. and Cruz, A. (2009) Noise pollution changes avian communities and species interactions. *Current Biology* 19(16), 1415–1419.

Harper, K.A., Macdonald, S.E., Burton, P.J., Chen, J., Brosofske, K.D. *et al.* (2005) Edge influence on forest structure and composition in fragmented landscapes. *Conservation Biology* 19(3), 768–782.

Heatwole, H. and Wilkinson, J. (eds) (2009) Amphibian Biology: Amphibian Decline, Diseases, Parasites, Maladies and Pollution. In: *Amphibia Biology*, Vol. 8. Surrey Beatty & Sons PTY Limited, Australia.

IFC (2012) Performance standard 6: Biodiversity conservation and sustainable management of living natural resources. International Finance Corporation.

Laurance, W.F., Clements, G.R., Sloan, S., O'Connell, C.S., Mueller, N.D, *et al.* (2014) A global strategy for road building. *Nature* 513(7517), 229–232. DOI: 10.1038/nature13717.

Menegon, M., Lyakurwa, J. and Loader, S. (2020) Version 1.0. *The Amphibians of the Tanzanian forests: A freely Downloadable visual guide*.

Mligo, C. (2019) Post fire regeneration of indigenous plant species in the Pugu Forest reserve, Tanzania. *Global Ecology and Conservation* 18, e00611.

Okita-Ouma, B., Lala, F., Koskei, M., Mwazo, A., Kibara, D. *et al.* (2016) *Movements of satellite-linked collared elephants and other wildlife in relation to the Standard Gauge Railway (SGR) and highways in Tsavo Ecosystem, Kenya (March–September 2016).* Save The Elephants and Kenya Wildlife Service, Nairobi.

Perkin, A.W. (2020) *Paragalago rondoensis*. DOI: 10.2305/IUCN.UK.2020-2.RLTS.T40652A17962115.en.

Rocky, J. and Mligo, C. (2012) Regeneration pattern and size-class distribution of indigenous woody species in exotic plantation in Pugu forest reserve, Tanzania. *International Journal of Biodiversity Conservation* 4, 1–14.

TRC (2021) Restoration and Monitoring Plan for Pugu-Kazimzumbwi and Ruvu South Forest Reserves. In: *A Report Prepared for TRC by Kilawe, C. and Emily, C.J.* Dar es Salaam, Tanzania.

Index

Printed and bound by CPI Group (UK) Ltd, Croydon, CR0 4YY

13/09/2026

14961107-0001